Environmental Education in Ind

Indonesia's wealth of natural resources is being exploited at breakneck speed, and environmental awareness and knowledge among the populace is limited. This book examines how young people learn about the environment to see how education can help to develop environmental awareness and avert vast environmental destruction, not only in Indonesia, but also in the Global South more generally.

Based on in-depth studies conducted in the cities of Yogyakarta and Surabaya, complemented with surveys of students in secondary schools, *Environmental Education in Indonesia* examines educational curricula, pedagogy and "green" activities to reveal what is currently being done in schools to educate children about the environment. The book investigates the shortcomings in environment education, including underqualified teachers, the civil service mentality, the still-pervasive chalk-and-talk pedagogy and the effect of the examination system. It also analyses the role of local government in supporting (or not) environmental education, and the contribution of environmental NGOs. The book establishes that young people are not currently being exposed to effective environmental education, and the authors propose that the best and most culturally appropriate way forward in Indonesia is to frame pro-environment behaviour and responsibility as a form of citizenship, and specifically that environmental education should be taught as a separate subject.

This book will be of great interest to students and scholars of contemporary Indonesia and Southeast Asia, education for sustainability and environmental education, as well as sustainability and sustainable development more generally.

Lyn Parker is Professor of Asian Studies in the School of Social Sciences at The University of Western Australia. She is the author/editor of numerous titles, including *Adolescents in Contemporary Indonesia* (with Pam Nilan, Routledge, 2013), *Contestations over Gender in Asia* (with Laura Dales, Routledge, 2015) and *Women and Work in Indonesia* (with Michele Ford, Routledge 2008).

Kelsie Prabawa-Sear has recently completed her PhD in Anthropology & Asian Studies in the School of Social Sciences at The University of Western Australia. Her research was on environmental education in Indonesia and, in particular, in schools in Java. She is now working as Manager of Environmental Education at Perth Zoo.

Routledge Explorations in Environmental Studies

Society, Environment and Human Security in the Arctic Barents Region
Edited by Kamrul Hossain and Dorothée Cambou

Environmental Performance Auditing in the Public Sector
Enabling Sustainable Development
Awadhesh Prasad

Poetics of the Earth
Natural History and Human History
Augustin Berque

Environmental Humanities and the Uncanny
Ecoculture, Literature and Religion
Rod Giblett

Ethical Responses to Nature's Call
Reticent Imperatives
James Magrini

Environmental Education in Indonesia
Creating Responsible Citizens in the Global South?
Lyn Parker and Kelsie Prabawa-Sear

Ecofeminism and the Indian Novel
Sangita Patil

The Role of Non-state Actors in the Green Transition
Building a Sustainable Future
Edited by Jens Hoff, Quentin Gausset and Simon Lex

www.routledge.com/Routledge-Explorations-in-Environmental-Studies/book-series/REES

Environmental Education in Indonesia

Creating Responsible Citizens in the Global South?

Lyn Parker and Kelsie Prabawa-Sear

Routledge
Taylor & Francis Group
LONDON AND NEW YORK

earthscan
from Routledge

First published 2020 by Routledge

2 Park Square, Milton Park, Abingdon, Oxon, OX14 4RN

605 Third Avenue, New York, NY 10017

Routledge is an imprint of the Taylor & Francis Group, an informa business

First issued in paperback 2020

British Library Cataloguing-in-Publication Data
A catalogue record for this book is available from the British Library

Library of Congress Cataloging-in-Publication Data
A catalog record has been requested for this book

ISBN: 978-0-367-02755-1 (hbk)
ISBN: 978-0-367-78424-9 (pbk)

Typeset in Goudy
by Wearset Ltd, Boldon, Tyne and Wear

To Clare and Rory, Shelley and Ben
and
Thel (Anang), Ruby, Kingston, Quinn and the Sudihartono and Sear families

Contents

List of figures   viii
List of tables   ix
Acknowledgements   x

1 Introduction   1

2 Theorising responsible environmental citizenship   18

3 Introducing environmental education   39

4 Introducing Indonesia   57

5 Education in Indonesia   79

6 Religious environmental education?   95

7 Is anyone responsible for the environment in Yogyakarta?   125

8 Hollow environmental education in Yogyakarta   146

9 A coordinated approach to environmental education in
   Surabaya   175

10 Environmentality in Surabaya?   207

11 Young people as environmental subjects? Identity,
   behaviour and responsibility   234

12 Conclusion, and a way forward   253

Index   266

Figures

1.1 The relationships between the environment, human society and the economy 6

6.1 A cartoon from *Becoming an Environmentalist – It's Easy! A Guide for Beginners* 96

6.2 Reading passage from Year 10 English language textbook 117

6.3 Comprehension questions in Year 10 English language textbook 118

7.1 Banner protesting the building of the Uttara apartments in Yogyakarta 127

8.1 Successful mushroom hut in Surabaya school 149

9.1 Doing fieldwork in schools in Surabaya 177

9.2 TENGO teaching students out of the Eco-mobile 188

9.3 Students collecting organic waste in the market 199

9.4 Students carrying the organic waste to be made into compost 200

10.1 Yel-Yel dance competition 209

10.2 Princess of the Environment 219

Tables

5.1	Size of the education system in Indonesia, 2015/16	81
6.1	Curriculum 2013: compulsory subjects	103
6.2	Curriculum 2013: subjects of choice	104
11.1	Characteristics of respondents	237
11.2	Environmental "care" activities conducted by senior high school students	239
11.3	Environmental "knowledge" activities conducted by senior high school students	240
11.4	Frequency of everyday pro-environment behaviours	241
11.5	Frequency of everyday pro-environment activities	242
11.6	In your opinion, what is the most important environmental issue where you currently live?	243
11.7	In your opinion, what is the most important environmental issue nationally?	243
11.8	In your opinion, what is the most important environmental issue internationally?	244
11.9	In your opinion, who is responsible for addressing this (most important) issue (locally)?	245
11.10	In your opinion, what can you do to address this (local) issue?	245
11.11	In your opinion, what are the barriers to addressing this (local) issue?	246
11.12	In your opinion, who is responsible for addressing this (most important national) issue?	247
11.13	In your opinion, what can you do to address this (national) issue?	247
11.14	In your opinion, what are the barriers to addressing this (national) issue?	248
11.15	In your opinion, who is responsible for addressing this (most important international) issue?	248
11.16	In your opinion, what can you do to address this (international) issue?	248
11.17	In your opinion, what are the barriers to addressing this (international) issue?	249

Acknowledgements

Our gratitude and thanks go to all of the students, teachers, NGO workers and volunteers in Yogyakarta and Surabaya who are trying their best to help the environment. Their participation not only provided the basic data for the field-work chapters but also enriched our understanding of the challenges facing those who are working against the odds for environmental sustainability.

This research was supported by Australia Research Council Discovery Grant: DP130100051 and was conducted as part of a team project entitled, "Fostering Pro-Environment Consciousness and Practice: Environmentalism, Environmentality and Environmental Education in Indonesia". Our sincere thanks go to Dr Greg Acciaioli, Professor Pam Nilan, Professor Yunita Winarto and Dr Suraya Afiff for their collegiality and teamwork. Fieldwork was supported by official permission from Ristek-Dikti.

And we would like to thank each other. Lyn thanks Kelsie for being such an amazing PhD student and colleague, and Kelsie writes: To Lyn, my mentor and supervisor, it has been an absolute pleasure working with and learning from you.

Every effort has been made to contact copyright-holders. Please advise the publisher of any errors or omissions, and these will be corrected in subsequent editions. We thank Taylor & Francis for permission to reproduce the following.

Lyn Parker (2017) Religious environmental education? The new school curriculum in Indonesia. *Environmental Education Research*, 23(9), 1249–1272. This is the authors' accepted manuscript of an article published as the version of record in © 2017 Taylor & Francis – https://doi.org/10.1080/13504622.2016.1150425.

Lyn Parker, Kelsie Prabawa-Sear, & Wahyu Kustiningsih (2018) How young people in Indonesia see themselves as environmentalists. *Indonesia and the Malay World*, © Editors, 46(136), 263–282. doi:10.1080/13639811.2018.1496630, on behalf of the Editors, *Indonesia and the Malay World*.

1 Introduction

This book examines the environmental education of young people in Indonesia, and focuses on efforts to educate them towards taking responsibility for the sustainability of the natural environment. Using the base of "what's happening now" in Indonesia, and taking into consideration the socio-cultural, economic and governance context of contemporary Indonesia, the book also suggests culturally sensitive ways forward, to transform young people into environmentally responsible citizens. In this sense, it is also an evidence-based public policy document.

The literature on environmental education (EE), and on environmentalism in general, is mostly about rich, Western, post-industrial, late capitalist countries where there are strong environmental movements and "green" political parties (Gough, 2003; Jickling & Wals, 2008).[1] Despite the international impacts of environmental problems and prolific use of the slogan "think global" in EE, the academic literature on EE in schools remains a Western, science-based discourse (Cole, 2007; Gough, 2003; Parker, 2016). The Global North is the "default position" in discourses of EE, and anything outside of that is still Other.

Despite the UN's Decade of Education for Sustainable Development (2005–2014), and the series of international conferences and protocols on climate change to which countries in the Global South are signatories, there is a real lacuna in our knowledge of environmental attitudes and knowledge, and pro-environment efforts, in non-Western cultures of the Global South. This book examines the situation in a non-Western, Global South country and argues that this very different socio-cultural and economic context makes a difference. It proposes that the best, most culturally appropriate way forward in Indonesia is to frame pro-environment behaviour and responsibility as a form of citizenship. The objective is the creation of practising pro-environment citizens, who share a collective environmentalist subjectivity.

Indonesia is a resource-rich, democratic, developing country; with 258 million people, it is the world's fourth largest country in terms of population (UNDESA, 2015, p. 14) and the largest Muslim-majority country; it is one of the most culturally and linguistically diverse countries on earth; and it has a magnificent wealth of biodiversity, both terrestrial and marine. Unfortunately, it is also a country of dire environmental problems: of untrammelled exploitation of forests and marine resources, of serious air and water pollution, of population growth and a large and growing middle class set on material prosperity. All this is matched with a low level of environmental consciousness among its population. While the government has made some efforts to address the problem, the research reported upon in this book makes clear that much more needs to be done.

In Indonesia, young people have an established historic role as "agents of change", both politically and socially. Their spirit and activism have been vital in ushering in each change of regime, beginning with the establishment of the independent nation-state and, most recently, in triggering the resignation of former President Suharto in 1998 and the re-establishment of democracy. They are "the hope of the nation", and are remarkably optimistic and positive (Nilan, Parker, Bennett, & Robinson, 2011; Parker & Nilan, 2013). They constitute a huge resource for socio-cultural change towards pro-environmental subjectivity and practice. Indonesia is an education "success story": in its short life as an independent, postcolonial nation-state, i.e. from 1945, it has gone from basically a country of nationwide illiteracy, without a mass, national education system, to a country where virtually all children attend primary school, the vast majority get nine years of schooling, and nearly 80 per cent attend senior high school. This amounts to an "education revolution". Further, Indonesia inherited the arbitrary borders of the Netherlands East Indies, and in a remarkable process of creating and harnessing nationalism, has successfully constructed itself as a functioning and unified nation-state.

Arguably, the principal mechanism by which it has achieved this is through the national education system: the deployment and teaching of a single national language in schools (in a country of hundreds of languages); the nationwide sharing of the experience of school education; the connection between school graduation and securing desirable jobs (although this is problematic in contemporary Indonesia); and the unifying struggle to achieve development and modernity. In Indonesia, schooling also involves the constant instilling and development of civic pride and national loyalty. Students are constantly exposed to Indonesia's national ideology, called Pancasila, in school lessons and school culture. Pancasila consists of five inter-connected "pillars": belief in one Great God, a just and civilised humanity, national unity, consensual and representative democracy, and social justice for all the people.

However, it has to be said that, until now, in this story of national development, "the environment" has barely appeared as a topic. In the discourse of national progress, the environment really only makes an appearance as the wealth of natural resources that it is Indonesia's prerogative to exploit to the maximum, to create prosperity for its citizens. In this book we call this "resource nationalism". This means that a transformation of the national discourse is required, if these natural resources are to be used wisely and sustainably. Given the ubiquity of schooling now, and its historic role in creating a patriotic citizenry, environmental education in schools appears as the most suitable vehicle for bringing about this much-needed transformation. In this book, we investigate schools' and others' attempts to bring young people to responsible environmental behaviour, because not only will today's young people inherit the problems wrought by irresponsible development, but also they represent the nation's best hope for staying their country's gung-ho destruction of the natural world.

A few notes of caution are warranted. First, there is potential here for unreasonable expectations. Collectively, young people have spearheaded social and political change, but one of the features of Indonesian societies is the strength of family and social norms that instantiate respect of children for their parents. Young people have a relatively powerless position in their families, and it is extremely difficult for children to suggest to their parents new ways of doing things, let alone to disobey their parents. There is something of a disconnect here in the historic public role of young people and their subordinate position within the family domain. Second, Indonesia starts its journey towards environmental sustainability a long way behind many countries of the Global North, where populations enjoy high levels of science knowledge and environmental understanding. For example, we have heard high school children in Indonesia explain that the "greenhouse effect" and global warming are caused by overuse of glass in houses ("glasshouses"); many farmers use red, blue and white chemicals on their crops, without knowing what elements or types of fertiliser, weedicide or pesticide they are applying, and they mix cocktails of these chemicals without wearing protection and using kitchen cooking utensils. International assessment tests of schoolchildren show that Indonesian students are woefully behind in science knowledge (OECD, 2016). Third, although this book suggests ways forward via the formal education system, there is great inertia in the enormous education system. It is not surprising, given that there are over 49 million students and ~3.5 million teachers in levels from kindergarten to senior high school (MOEC (Ministry of Education and Culture), 2016). There are entrenched reasons that teachers have no incentive to change their ways – particularly as many are civil servants first and educators second; and the capacity of teachers, in terms of their knowledge base and pedagogical capabilities, is limited. Fourth, environmental education cannot do the job alone. In many ways, it can be seen as a "safe" option, delaying or shifting responsibility for major structural changes that will only come about with political action. As Jucker said,

> The highly idealistic notion – which assumes that we just need to change the way we educate our kids and students in order to make sustainability fall into our lap – is both horribly naïve *and* utterly unfair to the younger generation.
>
> (Jucker, 2002, p. 9, emphasis in original)

It is important not to set up an oppositional dichotomy of young people versus the state, and/or versus a rapacious economic system. Young people too contribute to the consumption of material goods and hence natural resources, and young people in Indonesia are routine litterers. While many people in Global North countries would find it almost physically impossible to drop an empty plastic water bottle on the ground in the street, and would either cast about for a rubbish bin or carry it home, most young people in Indonesia would drop it without thought. This is all part of the low level of environmental awareness that characterises Indonesian society.

But this is not to demonise Indonesia and valorise the Global North. The model of economic development that has come to represent the desired goal of the post-colonial nation-state since the Second World War, derives from the Industrial Revolution of Euro-America and the Age of Empire (Escobar, 1995). This development model and its capitalist economic system is to blame for much of the world's environmental woes. And yet, not unreasonably, many post-colonial countries aspire to reach the same levels of prosperity and security that characterise the Global North. This introduces the Gordian knot of the global predicament today: disparate levels of responsibility for climate change and biodiversity loss; different levels of ability to pay for clean-up and switch to more sustainable economies; heightened concerns with national sovereignty as transnational companies and institutions extend and deepen their hold over the global economy; and undiminished commitments to economic growth and heightened prosperity. There is no prospect that a swash-buckling Alexander-like hero can slice through this knot. We must seek slower, wiser solutions.

The environment

Of course, everybody lives in an environment, and it affects their daily life in all sorts of ways: city dwellers may only have to decide whether or not to wear a coat or take an umbrella as they leave the house, but people in hunting and gathering societies rely for their survival on their successful utilisation of the natural environment in which they live. In the richer countries of the world, and in contemporary global discourse, "the environment" is externalised – as something apart from humans, as a bank of natural resources, sometimes as a threat (in the form of cyclones or earthquakes) and as something that can be manipulated and should be managed – hence climate change conventions, the declaration of national parks, etc. While most people in such countries assume that humans depend on the environment, opinions vary as to the extent to which humans can make "withdrawals" from that bank without thought for future generations; the extent to which continuing economic growth is desired over care for the sustainability of the environment; and the extent to which humans are perceived as an intrinsic part of nature (an eco-centric worldview), versus the anthropo-centric view that humans, as superior beings, are meant to have mastery (or stewardship) over nature, or indeed must "conquer" nature (see, for example, Schultz & Zelezny, 1999; Thompson & Barton, 1994).

The idea of a split between eco- and anthro-centric worldviews, of nature versus human society, of the natural sciences on the one hand and the humanities/social sciences on the other, has some validity because in some contexts it has real purchase. Many biologists would, for instance, favour the establishment of large, people-free protected areas such as national parks and wilderness areas, in both Global North and Global South contexts. On the other hand, many social scientists point out the "rich country" blindness of such actions and look for ways to simultaneously address the social justice issues that erupt with attempts to save or conserve people-less wilderness (see, for example, Guha, 1989; Nixon, 2011). In many real-world contexts, we find ourselves in quandaries over whether to prioritise the environment or society-driven demands (e.g. whether to go by public transport and take longer to commute, or spend the time more efficiently by taking the car). But the split is not necessarily that clear-cut. If public transport systems were adequate and efficient, there would be less of a quandary. If public policy and budgets prioritised the environment, individuals could sensibly take public transport. We have scientific and technical solutions to many of the world's "environmental" problems – which have been caused by humans – but lack the social understanding and political will to implement them. What is needed is a humanity-in-environment approach.

After all, we are in the Anthropocene Age. As Philips has written:

> Planet Earth is more than 4.5 billion years old; life has existed on it for more than 3.5 billion years, with humans on it for 2–3 million years, living with other life forms. But the Anthropocene Age is named for us. As its namers, Crutzen and Stoermer, put it: "It seems to us more than appropriate to emphasize the central role of mankind [*sic*] in geology and ecology by proposing to use the term 'anthropocene' for the current geological epoch."
>
> There has been overwhelming agreement with the thesis of this original scientific paper.
>
> (Philips, 2014, p. 978)[2]

As the draft Islamic Declaration on Climate Change states, "We have now become a force of nature."[3]

The environment is an empirical reality, which can be studied scientifically, but it is also a social construct. Different societies, different regimes and different organisations have their own perceptions of the environment and of environmental issues. Insofar as it has one, Indonesia's national discourse of the environment, as mentioned above, is one of abundant natural resources, such as forests, ripe for exploitation to enrich its people. Increasingly, there is a parallel but more muted discourse of global and local environmental issues, and Indonesia's international representatives sign commitments on behalf of the country to limit carbon emissions.[4] At the same time, wet-rice farmers in Java are primarily interested in their small environment of paddy field, water supply and weather; city dwellers mostly identify rubbish as the nation's number one environmental issue; and indigenous peoples are often engaged in site-specific fights to save their own enviro-economy, the forest. It is necessary to understand different people's different understandings of nature and the environment.

Environmental issues are almost by definition social issues, not least because "the environment" is a social construct. These different perceptions of the environment have real-world policy and on-the ground ramifications, as noted above. Environmental problems are mainly caused by human societies. "The environment" knows no political boundaries or jurisdictions: smoke from forest fires in Sumatra and Kalimantan (Indonesian Borneo) not only closes airports and schools in Sumatra and Kalimantan, sometimes for months at a time, but also damages the health of people in Singapore, Malaysia, Brunei and Thailand. However, responsibility for much of the burning can be sheeted home to business tycoons in Malaysia and Singapore, who invest in the palm oil industry – as well as to the government officials and politicians who should be controlling it but often stand to gain financially by not (Varkkey, 2015). Environmental problems have no time limits or statute of limitations – the ramifications of the Industrial Revolution that occurred first in Western Europe are still being felt in the rapid industrialisation of India and China and drastic global climate change, and international conventions struggle to deal with that legacy. The "past of slow violence is never past" (Nixon, 2011, p. 8).

The environment we have in mind in this book is the earth's life-support system. Following Griggs *et al.* (2013, p. 306), we can visualise this as shown in Figure 1.1.

This is the environment that is the subject of global concern, not only to environmental activists and those who are trying to live in more environmentally sensitive ways, but also to scientists and academics, policy-makers, public servants and the like: an environment that is degrading in quality because of human actions that are causing shrinking biodiversity and the deterioration of conditions that support life on earth. Nixon describes the "slow violence" of "[c]limate change, the thawing cryosphere, toxic drift, biomagnification,

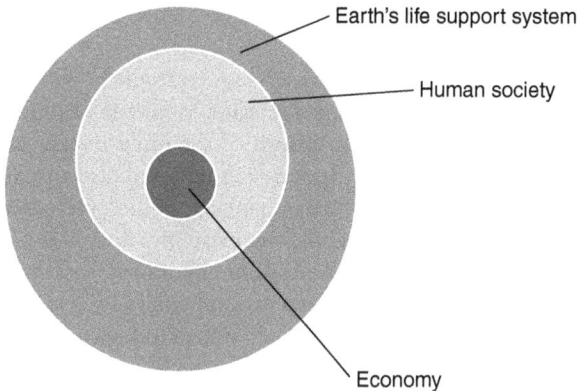

Figure 1.1 The relationships between the environment, human society and the economy.

Source: Griggs *et al.* (2013).

deforestation, the radioactive aftermath of wars, acidifying oceans, and a host of other slowly unfolding environmental catastrophes" (Nixon, 2011, p. 2). This violence is typically not perceived as violence at all, not least because those who are the most vulnerable to its ravages are the poor and marginal. These "long dyings" are largely invisible and uncounted.[5]

In the Global South, "the environment" is often disguised. Environmentalism often arises around a single issue, such as the building of a highway or a large dam, and is therefore local and often ephemeral (Kalland & Persoon, 1998). It might be labelled an issue of dispossession or social injustice. Sometimes a social conflict disguises the environmental issue at its heart, or, more precisely, conflict occurs over control of a natural resource, but may be fought in the name of religion, race or ethnicity. There is indeed an "environmentalism of the poor" (Erb, 2012; Martinez-Alier, 2002), but "the poor" lack access to big business media, so their protests often go unnoticed. For now, it is enough to emphasise that the environment is a social construct, and that social injustice is often environmental injustice. In Indonesia, many local environmental issues are also issues of social justice, and, when reported in the media, it is the social conflict, rather than the environmental damage, that is reported, e.g. when large dams, highways or ecotourism resorts displace local farmers in the name of Development (Colombijn, 1998). It is very rare for observers to link common local events, such as the closure of schools or airports due to smoke haze, to their real cause: deforestation of huge swathes of rainforests for the sake of a monoculture of palm oil and, ultimately, profit. This obfuscation not only inhibits holistic thinking and understanding of human–nature interactions, but also hides the identity of perpetrators and the systemic nature of social and environmental injustices.

Environmental responsibility

In the context of Indonesia, the phrase "social justice" has a great deal more pulling power than "environment" or "sustainability". Social justice is the fifth pillar of Pancasila, and justice (*keadilan*) is an Arabic-derived term that has many referents in the Qur'an. For these and other reasons described in the next chapter, this book borrows one of its philosophical foundations from the social justice theory of the feminist philosopher, Iris Marion Young (2006). In her paper of 2006, "Responsibility and Global Justice: A Social Connection Model", she posits that obligations of justice and responsibility derive from social connections.[6] This conceptualisation is useful because it explains why human beings are responsible to, and have obligations to, others beyond their family, community or even nation-state, but, at the same time, share responsibility, albeit unequally, to act collectively to restore justice. It is a theory that separates responsibility from blame (liability), looks forward rather than backward, and is concerned with action for global social justice.

Here we outline the theory, showing its relevance to environmental responsibility.

The social connection model of responsibility says that all people who contribute to structural social processes that produce injustice have responsibilities to remedy the injustice. This is not just about equal and universal human rights: it is about structural injustice brought about by transnational, institutional and personal relationships that are liable to cause conflict and inequalities of power. Young uses the transnational clothing industry as her example. The consumer who buys clothes in the Global North is, through her everyday act of purchase, engaging in a social process. Thereby, she is connected to and has responsibilities towards the exploited women who work long hours, in unhealthy and unsafe conditions, for below-minimum-wage pay in the Global South. This is an example of a transnational social structure that produces injustice and therefore obligations beyond the known local context and indeed beyond political borders. Similarly, we could posit a home-owner in a prosperous country buying a new wooden dining suite. By virtue of this social act of consumption, that person is involved in a global structure that does harm: it does violence not only to the rainforest and rainforest dwellers in, say, Borneo, but also to the ecosystems that have been disrupted to produce the oil to fuel the chainsaws, trucks and ships used to transport the timber; it exploits cheap labour in the furniture factory in the town in, say, north coast Java; or it involves the better-paid cabinet-maker in the rich country where the furniture is made or finished, the wholesalers and the retailers and finally the consumer. Thus, the social act of consumption entails a structural social process that does both social injustice and environmental harm. We would argue then, that it necessitates social, economic and environmental obligations and responsibilities beyond the national border of the country of consumption.

However, Young's social connections theory of global social justice cannot do all the work. Our extension of it to environmental justice and our comingling of harms to humans and harms to the environment presumes the absolute value of natural ecosystems. We think we need to accept the absolute value of the environment and its complex systems and assign it rights to exist per se. However, to pre-empt the findings of our fieldwork and analysis of the discourse in Indonesia, an eco-centric worldview is not prevalent in Indonesia. At base, the Indonesian national discourse states that the environment is the creation of God: God is the Creator. We can borrow from this religious discourse its sense that humans should act as the stewards of God, with the responsibility to preserve God's Creation. Probably at this point we will have raised some hackles: many secular readers would find it problematic to base an environmental programme on a religious, and further, creationist, base. But we would argue that in searching for culturally appropriate ways to bring Indonesians to environmental responsibility, and in order to minimise cognitive dissonance, we can deploy this religious discourse to motivate environmental action – and indeed we find that many environmentalists and activists engaged in EE in Indonesia feel that their work is a religious vocation.[7]

In introducing this element of religion to the Introduction we are alluding to the fact that the research on which this book is based is ethnographic, real world, field work. While there is some fine academic work on notions such as environmental citizenship and responsibility, this academic work is mainly theoretical and philosophical. Here we take an anthropological approach, which means we start with the fieldwork and work up from there. We echo MacGregor, who writes on feminist ecological citizenship, that

> empirical research [is] both necessary to the development of theoretical ideas and frustrating for the desire to arrive at pat conclusions. I [choose] to include the first-hand accounts […] of women activists in my research because of a dissatisfaction with the absence of "the empirical" in the writings of green theorists of citizenship and the over-reliance on women's experiences (as incontestable truth) in ecofeminist scholarship. While my effort to synthesize theory and practice makes for a much messier narrative than the ones now on offer, it is my hope that it will also provide a much more useful one.
>
> (MacGregor, 2009, pp. 292–293)

The theoretical framework of the book and concepts such as environmental awareness, responsibility and citizenship will be elaborated in Chapter 2.

The research

The first author, Lyn Parker, is a social and cultural anthropologist who has conducted fieldwork in Indonesia since conducting doctoral fieldwork in east Bali in 1980–1981.[8] The environment has rarely been the main subject of her research but has been ever-present, beginning with her Honours project on the effects of the then-new High-Yielding Varieties of rice and the Green Revolution in Indonesia. At the time of PhD fieldwork in Bali, beginning in 1980, villagers where she was studying most wanted the provision of electricity. (She thought that a clean water supply would have been more advantageous, given the high rates of gastro-intestinal diseases, dysentery, and so on.) She knew that they could use a combination of small hydro-power and solar-powered photovoltaic cells, but her efforts to advocate for this were stymied by a complex array of apathy, feelings that "it's the government's job", doubts over her capacity and ability, and fear of the new.

Over the decades, the dire state of the environment in Indonesia triggered questions for her about local environmental knowledge, attitudes towards the environment and environmental values among the populace. She wanted to foster environmental consciousness and pro-environment action in Indonesia. In 2011 she invited several researchers to a large team research project which aimed to identify how various types of education and environmentalism in different contexts in Indonesia could contribute to creating environmentally aware citizens in Indonesia. The team members were Indonesian and Australian scholars – anthropologists and sociologists – working in different parts of Indonesia: Central Kalimantan, Yogyakarta in Central Java, Surabaya in East Java, rural Indramayu in West Java, rural East Lombok, and so on. The team members have been researching in different educational contexts: senior high schools, universities, wet-rice and mixed farming communities, and protected areas such as national parks.[9] One of the members of the team was the co-author, Kelsie Prabawa-Sear. She was a PhD student on the team, and conducted long-term immersion fieldwork in Yogyakarta and Surabaya. It is her fieldwork that forms the basis of the four ethnographic chapters in this book, Chapters 7–10.

The research on which this book is based is anthropological research. As anthropologists, we are committed to entering the world of the "Other" and, coming out of that world, to making sense of it to outsiders, to explain it as meaningful in accord with its own logic, values and system of meaning. In this sense, we are dealing with multiple understandings of "the environment" – the global environment that we perceive needs protection and restoration, the national Indonesian context as well as local understandings of the environment. Of course, in such a large and complex nation-state as Indonesia, the latter is not a single thing, and ranges from the shared understandings of the global discourse (e.g. among some scientists and activists in Indonesia) to uncaring and apathetic attitudes coupled with rapacious and avaricious material consumption, to the parochial worldviews of small communities. In the middle are a plethora of interpretations and meanings, often contradictory and ambivalent. But the aim of the book is not just to "translate" Indonesian understandings of the environment to an English-speaking academic audience. It is also to establish some ways forward, some culturally appropriate approaches to educating student citizens in the interdependent relationships of humankind and the environment. As we see it, environmental awareness and knowledge are important but not sufficient precursors to changing environmental behaviours. If we are to change the sensibilities and everyday practices of millions of young people, creating a collective environmental subjectivity, it will be necessary to engage the larger context and its influential institutions – national- and district-level government ministries, religious authorities, policies and curricula, the media and schools. The aim is transformation in the direction of appreciation of the urgent need to protect the conditions that support the diversity of life on earth, manifest in everyday care of the environment.

Outline of the book

Chapter 2 explores the key concepts and theories used in the book. After discussing understandings of "responsibility", it outlines some of the major approaches to environmental responsibility. After consideration of liberal views of environmental responsibility, the chapter introduces the Foucauldian concept of governmentality, and its derivative, environmentality, and the term responsibilisation, as these terms are potentially relevant for one Indonesia context, the city of Surabaya, where environmentalism is to some extent compulsory. Then the chapter elaborates on Young's approach to "responsibility for justice", which is basically a moral approach to environmental responsibility, and shows how it is appropriate in the Indonesia context.

In Indonesia, an enduring aim of education is to create loyal citizens. Given the strength of citizenship education in schools in Indonesia since 1945, the wave of Islamisation in Indonesia since around 1990, and the strength of the discourse around morality (and especially character education) in Indonesia today, the book argues that environmentally responsible citizenship could resonate as a culturally appropriate discourse in Indonesia. The chapter therefore traces some of the major theories of environmental citizenship, as this is the frame that we propose will best get traction in the context of the Indonesian education system.

Chapter 3, "Introducing environmental education", distils the most relevant themes and issues that have characterised the sub-discipline of "environmental education" (EE) since the Tbilisi Declaration in 1977. It introduces the academic literature on EE and identifies some of the salient issues in EE, such as the choice of terms (EE, Education for Sustainable Development and Education for Sustainability) and the location of EE in schools. This book uses the term EE because it is known in Indonesia and because the term "sustainable development" often leads to the neglect of the sustainability of the environment in favour of sustained development. One issue of great significance is that, as schools are part of the system that created the world's environmental problems, we cannot expect them to deliver education that will critique and transform the larger structures of capitalism, inequality and injustice that produce environmental destruction. The question is: can schools produce responsible environmental citizens? Our response is simply that they must, because the problems are so urgent. The chapter then turns to the issue of pedagogy in EE, first looking at best practice in the Global North then at four problematic aspects of pedagogy in Indonesia and many other developing countries:

1   The continuing dominance of rote learning.
2   The focus on the transmission of facts.
3   The gap between environmental awareness and knowledge on the one hand and pro-environmental behaviour on the other.
4   The effect of learned helplessness and apathy.

Finally, the chapter proposes "critical ecopedagogy" as an ideal. In summary, the chapter echoes the call "for education to accept full responsibility in addressing global survival issues" (Pinar *et al.*, 1995, p. 841).

Chapter 4 introduces Indonesia. It surveys the demography of Indonesia, the economy and the broad socioeconomic context in which our study is embedded, politics and government, religion and culture, and the environment. Indonesia is the fourth largest country in the world in terms of population, and has a young, growing population. It is a developing country, with growing prosperity and a declining incidence of poverty, but with most people still vulnerable, and a shockingly high, and rising, level of inequality. Two salient trends are the growth of the middle class and a rapid rate of urbanisation. Indonesia is the largest Muslim country in the world and, since the 1980s, has experienced massive Islamisation, with rising religious intolerance and growing fundamentalism. The chapter shows how the demographic shifts and economic development have had a deleterious effect on the environment, and how democratisation and decentralisation have also not produced expected gains for the environment.

The second half of the chapter presents what is known about environmental awareness in Indonesia, underlining the observation that the populace, the public service and the government have very little knowledge, understanding, or even awareness of the dire environmental problems that Indonesia faces – let alone what to do about them. Finally, the chapter examines government attitudes and capacity for responsible environmental management, and environmental non-government organisations (ENGOs) and the role they have played thus far in plugging that hole, trying to spread environmental concern and enthusiasm among the populace.

Chapter 5 sets out the education system in Indonesia. Many of the problems with EE that appear in the ethnographic chapters (7–10), can be traced back to problems with the education system in general. However, to give credit where it's due, Indonesia can be characterised as an education "success story" because it has gone from a basically illiterate colony in 1942 to a country of almost universal literacy for those under 25 years of age. Gendered inequalities in access to schooling are almost gone, although there are still pockets of female disadvantage; there are significant and growing inequalities by socioeconomic status, province and remoteness, with areas in eastern Indonesia consistently disadvantaged and lacking basic infrastructure such as health facilities, roads and schools with adequate teaching staff.

After describing the structure of the system, the chapter outlines the main objectives of education in Indonesia, highlighting the continuing emphasis on nationalism and the objective to create loyal, and pious, citizens, as well as the abiding need to produce effective workers. These days, the attention is on the notoriously poor quality of the education that students receive. Indonesia consistently scores very poorly in international tests (such as PISA, Programme for International Student Assessment). The low level of subject knowledge and poor pedagogical capabilities of teachers are often blamed. There have been wide-ranging efforts to improve the quality of teaching, such as increasing the pay of teachers, and upgrading their professional qualifications, and the management of schools, with a shift to school-based management.

Chapter 6 is titled "Religious environmental education?". It examines the latest curriculum, Curriculum 2013, for senior high school, to see how "the environment" is taught. It is here that we see the effect of recent Islamisation upon the education system. The chapter first outlines early efforts to introduce EE to schools in Indonesia. It describes the impetus for the new Curriculum, noting that although educators were concerned with Indonesia's poor showing in international tests, the new curriculum actually gutted the curriculum of academic content and substituted that with a new emphasis on character, moral and religious education. The chapter goes through the curriculum in detail, to discern the way "the environment" is represented. The salient feature of the Science curriculum is the religious framing. Students are to realise and be thankful that God created the universe in such a way that it is suitable for humans to live in, that it is to some extent knowable by humans (e.g. through science), and that so much has been provided for human exploitation. The same framing occurs in Geography, but in Geography there are also messages about caring for the environment, using resources wisely and responsibly, and we get the first and only mention of environmental sustainability and sustainable development. True EE carries very little weight in the new Curriculum.

The chapter takes a detour to discuss how EE as a sub-discipline has regarded religion, and the relation between Islam and science. It then addresses the following themes in Curriculum 2013: creationism; instrumentalism and the way humans are presented as having been created as separate from the environment, which was created as natural resources for humans to exploit; divine and human agency and responsibility; the desired affects and values; environmental nationalism – i.e. the idea that God created Indonesia with rich natural resources for humans to exploit for prosperity; religious resources that are neglected in the Curriculum; and it then discusses some problems with religious EE. Finally it examines how the environment is presented in textbooks.

Chapter 7 asks, "Is anyone responsible for the environment in Yogyakarta?", and the answer is basically "No". This is the first of four ethnographic chapters, and the first of two chapters on the Central Javanese city of Yogyakarta. The chapter describes the governmental context of Yogyakarta and argues that it is inimical to the fostering of good EE: the Sultan and Mayor are committed to rampant, unsustainable development; there is a lack of government commitment to the environment, poor coordination between agencies; and government officials ostensibly responsible for the environment lack expertise and interest. The chapter describes the fieldwork and selection of schools. The second half of the chapter describes the Adiwiyata Programme, the government's flagship national environmental education project: the dubious reasons that schools sign up for it; the forced participation of schools and teachers; and the way the programme is run. Its obsession for documentation and numerical KPIs turns accountability into cheating; and its emphasis on prize-winning hijacks environmental aims, turning the programme into a mechanism for school marketing and status performance.

Chapter 8 is the second chapter on Yogyakarta and uses classic educational ethnography to show what is happening in schools in Yogyakarta – both in classes where teachers are using the Curriculum, and in and around schools, where students are doing Adiwiyata activities. We examine three classes which show the hollowness of EE in schools, and the critical importance of high-quality teachers. An abysmal craft lesson, the Mushroom Fiasco, shows that when teachers stray from the traditional pedagogy, they run into problems – mainly because of their own lack of knowledge. The Biology class is an example of a more confident teacher following the curriculum to allow students to leave the classroom and explore the natural world. The third class is the best lesson we saw during fieldwork: it shows a smart, knowledgeable Geography teacher encouraging students to ask questions that go way beyond the textbook topics. The second half of the chapter reports on a student-led environmental event, Rubbish Day, at one of the schools. We use that example to examine the (limited) possibilities for the amplification of "student voice" in EE in Indonesia. We discuss the discursive impact of the fact that many teachers are public servants first, and teachers second; and the power of the social value of *sungkan* (respectful politeness) among students. These combine to work against teachers innovating and investing time and energy in the improvement of their knowledge base and pedagogy, and against students exercising initiative, suggesting innovations or critiquing their lessons or teachers.

Chapter 9 describes "A coordinated approach to environmental education in Surabaya". After introducing the fieldwork, this chapter describes the "forced volunteering" (*paksarela*) approach to EE in Surabaya, beginning with the vital role played by the Mayor of Surabaya, Ibu Risma. Then we examine the cooperation among government agencies, an environmental NGO (hereafter TENGO) and schools in enforcing a city-wide approach to EE. On the face of it, this coordinated approach looks very much like Agrawal's "environmentality" in practice. While acknowledging that Surabaya is indeed becoming "clean and green", partly through the environmental actions of students, the chapter questions the effectiveness of this approach to EE in enabling young people to solve environmental problems and understand the complex interactions between socio-economo-political systems and the natural world. Finally, the chapter examines the free labour of children to gauge if the deployment of school children in environmental services should be seen as exploitation, the exacerbation of inequalities or as "responsibilising" children towards environmental citizenship.

Chapter 10 examines various EE projects and events in Surabaya to see how students are involved, what they learn from participation in competitive events, the variable involvement and expectations of teachers and TENGO staff, how some approaches fail, and how an EE trip to Perth failed to educate participants. Even in Surabaya there is evidence of meaningless performance of environmentalism without understanding, with ritualistic compliance with rules and commitment to competition but no commitment to solving real-world environmental problems. Finally, the chapter revisits the question of whether this forced volunteering of environmental work can be considered an effective form of environmentality.

Chapter 11 is titled "Young people as environmental subjects? Identity, behaviour and responsibility". The team of researchers in this project designed a survey that was administered to 1000 senior high school students in our target schools. All of the students had been exposed to some form of EE. Our survey showed that almost 82 per cent of students self-identified as environmentalists. However, when we asked about their perceptions of environmental problems in the world, their responses were not well-informed. Further, when asked about their pro-environmental behaviours, it became clear that, while theoretically students are happy to identify as "green", in practice they are not behaving as environmentally responsible citizens. Finally, when asked who they thought is responsible for caring for and cleaning up the environment, students over-whelmingly answered "society", showing no indication that they expected their government or industries to contribute, nor was there any sense that overween-ing consumption or development was to blame. Thus, young people have absorbed the neoliberal message of small government and have assigned respons-ibility to "society", i.e. those who are least aware, most ignorant, and most poorly equipped to meet the challenges of environmental destruction.

Chapter 12 proposes some ways forward for EE in Indonesia, taking into account findings from fieldwork and consideration of the literature. The authors make recommendations for the way forward in Indonesia. This chapter also con-siders the ramifications of the study's findings for other Global South countries, with a view to the practical exigencies of weak education systems and the imperi-alism entailed in the globalisation of EE. The authors advocate the development of culturally sensitive, locally relevant, environmental education programmes that lead young people to become environmentally responsible citizens.

Notes

1 In this book, the tricky terminology of First versus Third World, the West versus the Rest, developing versus developed, advanced versus emerging economies, Global South and Global North, is mainly dealt with by using the last set of terms. However, for the Australian authors – citizens of a rich, "Western" country, economically a member of the Global North but geographically situated in the southern hemisphere – that particular binary feels decidedly odd. The advantage of this set is that it does not predetermine that the goal of "developing" countries is to become like the already-"developed" countries, or that the latter stand as some sort of model or ideal. Of course there are many problems with using such blanket terms, not least of which are their homogenising effect – as though countries as different as Indonesia and, say, Chad were somehow similar – and the static and ahistorical images they conjure.

2 The footnote in the extract (5) reads:

> This is considered the first naming of the Anthropocene Age; see Paul Crutzen and E. F. Stoermer, "The 'Anthropocene,'" *Global Change Newsletter* 41 (2000): 17–18. In a historically specific paragraph, Crutzen and Stoermer date the Anthropocene to the industrial era:
>
> We propose the latter part of the 18th century, … [although] some may even want to include the entire holocene.… We choose this date because, during the past two centuries, the global effects of human activities have become clearly noticeable. This is the period when data retrieved from glacial ice cores show the beginning of a growth in the atmospheric concentrations of several "greenhouse gases", in particular CO_2 and CH_4. Such a starting date also coincides with James Watt's invention of the steam engine in 1784.
>
> (Crutzen and Stoermer, 2000, pp. 17–18)

3 Interestingly, the final Declaration rephrased this to: "Moreover, it is human-induced: we have now become a force dominating nature." ("Islamic Declaration on Global Climate Change", 2015).

4 For instance, in February 2017, Indonesia pledged to the UNEP to cut plastic waste in 25 coastal cities and reduce marine litter by 70 per cent in eight years (UNEP, 2017).

5   National accounts figures rarely factor in the costs of erosion or pollution, unless there is an event such as a flood or an oil spill, nor do they count the opportunity cost of species loss or failure to guard against and adapt to climate change. Typically the environment features in economic accounting as the cost of property and of clean-up of (often human-caused) environmental disasters. Attempts to "count" or "economise" the environment, for instance in triple bottom lines or through corporate social responsibility programmes, merely tinker at the edges.
6   This chapter is the heart of the book, *Responsibility for Justice* (Young, 2011). Other chapters provide context and later chapters elaborate on certain issues raised in the chapter.
7   See Nilan and Wibawanto (2015, pp. 66ff.), especially the story of Romo Yatno.
8   This research was eventually published as Parker (2003).
9   The project was funded by the Australia Research Council Discovery Grant DP130100051. Some findings of the project can be found in special issues of *Inside Indonesia* (127, 2017) and *Indonesia and the Malay World* (vol. 46, issue 136, 2018) as well as scattered journal articles (e.g. Nilan, 2017; Nilan & Wibawanto, 2015; Parker, 2016).

References

Cole, A. G. (2007). Expanding the field: Revisiting environmental education principles through multidisciplinary frameworks. *The Journal of Environmental Education, 38*(2), 35–44.

Colombijn, F. (1998). Global and local perspectives on Indonesia's environmental problems and the role of NGOs. *Bijdragen tot de Taal-, Land en Volkenkunde, 154*(2), 305–334.

Erb, M. (2012). The dissonance of conservation: Environmentalities and the environmentalisms of the poor in eastern Indonesia. *The Raffles Bulletin of Zoology, Supplement No. 25*, 11–23.

Escobar, A. (1995). *Encountering Development: The Making and Unmaking of the Third World*. Princeton, NJ: Princeton University Press.

Gough, N. (2003). Thinking globally in environmental education: Implications for internationalizing curriculum inquiry. In W. Pinar (Ed.), *International Handbook of Curriculum Research* (pp. 53–72). Mahwah, NJ: Laurence Erlbaum Associates.

Griggs, D., Stafford-Smith, M., Gaffney, O., Rockström, J., Öhman, M. C., Shyamsundar, P., ... Noble, I. (2013). Sustainable development goals for people and planet. *Nature, 495*(7441), 305–307.

Guha, R. (1989). Radical American environmentalism and wilderness preservation: A third world critique. *Environmental Ethics, 11*, 71–83.

Islamic Declaration on Global Climate Change (2015). Retrieved from http://islamic climatedeclaration.org/islamic-declaration-on-global-climate-change/.

Jickling, B., & Wals, A. E. J. (2008). Globalization and environmental education: looking beyond sustainable development. *Journal of Curriculum Studies, 40*(1), 1–21.

Jucker, R. (2002). "Sustainability? Never heard of it!" Some basics we shouldn't ignore when engaging in education for sustainability. *International Journal of Sustainability in Higher Education, 3*(1), 8–18.

Kalland, A., & Persoon, G. (1998). An anthropological perspective on environmental movements. In A. Kalland a. G. Persoon (Ed.), *Environmental Movements in Asia* (pp. 1–43). Richmond, Surrey: Curzon Press and Nordic Institute of Asian Studies.

MacGregor, S. (2009). The project of feminist ecological citizenship. In M. Reynolds, C. Blackmore, & M. J. Smith (Eds), *The Environmental Responsibility Reader* (pp. 292–301). London and New York; Milton Keynes: Zed Books Ltd. and The Open University.

Martinez-Alier, J. (2002). *The Environmentalism of the Poor: A Study of the Ecological Conflicts and Valuation.* Cheltenham, UK: Edward Elgar Publishing.

MOEC (Ministry of Education and Culture), Centre for Educational Data and Statistics and Culture (2016). *Indonesia. Educational Statistics in Brief 2015/2016.* Retrieved from http://publikasi.data.kemdikbud.go.id/uploadDir/isi_AA46E7FA-90A3-46D9-BDE6-CA6111248E94_.pdf.

Nilan, P. (2017). The ecological habitus of Indonesian student environmentalism. *Environmental Sociology.* doi:10.1080/23251042.2017.1320844.

Nilan, P., Parker, L., Bennett, L., & Robinson, K. (2011). Indonesian youth looking towards the future. *Journal of Youth Studies, 14*(6), 709–728.

Nilan, P., & Wibawanto, G. R. (2015). "Becoming" an environmentalist in Indonesia. *Geoforum, 62*(2), 61–69.

Nixon, R. (2011). *Slow Violence and the Environmentalism of the Poor.* Cambridge, MA: Harvard University Press.

OECD. (2016). *Program for International Student Assessment (PISA) Results from 2015: Indonesia.* Retrieved from www.oecd.org/education/pisa-2015-results-volume-i-97892 64266490-en.htm.

Parker, L. (2003). *From Subjects to Citizens: Balinese Villagers in the Indonesian Nation-state.* Copenhagen: Nordic Institute of Asian Studies Press.

Parker, L. (2016). Religious environmental education? The new school curriculum in Indonesia. *Environmental Education Research.* doi:10.1080/13504622.2016.1150425.

Parker, L., & Nilan, P. (2013). *Adolescents in Contemporary Indonesia.* New York: Routledge.

Philips, K. (2014). Doing interdisciplinary Asian studies in the age of the anthropocene. *The Journal of Asian Studies, 73*(4), 975–987.

Pinar, W. F., Reynolds, W. M., Slattery, P., & Taubman, P. M. (1995). Understanding curriculum as international text. In *Understanding Curriculum: An Introduction to the Study of Historical and Contemporary Curriculum Discourses* (pp. 792–843). New York: Peter Lang.

Schultz, P. W., & Zelezny, L. (1999). Values as predictors of environmental attitudes: Evidence for consistency across 14 countries. *Journal of Environmental Psychology, 19,* 255–265.

Thompson, S. C. G., & Barton, M. A. (1994). Ecocentric and anthropocentric attitudes toward the environment. *Journal of Environmental Psychology, 14,* 149–157.

UNDESA, UN Department of Economic and Social Affairs, Population Division. (2015). *World Population Prospects: The 2015 Revision, Key Findings and Advance Tables* (ESA/P/WP.241).

UNEP. (2017). Indonesia joins UN in a bid to eradicate ocean plastic. Retrieved from www.unep.org/asiapacific/indonesia-joins-un-bid-eradicate-ocean-plastic.

Varkkey, H. (2015). *The Haze Problem in Southeast Asia: Palm Oil and Patronage.* Florence: Taylor & Francis.

Young, I. M. (2006). Responsibility and global justice: A social connection model. *Social Philosophy & Policy Foundation, 23*(1), 102–130. doi:10.1017/S0265052506060043.

Young, I. M. (2011). *Responsibility for Justice.* Oxford and New York: Oxford University Press.

2 Theorising responsible environmental citizenship

This chapter explores some of the key concepts and theories used in the book.[1] We begin by outlining some of the major approaches to environmental responsibility in order to show some of the possibilities for analysis, the better to show why Young's approach of "responsibility for justice" is appropriate in the Indonesia context. After consideration of liberal views of "environmental responsibility", we introduce the Foucauldian concept of governmentality and the awful term "responsibilisation", as this may be applicable in one Indonesia context, the city of Surabaya. Then we elaborate upon Young's approach to responsibility for justice. This is basically a moral approach to environmental responsibility.

In Indonesian education, one constant has been that the aim of education is to create loyal citizens.[2] Given the strength of citizenship education in schools in Indonesia since 1945, the wave of Islamisation in Indonesia since around 1990, and the strength of the discourse around morality (and especially character education) in Indonesia today, we propose that environmentally responsible citizenship could resonate as a culturally appropriate discourse. We therefore trace some of the major theories of environmental citizenship, as this is the frame that we think will best get traction in the context of the Indonesian education system. While not ideal, we consider the environmental conditions in Indonesia so dire, and the absence of other desirable conditions so significant,[3] that we propose that responsible environmental citizenship is the way forward for EE in schools in Indonesia.

Environmental responsibility

Responsibility

The word "responsible" has a long history, going back to classical Latin, Anglo-Norman, Old French and Middle French. In legal history, in thirteenth-century Anglo-Norman it had the meaning of being required to answer, of being accountable. This meaning of responsible still holds: that one is answerable or liable to be called to account to another person for something. In the history of the Church, the liturgy required spoken or sung "responses", and therefore being "responsible" implied that one was capable of responding to a question, and, by extension, of fulfilling an obligation or duty. Today one of the salient meanings of "responsibility" is the sense of moral obligation: this often implies being in charge of or having a duty towards a person or thing; there is also the sense of being capable of rational, sensible conduct (OED) – of acting responsibly.

The relevant Indonesian term is the compound verb, *bertanggung jawab*, which has several similar meanings: to be responsible for, to be in charge of, to be liable for, to be accountable for and to report to; the word "*jawab*" (answer) in the compound seems to echo the old Christian meaning of response.

Responsibility is a relational term involving several components, which Schicktanz and Schweda (2012) call "relata". They identify these relata:

> i someone/something/a process (Agent/Cause)
> ii brings something about (Problem, e.g. environmental damage)
> iii which may or may not have been intended (Intention)
> iv someone (may or may not be (i)) is responsible (Responsible Agent)
> v in a particular time frame (Time)
> vi for responding (Response)
> vii on the basis of a normative Standard (measure of acceptable standard)
> viii with certain sanctions/rewards (Consequences)
>
> (Schicktanz & Schweda, 2012, p. 133)

Thus, responsibility is both retrospective (relata (i)–(iii)), looking for a causative agent, and prospective (relata (iv)–(viii)), looking for an agent who bears the burden of response or solution. While this list of eight relata is the most expanded of any we have seen, perhaps the most common elements in the idea of "responsibility" are (1) being blamed (liable) for a past phenomenon; (2) being held accountable to someone for carrying out an action; and (3) being charged with a task in the future. In all three, there is a transcendent authority – whether it be an impersonal standard of excellence or morality (which, in the case of (i), may not have prevailed at the time of past harm), or a person or position of authority.

Environmental problems are complex, often multi-causal, with many actors involved, over long time spans. And indeed the assignation of causes and causal agents is both difficult and highly political. One could take as example, a Problem that is ubiquitous in Jakarta: a river is blocked with rubbish and when the heavy monsoonal rains come, the whole area is flooded. We can assign relata for the Problem:

> i Agents/Causes: upstream deforestation; lack of clean water supply and sanitation services means people buy water in plastic bottles and excrete in the river; groundwater pumping (supplies >60 per cent of Jakarta's inhabitants with water) (Furlong & Kooy, 2017, p. 888);[4] lack of garbage collection by government services means upstream and nearby inhabitants throw waste into river; creation of non-biodegradable waste; encroaching seawater.

   ii Problem: flooding.

  iii Intention: unintended.

  iv Responsible Agents: Ministry of Environment and Forestry, logging companies, farmers, local governments, inhabitants.

   v Time: specified.

  vi Response: sustainable forestry and agricultural practices (might involve cleaning up corruption, sorting out land-ownership issues); local governments introduce clean water supply and sanitation services, garbage sorting and collection, renewable waste management; inhabitants made aware of the causes and educated towards responsible environmental behaviour.

 vii Standard: regular monitoring including visual inspection and testing of water against standards for factors such as acidity (pH), colour, dissolved oxygen, turbidity, biota (especially coliform bacteria), concentrations of metals, nitrates and phosphates, pesticides, and other substances.

viii Consequences: Fines, withdrawal of forestry licences, reputation damage and enhancement, electoral outcomes, awards for service, hygiene and cleanliness, a clean, free-flowing river and reduced flooding.

Liberal environmental responsibility

To begin, we borrow a simple "working definition" from Reynolds. Environmental responsibility involves:

> two complementary actions[:] … (a) caring for an environment comprising the natural world of life and the life support in which humans are an integral part, and (b) ensuring guidance and accountability for any harm or wrong done to the environment.
>
> (Reynolds, 2009, p. 3)

Environmental responsibility implies that people are aware of environmental problems, value the environment, and are conscious that their everyday actions have environmental consequences: that making the effort to refill a water bottle and carry it around all day, to turn off a TV that no one is watching, or to not drop rubbish in the street, will contribute positively to environmental sustainability. This presumes an understanding of different scales and the interactions among them. It implies a sensibility that small actions are connected to, and have impacts upon, large, global processes, which might be natural systems such as the water cycle, or economic systems such as a supply chain. It also implies a particular understanding of causation, within a concept of linear time, both of which appear to be common in European societies.[5] Anthropology has shown that many cultures have multiple and non-linear concepts of time. As Visvanathan says, "A tribesperson involved in shifting cultivation operates in a world of over twenty different kinds of time, which emanates from the way s/he deals with soil, seed, seasons, rituals, fast, feast, rest, work, domestic and communal space …" (Visvanathan, 2005, p. 90). For the first author, fieldwork in

Bali was confusing, not so much because of the many difficult calendars in use simultaneously (determining the timing of offerings, markets, temple anniversaries, rice planting, and the like) but because, in telling stories, it became apparent that the storyteller would be sliding from one time-frame to another, but without alerting the audience – e.g. a story that was ostensibly about a pre-colonial dynastic battle, in which agentic krises (daggers) would fly to attack the antagonist "king", would suddenly morph into a story of the Japanese Occupation with bullets emanating from Japanese rifles. Different understandings of time are only part of the problem. Understandings of causation are also culturally diverse. In Bali, for instance, all sorts of animate and inanimate phenomena are believed to have causative agency (Hobart, 1990). Work with primary school children in Bali seemed to support Sweeney's finding: the same difficulty with understanding and expressing causation.[6] Perhaps we should not expect students who are not exposed to the European tradition of philosophy and storytelling as part of their "cultural capital", to easily understand that Action A, such as buying plastic water bottles, will, over time, cause, or contribute to causing, Concomitant B, the clogging of a river. Multiple and intersecting causations, such as untoward upstream deforestation and unenforced government regulation, and further complications, such as the unsustainable use of petrochemicals, might also be incomprehensible. Environmental educators in Indonesia therefore need to be aware of the Eurocentrism of conventional approaches, and make adjustments in their pedagogy so that they can sensitise Indonesian learners to the significance of their everyday actions.

This sort of environmental responsibility implies intention, but it is a "looking back" intention related to the causes of the problem (a liability approach) rather than a forward-looking intention to fix the problem. It shows a great interest in perpetrators rather than solutions. There is research that shows that a sense of environmental responsibility is most significant when a person knows that their action will be efficacious – that is, that their action (such as taking the bus rather than driving a car, writing a letter or signing a petition urging some pro-environmental action) will contribute to an aggregate effect (Eden, 1993). A forward-looking sense of environmental responsibility should inspire learners to "make a difference" through their lifestyle.

Environmental responsibility depends upon a certain level of general knowledge – e.g. of understanding that dropping rubbish might be a health hazard for humans and other life forms – and of cultural context. While the authors could not explain scientifically why the weather today is the way it is, we accept the advice of scientific experts, and have a general knowledge of high and low air pressure, the water cycle, the effect of the local landscape on temperatures and rainfall and so on, even though we are not natural scientists. In Indonesia, we cannot say that people have this sort of general scientific knowledge base. In general, the level of science and mathematics knowledge is abysmal, even for people who have had 12 years of schooling. In the Introduction we gave some examples: students who think that that the greenhouse effect has something to do with the overuse of glass in houses; farmers who have no idea which chemicals they are using on their crops and soil. International tests, such as PISA, "show a poor performance for Indonesian 15-year-olds, with 76% and 67% respectively failing to reach basic proficiency" (OECD/Asian Development Bank, 2015, p. 139). However, the tests in 2015 show some improvement (OECD, 2016a). In Australia, we can say that farmers are smart and looking for ways to adapt to climate change and make their farms more environmentally sustainable. We cannot say that of many farmers in Indonesia.

If it is the usual, normal thing in a society to drop rubbish on the street or in the river, the average individual will not know that it is environmentally deleterious nor think to look for another way to dispose of the rubbish. In such a context, it would seem unfair to label dropping rubbish "irresponsible", because in that context it is not part of everyday knowledge and discourse. Making the effort to carry home one's rubbish or to look for a rubbish bin would be culturally unconventional. In a country such as Indonesia, it could be considered puzzling behaviour and be commented upon. However, in some locales, such as university campuses, as we will see, it might be considered a pro-active instance of environmentally responsible behaviour (ERB). In countries such as Indonesia, where the level of environmental awareness and knowledge is low, there is a crying need for decent EE. However, as we will see, there is also potential for clashing ontologies and epistemologies,[7] as well as a danger of environmentalist imperialism, so this needs to be done sensitively.

On the other hand, in those countries where ER is more normalised, not littering has become internalised by society at large and, for many, dropping rubbish, even a tiny sweet wrapper, on the street, would be not only unthinkable but also physically difficult to do. We have trained our bodies not to litter. In such contexts, environmentalists struggle to understand how other people cannot act responsibly, and generally put it down to ignorance, lack of information or lack of understanding of the significance of things: "People don't realise. If people realised, there would be an outcry" (reported by Eden, 1993, p. 1750). This means that the work of ENGOs (Environmental Non-Government Organisations) often blurs the line between advocacy and EE.

Many studies have established that environmental awareness and knowledge is not enough. ER also entails an ethos of care for the environment. In some countries, the charge has been laid that environmentalists are selectively "caring" – i.e. they are too preoccupied with saving wildernesses and wild places, iconic furry species and the like, and never mention saving spiders and snakes, parasites and microorganisms; others counter that such campaigns identify key issues and iconic species for sound marketing reasons.

Nevertheless, it is impossible to imagine environmental responsibility without some ethos of care for the environment. Some people might know that littering is a problem for the environment or human health, but not care that buying many new plastic water bottles each week, or that dropping a used plastic bottle in a river, is bad for the environment. Their attitude might be cavalier or lazy – why should I go to all the trouble of looking for a rubbish bin? – or assume that their wealth gives them the right to act however they like. Those who are not environmentally responsible might be judged "uncaring, apathetic, lazy, individualistic, greedy, and materialistic" (Eden, 1993, p. 1750) by those who are. This derogatory labelling can reinforce environmentalists' sense of their own agency, the validity of their moral stance and hence their group identity, but does not solve the problem of environmental irresponsibility.

Those who self-identify as environmentalists and are active as such often identify an ethos of environmental care and a sense of moral obligation as essential components of ER. The obligation is to other people, especially future people, and to the planet and the perpetuation of its natural systems. The ethics of these different commitments are debatable – do we owe a duty to unborn generations to preserve the diversity and quality of our planet's environment as it is now, or do we have to make substantial sacrifices so as to restore some ecological balance? Are some people, or some species, more important than others? Or do we acknowledge that human beings are just one of many life forms (or one of the components of the universe, sentient and non-sentient), and that we owe it to the planet, or the universe, to look after this one while we have it in our care (as the dominant, highly destructive, species)?

This last stance is that of Deep Ecology, which can stand here for an ecocentric approach to the environment. Deep Ecology is a philosophy and environmental movement, begun by Norwegian philosopher Arnold Naess (Naess, 1998). It takes as axiomatic that every living thing is intimately connected to everything else, and is valuable in its own right. Therefore Deep Ecology advocates the practice of non-violence to non-human species.[8] Deep Ecology rejects conventional anthro- (or human-) centric approaches, which believe that the world and all its creatures were made for humans; most anthropocentric approaches believe that humans, as superior beings, have mastery (and "softer" versions would say "stewardship") over nature, so the responsibility is how best to manage (or care for) nature. Many environmentalists might belong to the former camp philosophically, but struggle to translate its ideals into action in their daily workaday lives. Others would not agree that humans are superior beings or that the world was made for humans, would agree that humans should care for nature and try to manage the environment wisely, but would struggle to ascribe equal value to humans and gnats.

While personal morality and a caring ethos are cornerstones for some environmentalists, it is always important, as noted above, to consider context. Here we are thinking less of the general society and more of governmental support and infrastructure: if governments do not supply clean, piped water or regular rubbish collections, it is much more difficult for citizens to be environmentally responsible. The little exercise in unblocking the river above, shows the interconnectedness of the relata of responsibility. So governments, just as much as citizens, should be ER, not least in order to enable citizens to act on their awareness, knowledge and care for the environment.[9] We have seen that the perceived efficacy of ERB engenders further ERB, so with the support of government, a "virtuous circle" can be constructed.

The picture of ER outlined above is a liberal understanding of ER. Its presumed subject is a largely autonomous, rational individual, and an understanding of "environment" as something that can be affected by humans and their actions into the future. It is important to remember the diversity of understandings about the relationship between people and the environment – as noted above, some Indigenous peoples and environmentalists who stand toward the Deep Ecology end of a continuum may see that humans are embedded in the environment, while others may see the environment as a separate, manipulable entity.

Most of the academic literature on EE comes from the prosperous West and presumes such a subject. The history of the development of Western liberal democracy is undergirded with the development of a population of autonomous and responsible citizens. Representative government requires a population that is responsible, and to some extent educated. The development of mass education, besides being necessary for the machinery of capitalism, helped to create a responsible, educated citizenry. However, as we will see below, liberal citizenship does not explicitly focus on responsibility (Dobson, 2003, p. 2): the philosophical base is one of rights, not responsibilities. We use the term liberal environmental responsibility first to emphasise the components that characterise liberal environmental education – environmental awareness, knowledge, values and a care ethos – and second in order to highlight the contrast with the understanding of responsibility (actually responsibilisation) under conditions of environmentality.

Responsibilisation under environmentality

Here we shift to neoliberal discourses, and reference a different, critical literature around the concept of governmentality. "Governmentality" emerged in Foucault's later writings and was developed by other researchers, such as Dean (1999) and Rose (1999). Governmentality refers to something much more pervasive and all-encompassing than "governance" or "government": "it becomes an interconnection of the government of the self, the government of others and the government of the state, which Foucault (2007) termed the 'conduct of conduct'" (Fejes, 2010, p. 90). Where once there was legislation, the bureaucracy, the judiciary and penal system, repression and such, there is now "self-governance", by which Foucault means that, because there is an educated, autonomous subject citizen, and an enabling government apparatus, "governing is conducted by the citizens themselves" (Fejes, 2010, p. 92).

One relevant strand of governmentality is "green governmentality", which begins with the objectification of nature to render it manageable.

> Corporations, all levels of government, scientists, United Nations organizations, and global think tanks ... work to produce knowledge about the earth and its resources that cast it as manageable ... Within this discursive regime we must also situate the environmental organizations, which shape the truth about nature, and seek to regulate and ameliorate its (ab)use....

[S]ome previously state-based responsibilities have been shunted onto the market, quasi-private sector or the nebulous catchall, civil society. In this vein, we see the birth of public-private partnerships, where non-governmental organizations, corporations and the state work together to manage the "problems" of society.

(Rutherford, 2007, p. 295)

Foucault conceptualised power not just as the power to say no, to discipline or to repress, but as a positive, constructive force that is pervasive and naturalised through institutions:

What makes power hold good, what makes it accepted, is simply the fact that it doesn't only weigh on us as a force that says no, but that it traverses and produces things, it induces pleasure, forms knowledge, produces discourse.

(Foucault, 1980, p. 119)

Although notoriously uninterested in the natural world, Foucault's notion of "biopower", or the power to regulate and control entire human populations, from conception to death, can be extended through green governmentality to the management of all life on earth. A third component of his work is the way governmentality works upon the human subject to create normalised, desirable citizens. This, again, should not be seen necessarily as a negative or repressive process, but as the constructive creation of often self-disciplining, self-controlling subjectivities (Foucault, 1983).

Thus, in today's neoliberal world, the state is in some ways in retreat, but can simultaneously be seen to be powerful in its creation of complicit citizens:

Nikolas Rose (1999) has called the contemporary state the "enabling state". An important aspect of such a state is providing the opportunity (enabling) for citizens to make choices in accordance with her/his wishes and desires....

(Fejer, 2010, p. 92)

Fejer is writing of the neoliberal subject who is unemployed, but positioned as "employable", if s/he constructs her/himself as flexible, mobile and adaptable:

[I]t is still the individual who is positioned as responsible for becoming adaptable and flexible as a way to become/remain employable. One could say that there is a responsibilisation of the individual.... The individual needs to take responsibility for using the opportunities for lifelong learning, by means of education and in-service training, offered by the state and the market, thus transforming her/himself into an employable person. The role of the state is then more distanced than was previously the case.... Now, structures for supporting the individual in her/his own choice are created instead of collectively planning the future by means of legislative measures and regulations.

(Fejes, 2010, p. 95)

One could say that the role of the state is redefined from being a distributor of resources to offering services ..., or as Rose (1999) argues, there has been a shift from a social state to an enabling state, where the state should make it possible for the citizen to make active choices.

(Fejes, 2010, p. 100)

Scholars have pointed out that contemporary neoliberal discourse is particularly fond of "responsibilisation". With its ideal of the shrinking state, drive towards deregulation and privatisation, and belief in the market to govern all aspects of life, neoliberalism particularly fosters an emphasis on personal choice and "freedom", and therefore, ironically, individuals are "made responsible". In neoliberalism, personal choice and autonomy are the "means through which responsibility is enacted" (Trnka & Trundle, 2014, p. 3). Self-responsibility is an important aspect of the neoliberal subject-citizen (Miller & Rose, 2008; Rose, 2007).

Neoliberalism is not an ideology that sees individuals as socially situated – rather, they are seen as autonomous actors making choices that determine their lives. Thus, there is a discourse about "flexible" citizens, who adapt to changing conditions but also are responsible for their own success or failure in life; the "changing conditions" might be the neoliberal retreat from social welfare; and the newly flexible citizens might have to lose some autonomy as they become more responsible. An example might help to show the rather perverse and contradictory discourse and sometimes competing discourses.

Recently, the Australian government made a so-called "emergency" Intervention in Indigenous communities in the Northern Territory (NT). NT Government reports apparently revealed that liberal policies of self-determination were failing, and that Indigenous communities were rife with domestic violence and social dysfunction. The government stepped in, linking welfare payments to school attendance, income use, family life and employment. The idea was not to create new dependencies but to shift responsibility to Indigenous people themselves, based on a model of a school-attending, non-alcohol-drinking, disciplined, ideal citizen. Thus, there were no qualms from the government about "more government" or the curtailment of individual freedoms and autonomy when such curtailments were held to enable the development of more valued, more "responsible" forms of "autonomy", albeit within a dense web of governmental structures and regulations (Trnka & Trundle, 2014). A host of evaluation reports has shown that "the Intervention" not only curtailed autonomy and was perceived by recipients as "unfair, embarrassing and discriminatory" (Bray, Gray, Hand, & Katz, 2014, p. xxi) but also "encouraged increasing dependence upon the welfare system" (Bray *et al.*, 2014, p. xxii) while not improving domestic violence rates, school non-attendance rates or other key social dysfunction indices (e.g. suicide and incarceration rates).

Agrawal introduced the concept of "environmentality", focusing on Foucauldian governmentality rather than neoliberal responsibilisation (Agrawal, 2005). He argued that in northern India, participation in a plethora of government-imposed programs focused on environmental protection was the key variable in transforming people into "environmental subjects" in rural populations. Acciaioli criticised Agrawal's environmentality framework, arguing that its Foucauldian emphasis upon governmentality in the environmental context did not sufficiently consider local peoples' agency in their decisions to accept or reject such attitudes (Acciaioli, 2008). In his account, Lindu people in Sulawesi, Indonesia, accepted conservation initiatives only when they could harness them to their own local projects of controlling migrants. Cepek also found that the Cofán people of Colombia only accepted scientific conservation when they received political and economic resources from such interventions (Cepek, 2011). These studies demonstrate that any inculcation of environmentalist attitudes must take into account not only objective ecological problems and government programs to deal with these, but also the interests and initiatives of local people. In the Indonesia case, there was not the dense net of environmental protection regulations and subjection to government discourse that pertained in northern India; nor were local interests homologous with environmental conservation objectives.

In short, for Indonesia generally, the context for environmentality approaches to EE is not amenable. Indonesia is a "soft" state, with a massive population, an archipelagic geography and a sprawling bureaucracy.[10] It is hard to imagine that the civil service and education system could combine nationally to produce strong governmentality directed toward environmental sustainability and the construction of environmental subjects. However, decentralisation and a forceful, environmentally committed mayor might present the exception that proves the rule: in Chapters 9 and 10 we will consider the city of Surabaya. There is a lesson in the environmentality approach: like conservation and sustainable development projects, EE must build on understanding of how local communities deal with their natural environment, the socio-political context and local economic imperatives, including livelihoods.

Responsibility for justice

We stated in Chapter 1 that Young's social connections theory of global justice would form the philosophical foundation for this book. The social connection model of responsibility says that all people who contribute to structural social processes that produce injustice have responsibilities to remedy the injustice. Although it is a theory of social justice, it can be borrowed to address questions of responsibility for environmental harms and injustice. It is a theory that is concerned about structural injustice brought about by transnational, institutional and personal relationships that are liable to cause conflict and inequalities of power.

The "structure" is not a stationary entity, facilitating or restricting action: it is a process of repeat, large-scale action or everyday practice (called "structuration" by Giddens (1984)). The everyday act of consumption is shaped by complex cultural and economic processes, e.g. of "fast fashion", overconsumption, status competition, gender norms and so on – at both ends of the production line and along the way. Scholars should enquire into these structures: why are women and not men employed in clothing sweatshops and the clothing departments of fashion stores, but only men are employed to chainsaw trees, staff the sawmills, drive the trucks and crew the ships that transport the goods across the oceans? Why did that consumer buy yet another new dress or replace their perfectly good old dining suite? Why are virgin rainforest trees and not sustainably grown plantation timbers used? Why are cabinetmakers in Java paid so much less than those in the Global North? People act according to the norms and conventions they have internalised through socialisation within the family, and later through schooling and social life in communities, engagement with mass media and popular culture, and eventually through conditions of employment and involvement with large entities such as legal systems, transnational companies and the nation-state. This does not mean that they are automatons or puppets. As Giddens, Butler, Scott and many others have noted, daily acts bring opportunities for "ad libbing" – for innovation and revision, for questioning and assessment – such that a better word than "structures" is "structuration", in order the better to suggest a process rather than an immovable object.

Structural injustices are moral wrongs that occur when large categories of people are rendered powerless, unequal or subordinate, unable to develop or exercise their capacities by large processes that simultaneously enable others to dominate or exploit opportunities or act in pursuit of their own interests. Environmental injustices are often also social injustices; but some environmental harms are only harms to natural ecosystems. All participants in such processes are connected by virtue of their participation, whether they know it or not, and whether they intend it or not. Thus, people across the world are connected and have different responsibilities with regard to structural injustices and environmental harms.

According to Young, the social connection theory of responsibility does not aim to isolate and punish individual wrong-doers like a conventional justice system. It acknowledges that individual perpetrators may be personally liable for specific wrongful actions, such as cutting down trees, but it is the large structural processes that produce harms and injustices – the loss of oxygen-creating, biodiversity-producing, erosion-preventing forests, the dispossession of marginalised people who once called the rainforest home, of long supply lines from source to consumption that in turn create pollution and disrupt natural ecosystems. Individuals have mediated relationships to the production and reproduction of structural injustices, and this means that background conditions – the rules and conventions of society, the roles of institutions, the everyday practices – should come in for scrutiny in order to ask about responsibility for harms.

While all "responsibility" theories look both forward and backward, as we have discussed, the liability model is most concerned with identifying perpetrators and apportioning blame or guilt. The social connection model is more forward-looking, being concerned with the ongoing process and with devising interventions that will disrupt the harm-causing process. The point is to "enjoin those who participate by their actions in the process of collective action to change it" (Young, 2006, p. 122). This can be the heart of environmental education programmes. There is an emphasis on shared, collective action by those involved in an injustice-causing social process in order to reform unjust structures. This usually means taking political responsibility, because power is involved; and both those who are perpetrators and those who are victims of injustices are implicated in such political reform. Since most of us are involved in multiple structural processes, and are positioned differently in each, it is probably not possible to discharge all our responsibilities. We must use criteria – such as which action would be most effective, which positioning gives us most leverage, which injustice is most urgent – in order to best use opportunities to ameliorate harms and change structural processes. Young suggests we look at the parameters of power, privilege, interest and collective ability (Young, 2006, pp. 127–130).

There are points of issue arising from this theory, and Young herself addresses some of these in the later chapters of her book (Young, 2011). In a context such as Indonesia, many harms are both invisible and unknown. As discussed above, the level of general environmental awareness and knowledge is low, and that makes it difficult when one is trying to alert students, for instance, to environmental injustices. It is obvious that environmental responsibility must encompass social responsibility, and that both are types of moral responsibility. While Young's formulation is positive and constructive, there are still questions. How important is responsibility for historical injustices (of colonialism, of dispossession)? How do aware people share responsibility? How can we act responsibly (ethically) when larger harms and unfair processes are invisible and unknown?

And yet there are good reasons for optimism, and promising entry points for the mobilisation of young people to take responsibility for environmental justice. In Indonesia, young people are optimistic and idealistic, and they are keen to work for social justice (Nilan, Parker, Bennett, & Robinson, 2011, pp. 724–725; Parker & Nilan, 2013). Conflicts over natural resources are often presented as ethnic and religious conflicts, so while environmental awareness is low (and will be discussed in detail in Chapter 11), social justice issues and minority disadvantage are on everyone's radar. Access to schooling has improved dramatically, and this is the best-educated generation in Indonesia. While the level of science knowledge in the Indonesian community is abysmally low, student scores in science in the international PISA tests are at least improving (OECD, 2016a, 2016b). Interestingly, there is not a discourse of embittered anti-colonialism or environmental finger-pointing at "advanced, post-industrial" countries in Indonesia. It is hard to know if this is because people do not know that the Industrial Revolution was, and still is, so destructive environmentally, or because they do not understand the rapacity of the international capitalist system (Klein, 2014).[11]

Another promising point is that in Indonesia,

> Collective rather than individual concerns are still salient for young Indonesians....
>
> In Indonesian cultures, emphasis on family and kin interdependence and relatedness still prevails.
>
> (Nilan *et al.*, 2011, p. 715)

"Collective" probably most often refers to family and extended family, but sometimes also it references clans, lineages, tribes and ethnic identities – there are many different kinship systems in Indonesia. On a global scale, and more metaphorically, Haraway (2015, p. 161) exhorts us to "make kin", to work towards "a multispecies ecojustice". Since we want to be talking about collective responsibilities, multiple complex relationships between things and species and humans, imagined relationships and obligations, and connections – E.M. Forster's "only connect" – we must also talk about the nation-state and the nature of people's belonging to Indonesia. So we next turn to understandings of citizenship.

Environmental citizenship

Environment Canada has helpfully put out a rousing call for environmental citizenship:

> Each of us has an effect on the environment every day; the key is to make this impact a positive one. We must all take responsibility for our own actions, whether as individuals, or as members of a community or an organization. Let's work together and become good Environmental Citizens! If you don't, who will? (Environment Canada, 2004).
>
> (Bell, 2005, p. 179)

We argue that in the context of Indonesia, the concept of responsible environmental citizenship will be useful in EE. As the quote from Canada shows, the concepts of responsibility and citizenship are linked, and this is particularly the case in Indonesia. In Indonesian schools, it is common to see posters of school rules on notice-boards where one side of the poster has a list of (a few) rights (*hak*) and the other side a list of (many) responsibilities or obligations (*kewajiban*). Often the list of obligations is accompanied with a list of the penalties incurred by infringements (e.g. credit points deducted, detentions, meetings with parents and, the most extreme penalty, expulsion). School children are heavily socialised into a regime of "responsibilities", such as sweeping the classroom and picking up litter, and these are typically organised in teams, that take turns, and the roster is posted in classrooms. We argue that EE that appeals to students' sense of responsibility, as members of the school community, and as future citizens of larger communities, will be an effective starting block for sensitising young Indonesians to environmental problems, and for mobilising them to pro-environment practice.

Citizenship is a large topic in political philosophy and here we will only point to some ideas and issues that are pertinent to this book. Conventionally, citizenship is understood as a legal status, an administrative category and a political practice that accounts for one's membership, rights and obligations to a political community (Stokes, 2008). It is conventional to describe two main approaches to citizenship in political philosophy: the liberal and civic republican approaches. Western liberal theorists emphasise rights, and posit that citizens are entitled to universal rights granted and guaranteed by the state. Theoretically, all citizens have the right to public participation – such as in democratic and representative government.

> Liberal environmental citizenship emphasises the importance of rights in safeguarding the environment. Bell (2005) has argued, for instance, that insofar as the environment (e.g. water) is central to human well-being, people have a right to a life-giving environment (e.g. a right to water) and a corresponding duty to comply with laws designed to protect this right.
>
> (Baldwin, 2012, pp. 626–627)

For our case of Indonesia, with its blocked and polluted rivers, and often unbreathable air, one could argue that the state is failing to guarantee its citizens' right to water and air – the right to a life-sustaining environment.

Civic republican theories of citizenship downplay rights and instead focus on a common political culture, where citizens have the obligation or responsibility to participate in communal affairs. Emerging out of debate among common interest groups, which gives way to collective agreement, a notion of the common good prevails.

> [C]ivic republican environmental citizenship emphasises the citizenly duty to uphold certain virtues in the interest of environmental and common well-being. Barry (2006), for instance, argues that liberalism has failed to safeguard the environment through its system of substantive and procedural rights, and thus advocates a system of compulsory actions designed to safeguard the life-giving capacity of the environment to be taken by individuals and enforced by the state.
>
> (Baldwin, 2012, p. 627)

Of course there are many other perspectives that emphasise different aspects of the state–citizen relationship – some downplay the individual and play up the idea that individuals are embedded in groups, that the collective good is prioritised over the pursuit of individual interests. This sort of citizenship is applicable in community-based natural resource management projects, where local or Indigenous knowledge and traditions of a shared commons still prevail or are resuscitated. The discourse of citizenship has evolved and extended into areas beyond the legal and political, to include civil, cultural, social and environmental areas (e.g. Marshall, 1950; Ong, 1999; Rosaldo, 1994). Citizenship is more than a legal status or administrative category and includes political practice, a mode of participation in public life, recognition of colonised and marginal groups, and affective belonging.

Theorists of environmental citizenship have often followed the distinction between liberal and civic republican citizenship in exploring how "the environment" can be incorporated into the "institutions and practices" of citizenship (e.g. Dobson & Bell, 2006; Latta & Wittman, 2012, p. 4). An important departure was made by Dobson in his 2003 chapter on "ecological citizenship", which he distinguished from environmental citizenship as complementary. He pointed to the failure of the liberal environmental citizenship model to uphold environmental rights against the state. This is evident in Indonesia, where the people's right to clean air and water is not upheld in practice by successive and multilevel governments. He accepts that the liberal rights-based movement has played a significant role in the environmental movement (everyone has the right to an environment adequate for their health and well-being) that tapped into the civil rights movement. But he builds on this: first by accepting that citizenship must extend to environmental issues, then by declaring that a global or cosmopolitan scale is necessary, he rejects the "reciprocal" rights and obligations of liberal cosmopolitan citizenship, and stresses unequal obligations and responsibilities rather than rights. He argues that we can have obligations for the effects of our actions even when those affected cannot easily be included in our immediate political community – in other words, that environmental rights, responsibilities and impacts are not always reciprocal, and they are potentially global.

> Ecological citizenship deals in the currency of non-contractual responsibility, it inhabits the private as well as the public sphere, it refers to the source rather than the nature of responsibility to determine what count as citizenship virtues, it works with the language of virtue, and it is explicitly non-territorial.
>
> (Dobson, 2003, Chapter 3, p. 6/47)

Harking back to Aristotle, he identifies that virtue is central to citizenship and that one of the chief virtues of ecological citizenship, if not the first virtue, must be justice. He calls this a post-cosmopolitan theory of justice,[12] and deploys the concept of the ecological footprint to argue that some members of some countries occupy more than their fair share of ecological "space", therefore having a damaging effect on the life chances of some other members of their own country as well as members of less ecologically expansive countries. His theory is homologous with that of Young, although coming from a different tradition and using different language. He also departs from liberal citizenship in nominating the private sphere as a legitimate site for citizenship activity (thus breaching one of the more sacred lines of liberal citizenship theory), noting that it is the private sphere that generates the ecological footprint that gives rise to the obligations of ecological citizenship. While this is arguable – we do not want to let global capitalism off the hook – it is not worth arguing about proportional responsibility between public and private: the more important point is the feminist one of dissolving the private and public spheres.

Dobson's theory of ecological citizenship is one of morality and justice (Dobson, 2003, pp. 259–260), and differs from environmental citizenship and other "emancipation movements" in its notions of responsibility rather than rights. A relevant question for us in the Indonesia context – a question that arises from both Young's and Dobson's theories – is the question of national citizenship. If ecological citizenship and justice responsibilities are non-territorial and extend beyond our immediate political community, what role is there for the nation-state? Here we refer back to McGregor and her point that she presents her theorising on feminist ecological citizenship in tandem with empirical research from "real life", producing "messier", but hopefully more useful, narratives (MacGregor, 2009, pp. 292–293). This book might exhibit "methodological nationalism" (Wimmer & Schiller, 2003) in its focus on citizenship in the nation-state of Indonesia, but the reason is that it emanates from "real life" – the reality of the education system in Indonesia. As we will see in following chapters, in the Indonesia context, education is inextricably entwined with citizenship in the nation-state. The challenge will be to present the responsibilities of (post-cosmopolitan) ecological citizenship and justice within the frame of the Indonesian nation-state. The obligations incurred by injustice are obligations to strangers, both within and without the nation's borders, and both past and future generations. We have to "make kin" of contemporary and future strangers.

Notes

1 It is not essential for those wanting only to know about the situation in Indonesia to read this chapter, but those who want to understand more deeply the reasons for our approach should read this chapter.

2 We elaborate on the education system in Chapter 5.

3 The "absences" include a well-educated teacher cohort, a scientifically literate population, a generalised environmental sensibility among the population, a "green" political party or at least pro-environmental presence in political life, and an environmentally well-informed media. Unfortunately, the latter element is not yet researched for Indonesia.

4 The water supply and sewerage system of Jakarta is notorious. Flooding occurs every wet season, causing the city to become even more dysfunctional than usual. Only 2 per cent of the population can access the centralised sewage collection system, and more than six million people use wells to access (contaminated) groundwater (Furlong & Kooy, 2017, p. 895). Those who live in high-rise apartments, businesses and larger factories access clean water from the deep aquifer. This example shows how environmental resources are unequally shared.

5 Perhaps controversially, the socio-linguist Sweeney suggests that in the Malay world, the oral narrative tradition shows a strong "adding-on" mode of communication (e.g. "and then" is a common way to establish continuity) and a lack of understanding of causation. Words of causation, such as "*apa sebab*" (why, what is the reason) and "*karena*" (because) are not commonly used (Sweeney, 1987, chapters 7 and 8), and

when his university students attempted to use words of causation such as "*oleh sebab*" (because of), "*jadi*" (so) and "*karena*" (because), they produced fractured sentences that did not subordinate or link logically one clause to another. The "and-then" mode was not suited to an analytical approach: the student wanted to memorise intact chunks and string them along in the correct order, not to analyse the meaning prior to construction and subordinate one sentence to another or compare or juxtapose sentences or clauses.

6 In doctoral fieldwork in schools in Bali, I conducted experiments in "creative writing" (Parker, 2003, pp. 240–241).

This opening sentence from a sixth grade student's composition about "The Island of Bali" is typical:

> The island of Bali is the island of the gods, what is the reason it is called the island of the gods, because of the many tourists from overseas as well as from within the country.
>
> (Composition of 4 February 1981)

... All the children who used the phrase "island of the gods" had trouble explaining it, yet 21 of the 24 pupils in the class used it. They knew the formulation and knew they could safely use it in their compositions. The pupils did not understand its meaning (i.e. the pervasive influence of religion in all aspects of life in Bali) nor the nature of its connection with the tourist trade. Also, they either did not realise that they did not understand it or they considered that it did not matter. As I expected, I found the phrase in the Social Science textbook, albeit the textbook for fourth grade:

> Children, the island of Bali is also called the island of the Gods. As a tourist place because the views of nature are beautiful. [It is] famous throughout the world.
>
> (Mugiyana *et al.* 1975, vol. 4A, p. 11; Parker's translation)

Sweeney linked the absence of causative thinking to the aim of learning in Malay universities:

> The aim of a pupil in an oral milieu is to acquire the sum total of his [*sic*] teacher's knowledge and to preserve it intact in his mind. His task is not to confront or argue with his teacher, for such activities are incompatible with the oral transmission of knowledge.
>
> (Sweeney, 1987, p. 269)

7 Lowe's work on conservation discourses in the Togean Islands of Central Sulawesi is a highly perceptive analysis of the attempts by Indonesian scientists to establish a national park that satisfied the "pure" conservation ethos of international ENGOs, the Indonesian bureaucratic-political context and the understandings of the local Sama people (Lowe, 2006). Fischer (2005), and other authors in the edited book, *Science and Citizens* (Leach, Scoones, & Wynne, 2005), provide good examples of the clash of ontologies and epistemologies when science is introduced to those "without knowledge".

8 While this moral concern and commitment is laudable, it is also important to remember the critique of scholars, such as Fletcher (2009), Guha (1989) and Lohmann (1993), who mount moral as well as discursive arguments against the value of wilderness, protected areas and Deep Ecology. We have here competing or contested moralities.

9 Business should also have an important role to play, and academics have studied "corporate social responsibility" in relation to environmental responsibility (e.g. Sevick Bortree *et al.*, 2013). Another interesting angle on corporate environmental responsibility is the work of ecological economists who look at chains of upstream responsibility, e.g. sourcing and procurement practices, transport and infrastructure;

downstream responsibility, e.g. uses to which products are put, effects of products on human health and environmental impact; and "shared responsibility" (e.g. Lenzen & Murray, 2010).

10 The terms "hard state" and "soft state" were introduced by the economist Gunnar Myrdal (1968). His idea was that in developing, postcolonial nations the traditional authorities and structures had been undermined by colonisation, leaving behind "soft states" that were characterised by social indiscipline and ineffective bureaucracies that were susceptible to corruption. Ironically, in view of Agrawal's claim, Myrdal's classic case was India, where, he argued, British rule had failed to establish long-lasting structures of authority and law and order to replace the pre-colonial structures that they had demolished.

11 In some countries, such as some of the countries of Latin America, Malaysia and India, the postcolonial discourse of bitter blame-laying at the feet of colonial masters is much stronger than in Indonesia. In contemporary Indonesia, anti-Dutch feeling is not noticeable. However, there is a strong anti-Western discourse. This often takes the form of a moral discourse, where Indonesia takes the moral high ground as an "Eastern" country, and critiques the morally degenerate "West", for instance, in magazines for young women, sermons from mosques, and teachers' discourse in schools (Bellows, 2003; Harding, 2008; Parker, 2013).

12 Baldwin explains the difference between a cosmopolitan and post-cosmopolitan view:

> If cosmopolitanism holds that global citizens have reciprocal obligations, non-reciprocity is the idea that not all members of the global community are equal in their capacity to fulfil these obligations and so should not be held to the universal principle of mutual obligation. In ecological terms, this means that those with disproportionately large ecological footprints have an obligation to those whose ecological spaces are diminished by them.
>
> (Baldwin, 2012, p. 627)

References

Acciaioli, G. (2008). Environmentality reconsidered: Indigenous to Lindu conservation strategies and the reclaiming of the commons in Central Sulawesi, Indonesia. In M. Galvin & T. Haller (Eds), *People, Protected Areas & Global Change: Participatory Conservation in Latin America, Africa, Asia & Europe* (pp. 401–430). Bern: University of Bern (NCCR North-South).

Agrawal, A. (2005). *Environmentality: Technologies of Government and the Making of Subjects*. Durham: Duke University Press.

Baldwin, A. (2012). Orientalising environmental citizenship: Climate change, migration and the potentiality of race. *Citizenship Studies*, 15(5), 625–640.

Barry, J. (2006). Resistance is fertile: From environmental to sustainability citizenship. In A. Dobson & D. Bell (Eds), *Environmental Citizenship* (pp. 21–48). Cambridge, MA: MIT Press.

Bell, D. (2005). Liberal environmental citizenship. *Environmental Politics*, 14(2), 179–194.

Bellows, L. J. (2003). "Like the West": New sexual practices and modern threats to Balinese-ness. *Review of Indonesian and Malaysian Affairs*, 37(1), 71–104.

Bray, J. R., Gray, M., Hand, K., & Katz, I. (2014). *Evaluating New Income Management in the Northern Territory: Final Evaluation Report*. Retrieved from Social Policy Research Centre, UNSW, Sydney, Australia.

Cepek, M. L. (2011). Foucault in the forest: Questioning Environmentality in Amazonia. *American Ethnologist*, 38(3), 501–515.

Dean, M. (1999). *Governmentality: Power and Rule in Modern Society*. London: Sage Publications.

Dobson, A. (2003). *Citizenship and the Environment*. Oxford: Oxford University Press.

Dobson, A., & Bell, D. (Eds). (2006). *Environmental Citizenship*. Cambridge, MA: MIT Press.

Eden, S. E. (1993). Individual environmental responsibility and its role in public environmentalism. *Environment and Planning A, 25*(12), 1743–1758.

Environment Canada. (2004). An environmental citizen … who me? Retrieved from www.ns.ec.gc.ca/udo/who.html (accessed 10 January 2005).

Fejes, A. (2010). Discourses on employability: Constituting the responsible citizen. *Studies in Continuing Education, 32*(2), 89–102.

Fischer, F. (2005). Are scientists irrational? Risk assessment in practical reason. In M. Leach, I. Scoones, & B. Wynne (Eds), *Science and Citizens: Globalization and the Challenge of Engagement* (pp. 54–65). London and New York: Zed Books.

Fletcher, R. (2009). Against wilderness. *Green Theory & Praxis: The Journal of Ecopedagogy, 5*(1), 169–179. doi:0.3903/gtp.2009.1.12.

Foucault, M. (1980). *Power/Knowledge. Selected Interviews and Other Writings 1972–1977* (C. Gordon, L. Marshall, J. Mepham, & K. Soper, Trans.). Brighton, Sussex: The Harvester Press.

Foucault, M. (1983). Afterword: The subject and power. In H. L. Dreyfus & P. Rabinow (Eds), *Michel Foucault: Beyond Structuralism and Hermeneutics* (2nd edn, pp. 208–226). Chicago: University of Chicago Press.

Furlong, K., & Kooy, M. (2017). Worlding water supply: Thinking beyond the network in Jakarta. *International Journal of Regional and Urban Research, 41*(6), 888–903. doi:10.1111/1468-2427.12582.

Giddens, A. (1984). *The Constitution of Society*. Berkeley: University of California Press.

Guha, R. (1989). Radical American environmentalism and wilderness preservation: A third world critique. *Environmental Ethics, 11*, 71–83.

Haraway, D. (2015). Anthropocene, capitalocene, plantationocene, cthulucene: Making kin. *Environmental Humanities, 6*, 159–165.

Harding, C. (2008). The influence of the "Decadent West": Discourses of the mass media on youth sexuality in Indonesia. *Intersections: Gender, History & Culture in the Asian Context, 18*.

Hobart, M. (1990). The patience of plants: A note on agency in Bali. *Review of Indonesian and Malaysian Affairs, 24*, 90–135.

Klein, N. (2014). *This Changes Everything: Capitalism vs. The Climate*. London: Penguin Books.

Latta, A., & Wittman, H. (2012). Citizens, society and nature: Sites of inquiry, points of departure. In A. Latta & H. Wittman (Eds), *Environment and Citizenship in Latin America: Natures, Subjects and Struggles* (pp. 1–23). New York: Bergahn Books.

Leach, M., Scoones, I., & Wynne, B. (Eds). (2005). *Science and Citizens: Globalization and the Challenge of Engagement*. London and New York: Zed Books.

Lenzen, M., & Murray, J. (2010). Conceptualising environmental responsibility. *Ecological Economics, 70*, 261–270.

Lohmann, L. (1993). *Green Orientalism*. The Corner House. Retrieved from www.thecornerhouse.org.uk/item.shtml?x=52179.

Lowe, C. (2006). *Wild Profusion. Biodiversity Conservation in an Indonesian Archipelago*. Princeton and Oxford: Princeton University Press.

MacGregor, S. (2009). The project of feminist ecological citizenship. In M. Reynolds, C. Blackmore, & M. J. Smith (Eds), *The Environmental Responsibility Reader* (pp. 292–301). London and New York; Milton Keynes: Zed Books Ltd. and The Open University.

Marshall, T. H. (1950). *Citizenship and Social Class: And Other Essays*. Cambridge, UK: Cambridge University Press.

Miller, P., & Rose, N. (2008). *Governing the Present: Administering Economic, Social and Personal Life*. Malden, MA: Polity Press.

Mugiyana, B. A. (1975) *Ilmu Pengetahuan Sosial untuk Sekolah Dasar* [Social Science for Primary School], vols. 1A-6B. Solo: Tiga Serangkai.

Myrdal, G. (1968). *Asian Drama. An Inquiry into the Poverty of Nations*. New York: Pantheon.

Naess, A. (1998). The deep ecological movement: Some philosophical perspectives. In S. J. Armstrong & R. G. Botzler (Eds), *Environmental Ethics: Divergence and Convergence* (2nd edn, pp. 437–449). Boston, MA: McGraw-Hill.

Nilan, P., Parker, L., Bennett, L., & Robinson, K. (2011). Indonesian youth looking towards the future. *Journal of Youth Studies*, *14*(6), 709–728.

OECD. (2016a). *Program for International Student Assessment (PISA) Results from 2015: Indonesia*. Retrieved from www.oecd.org/education/pisa-2015-results-volume-i-978926 4266490-en.htm.

OECD. (2016b). Summary description of the seven levels of proficiency in science in PISA 2015. Retrieved from www.oecd.org/pisa/test/summary-description-seven-levels-of-proficiency-science-pisa-2015.htm.

OECD/Asian Development Bank. (2015). *Education in Indonesia: Rising to the Challenge*. Retrieved from OECD, Paris.

Ong, A. (1999). Cultural citizenship as subject making: Immigrants negotiate racial and cultural boundaries in the United States. In R. D. Torres, L. F. Miron, & J. X. Inda (Eds), *Race, Identity and Citizenship: A Reader* (pp. 262–293). Malden and Oxford: Blackwell Publishing.

Parker, L. (2003). *From Subjects to Citizens: Balinese Villagers in the Indonesian Nation-state*. Copenhagen: Nordic Institute of Asian Studies Press.

Parker, L. (2013). The moral panic about the socializing of young people in Minangkabau. *Wacana, Journal of the Humanities of Indonesia*, *15*(1), 19–40.

Parker, L., & Nilan, P. (2013). *Adolescents in Contemporary Indonesia*. New York: Routledge.

Reynolds, M. (2009). Introduction to environmental responsibility. In M. Reynolds, C. Blackmore, & M. J. Smith (Eds), *The Environmental Responsibility Reader* (pp. 1–6). London: Zed Books.

Rosaldo, R. (1994). Cultural citizenship in San Jose, California. *PoLAR: Political and Legal Anthropology Review*, *17*(2), 57–64.

Rose, N. (1999). *Powers of Freedom: Reframing Political Thought*. Cambridge: Cambridge University Press.

Rose, N. (2007). *The Politics of Life Itself: Biomedicine, Power and Subjectivity in the Twenty-first Century*. Princeton, NJ: Princeton University Press.

Rutherford, S. (2007). Green governmentality: Insights and opportunities in the study of nature's rule. *Progress in Human Geography*, *31*(3), 291–307.

Schicktanz, S., & Schweda, M. (2012). The diversity of responsibility: The value of explication and pluralization. *Medicine Studies*, *3*(3), 131–145.

Sevick Bortree, D., Ahern, L., Nutter Smith, A., & Dou, X. (2013). Framing environmental responsibility: 30 years of CSR messages in National Geographic Magazine. *Public Relations Review, 39,* 491–496.

Stokes, G. (2008). Towards a conceptual framework for citizenship. In A. Azra & W. Hudson (Eds), *Islam beyond Conflict: Indonesian Islam and Western Political Theory* (pp. 85–92). Aldershot: Ashgate.

Sweeney, A. (1987). *A Full Hearing: Orality and Literacy in the Malay World.* Berkeley: University of California Press.

Trnka, S., & Trundle, C. (2014). Competing responsibilities: Moving beyond neoliberal responsibilisation. *Anthropological Forum, 24*(2), 136–153.

Visvanathan, S. (2005). Knowledge, justice and democracy. In M. Leach, I. Scoones, & B. Wynne (Eds), *Science and Citizens: Globalization and the Challenge of Engagement* (pp. 83–96). London and New York: Zed Books.

Wimmer, A., & Schiller, N. G. (2003). Methodological nationalism, the social sciences, and the study of migration: An essay in historical epistemology. *Transnational Migration: International Perspectives, 37*(3), 576–610.

Young, I. M. (2006). Responsibility and global justice: A social connection model. *Social Philosophy & Policy Foundation, 23*(1), 102–130. doi:10.1017/S0265052506060043.

Young, I. M. (2011). *Responsibility for Justice.* Oxford and New York: Oxford University Press.

3 Introducing environmental education

> The international community is in wide agreement that education has an enormously important role to play in motivating and empowering citizens to participate in environmental improvement and protection.
>
> (Fien & Corcoran, 1996, p. 227)

This chapter provides an overview of the sub-discipline "Environmental Education" (EE). Despite contestations, and its rather amorphous character, Stevenson *et al.* have usefully identified five characteristics of EE: it embraces normative (or value-laden) questions; it is interdisciplinary (people-society-environment); it is concerned not only with knowledge and understanding and attitudes and values, but also with developing the agency of learners in participating and taking action on environmental issues; it takes place in formal, non-formal and informal education settings; and it has both a global and local orientation (Stevenson, Wals, Dillon, & Brody, 2013, p. 2).

While EE is always evolving, and contested, here we try to distil the most relevant themes and trends that characterise it. Despite the UN's Decade of Education for Sustainable Development (2005–2014), there is a real lacuna in our knowledge of environmental attitudes and knowledge, and of efforts toward pro-environment education, in non-Western, Global South countries (N. Gough, 2008 [2003]). Apologies in advance for the Eurocentrism and "rich country" biases in this characterisation of EE.

The chapter introduces the academic literature on EE and identifies some of the salient issues in EE: the gap between theory and practice; the processes involved in effective EE; the location of EE in formal school curricula; issues about pedagogy and the type of knowledge being transmitted; how knowledge is not enough but is experiential EE the answer?

Further, the chapter examines the question of the validity of EE in schools. Many argue that as schools are part of the system that created the world's environmental problems, we cannot expect them to deliver education that must transform the larger structures of inequality and injustice that produce environmental destruction. The question is: can schools produce responsible environmental citizens? Our response is simply that they must, because the problems are so urgent. We echo Pinar *et al.*'s (1995, p. 841) call "for education to accept full responsibility in addressing global survival issues".

EE is widely acknowledged to have begun with the Tbilisi Declaration of 1977. The Declaration was written after the first UNESCO Intergovernmental Conference on Environmental Education in Tbilisi in 1977, and UNESCO has played a leading role in EE ever since. The Declaration defined EE as education that develops human awareness of the interaction of the physical, socio-economic, cultural and human biological systems; that enables students to develop the "knowledge, values, attitudes, and practical skills to participate in a responsible and effective way in anticipating and solving environmental problems, and the management of the quality of the environment" (UNESCO, 1977, p. 25). It is remarkable that this definition was so advanced for its time, particularly in its insistence on attention to the interaction between the social and natural systems, and is still inspirational today.

However, there were precedents going back to the 1920s. One of the useful works that preceded the Declaration was the PhD thesis of Arthur Maurice Lucas (1972). He identified three types of EE: education *about* the environment which focuses on knowledge, EE *for* the environment, which focuses on maintaining or improving the environment, and education *in* the environment, which can be glossed as outdoor education (Lucas, 1972, p. 98ff.). In this book, we use this handy prepositional framing of education *for* nature (Reid in Stevenson & Evans, 2011, p. 25), and the synonym, "pro-environment", as shorthand for a socially critical approach to transformative education towards global environmental health, diversity and sustainability. Nineteen-seventy-two was also the year of the first major international conference that recognised the environmental crisis, the United Nations Conference on the Human Environment in Stockholm. Its outcomes included Principle 19 that pointed to the importance of environmental education (Dyment, Hill, & Emery, 2015, p. 1106). Since then,

> all … major meetings and conferences related to the environmental crisis and sustainable development have advocated for the important role that education has in developing the capacities of people who are active and engaged and have the values, knowledge and skills to participate in making the major personal and structural changes for a transition to sustainability.
>
> (Dyment *et al.*, 2015, p. 1106)

Which term – environmental education, education for sustainable development or education for sustainability?

The international literature on EE sometimes speaks of a divide between EE and Education for Sustainable Development (ESD) (for example, Jickling & Wals, 2008). The ESD discourse is not as old as EE, and has its origins in the Brundtlandt Report, entitled *Our Common Future*, of 1987. "Sustainable development is development that meets the needs of the present without compromising the ability of future generations to meet their own needs" (World Commission on Environment and Development, 1987, p. 41). The concept of SD is much contested in the environmental movement: some see it as an oxymoron (e.g. Sachs, 1999), in which, really, economic growth is prioritised and the sustainability of "the environment" comes a very poor second (D. Reid, 1999, p. 59ff); others see it as vague, or worse – meaningless waffle – and therefore as something to which government, business and indeed everyone can safely subscribe. Others are more positive, and point out its inclusivity, i.e. it enables discussion without apportioning blame, and attempts to build consensus among a range of diverse and even oppositional forces; and others acknowledge that at least it is a discourse that includes the environment. However, it has become the dominant discourse globally, such that we have had a UN Decade of Education for Sustainable Development (UNDESD, 2005–2014) and now we have the Sustainable Development Goals (2016–2030).

The UNDESD highlighted the role of education in bringing about SD. ESD has been much criticised, sometimes for the same reasons as SD itself (e.g. Jickling & Wals, 2008), and sometimes on educational grounds (e.g. Jickling, 1992; Jickling & Wals, 2008; Sauvé, 1996), but the UN Decade was useful in focusing government attention on the issue, and provided governments with a comprehensive framework (UNESCO, 2006). It advocated educational reform at all levels of education, including teacher education, emphasised experiential learning, and, with its focus on economy, society and environment, did a good job of shifting traditional EE away from science and towards understanding the importance of socioeconomic processes and human values:

> ESD is fundamentally about values, with respect at the centre: respect for others, including those of present and future generations, for difference and diversity, for the environment, for the resources of the planet we inhabit. Education enables us to understand ourselves and others and our links with the wider natural and social environment, and this understanding serves as a durable basis for building respect. Along with a sense of justice, responsibility, exploration and dialogue, ESD aims to move us to adopting behaviours and practices that enable all to live a full life without being deprived of basics.
>
> (UNESCO, 2006, Executive Summary)

Nevertheless, in many countries, including Indonesia, the idea that any knowledge or learning to do with the environment should be in a Science subject, prevails. EE is thought to be *about* the environment, not *for* the environment. Typically, teachers in non-Science subjects claim that they do not know anything about the environment and so cannot teach EE. As we will see, in Indonesian schools, understanding that the content of EE is about the interrelations among society, economy and the environment is embryonic, at best.

Education for Sustainability (EfS) grew out of ESD, reflecting some of the reservations of environmentalists (see, for example, Berryman & Sauvé, 2016).

While dropping "development" from the title seems to solve many of the problems, Jickling and Wals (2008) point out that "sustainability" has become one of those double-speak terms that can evoke completely opposing points of view. They point to a conservation issue in Northern Canada where environmentalists used the term to mean conservation of the ecology of a watershed and those pushing for the building of a road and mine used the term to talk of jobs and economic "sustainability" (Jickling & Wals, 2008, p. 14). Sustainability has become a motherhood concept, that no one can argue against, and has often come to mean "long-lived" or "sustained", as in the phrase "sustainable growth", rather than conserved or restored environment.

In this book, we use the term environmental education (EE), rather than alternatives such as ESD or EfS, partly because EE is the term that is most commonly used in Indonesia (*pendidikan lingkungan hidup*) (Nomura, 2009), and partly because we agree with those critics who see ESD as inextricably entangled in the global neoliberal economy (e.g. Huckle & Wals, 2015).[1] Education for Sustainability seems ethically valid, but feels vague and vacuous, suffers from the misunderstandings noted above, and has little currency in Indonesia.

The objectives of EE versus the objectives of schooling

There is a potential contradiction between the goals and methods of EE and the aims and roles of schooling. While ideas about the goals of EE are diverse, the arguments outlined above make it clear that EE must critique the existing global socio-econo-political structures and processes, and prepare students to act for their transformation, in order to preserve and recuperate as much of the natural world as possible (Gruenewald, 2004). Such an education contradicts the present, largely uncritical role and functioning of many education systems, with their conservative curricula and pedagogies, their need for order and organisation, their orientation to the neoliberal, global market, and their tendency to reproduce social (gender, class, ethnic/racial, religious) hierarchies and inequalities (Stevenson 2007, p. 140). Some EE theorists argue that as schools are an inherent part of the global capitalist system that created the world's environmental problems, we cannot expect them to deliver EE goals.

This is partly a debate about what education is and what education ought to be. Certainly, in Indonesia, as we will see, education is directed towards two main goals: producing good, loyal citizens, and improving the quality of the "human resources" of the country (that is, producing good workers). But when we go inside classrooms and see what education currently *is*, we see what Jickling and Wals (2008, p. 7) call "transmissive education" – i.e. "the transmission of facts, skills, and values to students". We agree with them that education should be "transformative" rather than transmissive, enabling learners to co-construct with their teachers and peers a critique of the world as it currently is and plans for developing an ideal world in the future. Transformative education entails deep, open-ended learning, systemic knowledge and holistic understanding, combined with space for creative "autonomy and self-determination"

(Jickling & Wals 2008, p. 7). In this sort of education, children should be active, critical, but also constructive, learners – not sponges. EE fits beautifully into such a concept of transformative education. Through EE, they should be enabled to appreciate nature, to judiciously accept, question and transcend social norms and traditions, critique global corporate greed, find information, imagine ideal futures, solve problems and act on ideas for solutions.

> [A] function of environmental education is to enable students to become critically aware of how they perceive the world with a view to fostering citizen engagement with social and environmental issues and participation in decision-making processes.
>
> (Jickling & Wals, 2008, p. 7)

We agree that schools' usual obedience to and complicity with global forces, in which the market and economic growth are hegemonic, is a serious problem. We already have many technological solutions to environmental problems and quite sophisticated plans for new ways of living. What we lack is political commitment to implement those solutions and plans. Instead we have ever more dangerous, expensive, dirty, and destructive ways to keep going: "bitumen from the Alberta tar sands, oil from deepwater drilling, gas from hydraulic fracturing (fracking), coal from detonated mountains, and so on" (Klein, 2014, p. 2). The biggest obstacles to dealing with environmental problems are therefore political and economic: the actions we must take threaten deregulated capitalism, and our leaders are beholden to elite minorities, or indeed constitute the elite minority, which control our economy, our political processes and most of the media (Klein, 2014, p. 18). EE must, therefore, be political.

Moving back to the aims of education, and of EE in particular, we seem to be presented with a serious, if not terminal, question: if the current reality is of education systems that are transmissive, that are hopelessly embedded in the global capitalist system, and that reproduce social inequality, is it worth trying to engage in transformative EE in formal education contexts? Our answer is simply that we have to. The environmental problems are so serious and, in Indonesia, so unrealised, that not doing anything is not an option. The only questions are around "how to". We must do the best we can, given the conditions. One good thing is that in Indonesia we have a democratic government; another is that we have an education system that, at primary level at least, reaches virtually all children. We also have a young generation that wants to be engaged, and actively produce their version of "the good life". We certainly need to alert them to their dire environmental predicament, but we don't want to obliterate their "youthful optimism" (Nilan *et al.*, 2011, p. 725).

Content and location of EE

Only recently has EE been formally taught outside of the subjects known as science – geology and earth sciences, biology, hydrology, botany, natural history, astronomy, physics and chemistry. Since the 1970s, a host of more inter-disciplinary subjects, such as ecology and environmental science, has flourished.[2] Some of the social sciences have spawned sub-disciplines such as environmental anthropology and history, political and cultural ecology. However, it is still unusual for universities to offer majors that genuinely cross the natural/social science divide. This is also true of schools, though schools seem to have been more open and flexible, and it is now not unusual for school curricula to embed what we can call EE in many school subjects, if not across the disciplinary board.[3] This is truer for primary or elementary schools than for high schools and especially senior high schools.

Any education that includes subjects such as chemistry, physics and biology must be *about* nature but it is widely recognised that "the naturalist, apolitical and scientific work carried out under the banner of 'environmental education' in the 1980s and the early 1990s" (Jóhannesson *et al.*, 2011, 377) is not enough. As we argued above, many environmentalists argue that "education for sustainability needs more of a socio-political perspective than traditional environmental education" (Jóhannesson *et al.*, 2011, p. 377).

Nevertheless, the disciplinary "location" of EE remains a problem. The Framework for the implementation of the UN's Decade of Education for Sustainable Development (2005–2014) recommended that "sustainability be embedded across the curriculum" with an emphasis on "society, environment and economy with culture as an underlying dimension" (UNESCO, 2006). Further, "ESD is fundamentally about values, with respect at the centre: respect for others, including those of present and future generations, for difference and diversity, for the environment, for the resources of the planet we inhabit".

Pedagogy in EE

There is no widespread agreement upon pedagogy for EE. However, there is much discussion in the literature about how EE can be most effectively delivered in schools. It is generally acknowledged that EE is most effective when delivered as part of a whole school, cross-curricular approach that is *in* the environment (rather than *about* the environment – taught in a classroom or textbook), *for* the environment (action based), and experiential (hands on). Stern, Powell, and Hill (2013) list the characteristics that they consider to be generally agreed upon as part of "best practice" pedagogy in EE:[4] active participation, hands-on observation and discovery, place-based learning, project-based learning, cooperative/group learning, play-based learning, outdoor instruction, investigation, guided inquiry, pure inquiry, data collection, immersive field investigation, relevance, reflection, issue-based learning, learner-centred instruction, multimodal delivery of content, and multiple points of view.

Nevertheless, there is plenty of literature that criticises the way EE has been and is delivered in schools (e.g. Hart, 2008; Jucker, 2002; Stevenson, 2007). Hart is a critic of EE that favours imparting of knowledge and predetermined outcomes. "We want to speak and write in ways that make people think about human–environment relationships, yet we educate them as if they cannot think" (Hart, 2008, p. 32). Stevenson (2007) is another voice in the criticism of traditional pedagogies used in EE. Stevenson refers to the work of Young (1980) and Simon (1973) in describing the teacher as a *dispenser of knowledge* and (frequently) the only participant who actively engages in higher-order thinking processes. He argues that, characteristically, student thinking is confined to applying factual information to familiar "well-structured" problems that have a pre-determined, single correct solution, and that tests of students' thinking occur in written examinations on theoretical material which is usually far removed from the realm of the students' present or future life experiences (Stevenson, 2007).

It should be noted that many of the criticisms of the delivery and effectiveness of EE in schools align with an education ideology that can be called progressive education. Borrowing from Mead (2017), we can simplify and say that in education systems in the Global North there are two major opposing approaches to education: traditionalism and progressivism. Traditional approaches aim to equip all students with knowledge, through direct top-down instruction and a "disciplinary culture in which obedience is both prized and awarded" (Mead, 2017, p. 37). The aim is that all students will meet "measurable academic standards" through drilling, rote learning, intensive testing and preparation for tests, regimentation and strict discipline (Mead, 2017, p. 37). Progressivism, inspired partly by the American educator Dewey, involves a teacher developing a curriculum which is considered a "staging ground" from which students explore the world, often with hands-on activities, learning through play, student-centred learning, group work, problem-solving and open-ended discoveries in- and outside the classroom. There is a focus on developing individual curiosity, capacities and self-expression. Dewey was also very concerned that students develop the values and capacities that enable democracy to flourish: independent and critical thinking, the questioning of injustice, commitment to tolerance and respect, as well as what is known in the EE literature as "student voice" (Dewey, 1916 (1925 printing)).

There are varied arguments as to why student voice should be heard and considered in education. These arguments include issues of children's rights (Rudduck & Flutter, 2000), student empowerment and political agency (O'Boyle, 2013; Rudduck, 2002; Rudduck & Flutter, 2000), the value of the student perspective in education evaluation and improvement (Cook-Sather, 2002; A. Gough, 1999; McIntyre, Pedder, & Rudduck, 2005), educational benefits for students (Busher, 2012; Rudduck, 2002) and student voice as a transformative agent (Beattie, 2012).

O'Boyle (2013) reports on an English example where "young people's talk about themselves and their educational experiences" is not "valued in public discourse about education" (O'Boyle, 2013, p. 136). She argues that the cultural narratives that surround the education of young people seem to be founded on an underlying assumption of predetermination – i.e. a focus on their future state rather than their current existence. Further, education systems seem to instantiate both a need for society to homogenise and the objectification of young people – all of which seems to eradicate any real need to consult with them (O'Boyle, 2013, p. 136). She considers that a lack of student voice in education is a considerable barrier to young people becoming active and critical citizens.

EE that includes a student voice serves a broader purpose than merely allowing self-expression. It can also present young people's views to an audience of decision makers and those in positions of power and function as an avenue to promote young people's political agency, assisting them in creating a new order of experience for them as active participants in the broader society (O'Boyle, 2013; Rudduck & Flutter, 2000). It can help them move from silence and invisibility to influence and visibility.

Education that transforms thinking and behaviour about the environment requires new ways of teaching. Here we identify four problematic aspects of pedagogy that prevail in Indonesia and many other developing countries:

1   The continuing dominance of rote learning.
2   The focus on the transmission of facts.
3   The gap between environmental awareness and knowledge on the one hand and pro-environmental behaviour on the other.
4   The effect of learned helplessness, apathy.

Following this, we suggest the model that we call *critical eco-pedagogy*, which could theoretically address these problems.[5]

The first problem is that teaching for rote learning is still the basic pedagogy. In Indonesia, despite changes in curricula, and a theoretical shift towards a new focus on student-centred learning, in reality much pedagogy is still of the chalk-and-talk variety. Rote learning of facts is still the dominant model (Bjork, 2005, 2013). A typical lesson consists of a teacher reading a lesson from the textbook, or asking students to take turns reading the lesson out aloud, then asking students simple comprehension questions, often of the complete-the-phrase variety (Parker, 2003, Chapter 10). Rote learning of facts is tested in examinations, which are usually in the multiple choice format. Rote learning has its place, and as long as students do not just cram the night before an exam, they should learn some facts. However, behind rote learning is the unquestioned and unquestionable position of the all-knowing authority, whether it be the textbook or the teacher; and the assumption that knowledge should flow down from that authority to students. The problem then is that there is no legitimate position from which students (or teachers) can ask questions or critique what they are being taught. Students are simply expected to absorb information, as passive sponges. They are not given the skills to discover knowledge or answers for themselves, apart from what can be learned by rote, and are not given the confidence or the authority to question, counter or probe.

The second problem is connected to the first: the focus is on the transmission of facts. A related problem is that "facts" are not questioned – their selection, the conditions of their production, their reliability or generalisability. In secondary school, teachers are encouraged to prioritise knowledge transmission as it fits more easily into curricula and supports exam-focused outcomes (Jiang, 2004; Maulidya, Mudzakir, & Sanjaya, 2014; Steele, 2011). However, EE is about much more than facts: understanding complex processes is important, as are values of respect, care and responsibility, the ability to critique existing power structures, injustices and harms, and the capacity to solve problems and build anew.

The third problem is not unique to environmental education, but is a problem that dogs environmentalism in general: it is the gap between possessing environmental awareness and knowledge and behaving pro-environmentally. This gap was famously treated in the "Mind the Gap" paper by Kollmuss and Agyeman (2002), in which they identified factors that cause people to behave pro-environmentally, such as "demographic factors, external factors (e.g. institutional, economic, social and cultural) and internal factors (e.g. motivation, pro-environmental knowledge, awareness, values, attitudes, emotion, locus of control, responsibilities and priorities)" (Kollmuss & Agyeman, 2002, p. 239). They then proposed their own model of "environmental knowledge, values, and attitudes, together with emotional involvement" which together made up a "complex" they called "*pro-environmental consciousness*" (Kollmuss & Agyeman, 2002, pp. 256–257). They noted the barriers that intervene between knowledge and attitudes on the one hand and behaviour on the other, and state that there is no direct or automatic take-up of pro-environmental behaviours even if all the factors are lined up. There is a huge literature on the psychology (especially motivation) of environmental behaviours,[6] and some significant academic debates have been waged, e.g. the debate about the importance of significant life experiences in two special issues of *Environmental Education Research* in 1998, vol. 4(4) and 1999, vol. 5(4). Unfortunately, in all of this, there has been little input from the Global South.[7]

We propose the fourth problem with some reservations, because while it is commonly discussed for EE in Global North countries, it was not identified by our fieldwork in schools in Indonesia. It is that students (and their teachers) can feel helpless and hopeless once they realise the full extent of environmental degradation. Nagel interviewed Year 7 children in both Canada and Australia, and found "great concern and duress with their perceptions of the state of the environment and the future", e.g. one student said, "It won't get better. It can't get better. It's just getting worse. (Paul – Canada)" (Nagel, 2005, p. 75). Another, presumably unknowingly, expressed something of the Brundtland definition of SD, when he said,

> I think that it's just wrong that we hurt our children and their children and they did nothing. I think well that's just like what we're doing killing innocent people that we don't even know yet and they haven't really done anything to hurt the environment.
>
> (Jake – Australia)

While such negative expressions do not necessarily extend to "learned helplessness", developing into apathy, in the face of global environmental problems, others have also noted similar reactions. Payne (2001), in important work on identity and EE, cited students such as "Ned", who expressed guilt with regard to their own contribution to global ecological crisis, and "Kim" who agonised about her own lack of pro-environment behaviour:

> "Some of my inaction can be attributed to a sense of helplessness … and, the way in which I am made to feel part of the problem but not part of the solution."
>
> (Payne, 2001, p. 69)

Interestingly, Kim's words are echoed in the objectives of the 2013 Indonesian School Curriculum: students are to become "a part of the solution to various problems … in the social and natural environments" (Mendikbud, 2013, p. 7).

Nevertheless, such expressions of helplessness and apathy seem unlikely in Indonesia, where environmental consciousness in the community at large is low, and the mediascape is a long way from being saturated with environmental "doom and gloom". However, it might be noted, by way of a preview, that our researcher, Pam Nilan, reported – during fieldwork in a university campus in South Sumatra – that while the air on campus was unbreathable from illegal forest fires, water failed to come out of taps, and roads were unusable because they were being destroyed by trucks used for coal, oil and palm oil, those university students who considered themselves environmentalists, only seemed to concern themselves with litter (email communication, 7 November 2014). The explanation seemed to be that the forest fires, water supply and transport infrastructure problems were "too big to tackle", and involved too many powerful vested interests, so environmentally aware students were reduced to picking up rubbish (Nilan, 2018).

Critical ecopedagogy

There is a large body of literature on pedagogy for EE/ESD/EfS. Our model, which we call *critical ecopedagogy*, is based on the approach of Khan, and Misiaszek, which Misiaszek calls "ecopedagogy".[8] Ecopedagogy is "a critical environmental pedagogy that focuses on understanding the connections between social conflict and environmentally harmful acts carried out by humans" (Misiaszek, 2016, p. 587). His approach is derived from the work on education by the Brazilian education theorist Paolo Freire, known as critical pedagogy – notably Freire's book, *Pedagogy of the Oppressed* (Freire, 1996). However, it must be noted that Freire only became environmentally "critical" in later life. Critical pedagogy seeks to reveal how schooling benefits dominant groups and disadvantages the subordinate, helping to sustain and intensify oppressions. In this literature, words such as "social conflict" and "oppression" refer not necessarily to violent conflict but to significant inequalities and power relations which historically oppress some groups and work to the advantage of the powerful. Misiaszek opens the analysis to environmental harms so, basically, ecopedagogy is "the teaching and learning of connections between environmental and social problems" (Misiaszek, 2015, p. 280).

Ecopedagogy aims "to promote transformative action by helping to reveal socio-environmental connections that oppress individuals and societies" (Misiaszek, 2015, 280). The beauty of this approach is that it is a pedagogy of hope and thereby addresses problem four identified above:

> Freirean pedagogy refers to teaching which focuses on determining how someone wants the world to be, the gaps between this constructed reality and the perceptions of current reality … and what is necessary to eliminate these gaps. In ecopedagogy, this type of education is used to lay out current socio-environmental realities and possible realities in order to consider potential changes to existing societal, political, and economic structures and eliminate the gap between them. A fundamental philosophy of Freirean pedagogy is that if humans have constructed the current world, it is possible to change it. In this respect, it is a pedagogy of hope as opposed to pedagogies that promote fatalism in which deeper, structural change cannot occur.
>
> (Misiaszek, 2015, p. 283)

Misiaszek insists that ecopedagogy must be informed by this larger theoretical framing (Misiaszek, 2015, p. 282). His justification is that large theoretical framing can assist in giving insights into different cultural understandings and different positionings, e.g. a feminist or developing country perspective. Translated into practice, critical ecopedagogy gives opportunities for critical deconstruction (how the world currently is), for idealism (how the world could be better), and for active learning and problem-solving (how to reach the ideal). It is inherently political, because it is concerned with vested interests. As he notes, "it is necessary to … critique the benefits of environmentally harmful actions … [because] [a]ll environmentally harmful actions are likely to have some human benefit for someone" (Misiaszek, 2016, p. 590). In this regard it is a sort of political ecology for education.

In practice, ideally, critical ecopedagogy is an action-based, student-driven programme that involves students in place-based, experiential learning through which they can develop an environmentally sensitive view of their environment; critical analytical skills for making the connections between social and environmental problems; problem-solving skills; and capacities for finding information and solutions, making connections and developing relationships, negotiating with powerful others, and envisioning and constructing an ideal environment. Ideally, this would enable them to develop a closer experiential connection to the natural environment, but this might not be practical, depending on location. Thus, students would be more engaged in their learning, improve their academic learning beyond EE, and be happier and healthier students (Bowler *et al.*, 2010; National Environmental Education Foundation, 2013; Powers, 2004). It is important that they are not only "doing" and "experiencing", but also that they are learning about the interconnections among human society, the economic system and the environment, and developing capacities to think critically and constructively about how to solve problems of environmental harm, social oppression, marginalisation and inequality. If this can be in collaboration with not only peers but also "powerful others" in the community, who can help to bring about change in the community, that would be ideal.

Clearly, ideal critical ecopedagogy is a long way removed from the four problem scenarios outlined earlier in this section. It behoves us to take on board the caveats noted by scholars such as Barrett *et al.* (2005). They note the importance of practitioners attending to the difference between "token" and "authentic" participation, distinguishing between one-off actions and ongoing behaviour, and refraining from limiting their focus to science, the natural environment and lifestyle environmentalism (Barrett *et al.*, 2005, p. 507). They suggest that failure to attend to these concerns means that action-oriented and experiential educational initiatives may actually "undermine students' sense of agency, support student passivity and simple solutions … and gloss over the complexity of causes of environmental problems, including their intersections with social and economic systems, and ultimately, politics and power" (Barrett *et al.*, 2005, p. 507). More profoundly, and with a view to the problematic pedagogy prevailing in Indonesia (as described in Chapter 5 and in the ethnographic chapters, Chapters 8 and 10), engaging in such learning activities in schools challenges dominant conceptions about the organisation and transmission of knowledge, creating for most teachers and students contradictions with standard approaches to teaching and learning (Barrett *et al.*, 2005, pp. 511ff.). Based on their evaluation of action-oriented approaches in Canada, Barrett *et al.* (2005, p. 512) argue that "a key to success in action-oriented experiential education is a significant shift in teacher-student-teacher interactions, both inside and outside classroom walls." They acknowledge the difficulties experienced by both teachers and students in redefining and assuming unfamiliar roles that challenge the traditions of Canadian education and argue that the roles of "powerful teacher and submissive student carry both the authority of law and the weight of tradition, despite educational theory and educational practice purporting a counter theme of independent learning and critical and creative thinking" (Barrett *et al.*, 2005, p. 514).

Since critical ecopedagogy approaches are, by definition, implemented in the real-world locale of schools, teachers and students alike should be alive to the possibility of failure.[9] That is, engagement with local government and service providers would not necessarily lead to more environmentally friendly services or significant changes to locally oppressive structures. This means that inner-city students might not develop a closer experiential connection to the natural environment, that garbage might not be separated for recycling, and that slum dwellers might still have their shanty towns bulldozed.

Conclusion

We have borrowed from some scholars to float the idea that EE cannot escape the fact that it was born of the global economic system that caused global environmental damage. We argue instead that while it should acknowledge its past, it must transform itself. EE should be critical of the economic structures, as well as the international political relations and social values that underlie it. EE should also be transformative, to provide students with (a) real-world experiences of the natural world; (b) an understanding of human–nature interactions and the role of the economy in exploiting the environment; (c) critical thinking skills and a sense of responsibility to enable them to perceive social and environmental injustice; and (d) the knowledge, skills and capacities to engage in future imagining, problem-solving, creative thinking and collaborations that aim to conserve, sustain and enrich the natural environment. In short, EE must extend beyond the classroom, into the community, and it must be political.

Notes

1 Later, in Chapter 6, we discuss the role of UNESCO in leading ESD in Indonesia.
2 But see Korfiatus (2005) for an analysis of the vexed relationship between EE and ecology.
3 See the following references for country-specific evaluations: for India, Almeida & Cutter-Mackenzie, 2011; for an evaluation of the implementation of EE as a cross-curricula national priority in Tasmania, Australia, see Dyment *et al.*, 2015; for Jamaica, Ferguson, 2008; for a report on initiatives in the Asia-Pacific, Fien & Corcoran, 1996; for Iceland, Jóhannesson *et al.*, 2011; for Cyprus, Kadji-Beltran, 2002; for the UK, Scott, 2011.
4 These characteristics were collated from numerous sources including the Excellence in Environmental Education: Guidelines for Learning (K-12) developed by the North American Association for Environmental Education (2012). The following references present what is considered the most effective approach to school-based delivery of EE both in the classroom and as extra-curricular activities in the Global North: Cutter-Mackenzie, 2010; Ferreira, Ryan, & Tilbury, 2006; Powers, 2004; Steele, 2011; Stevenson, 2007; Tilbury, Coleman, & Garlick, 2005. There is little literature available on what is considered "best practice EE" in Southern education systems.
5 There is now an "ecopedagogy" literature, though not, as Kahn claims, an ecopedagogy "movement", e.g. *Green Theory & Praxis: The Journal of Ecopedagogy*; see also Hung (2014); Kahn (2010); Payne (2014, 2015), Payne (2017) and, for critique Reid and Payne (2011).
6 Apart from all the original studies of the differential impact of factors, such as awareness, knowledge, values, attitudes, intentions, the relative importance of personal versus contextual or situational influences, there have also been a series of meta-analyses of the body of literature: the first was Hines, Hungerford, and Tomera (1986/87) and a 20-years-later study was Bamberg and Möser (2007).
7 There was one article in the 1998 Special Issue which looked at SLE in nine countries, including a couple of Global South countries (Palmer *et al.*, 1998). The 1999 Special Issue had no such articles, although there was one short paper about El Salvador in an intervening issue (Sward, 1999). More recently, Li and Chen (2015) have looked at SLE in China – though its status as a Global South country is questionable – and Reibelt *et al.* (2017) at conservationists in Malagasy, and both revealed that childhood experiences were not seminal.

8 We retain the word "critical" both to keep the Freire genealogy at the forefront, and to refer to other literatures that have a more post-structural genealogy but still deploy a critical approach that keeps power relations at the forefront. For instance, the work of education scholars such as Apple (2004) and Giroux (1992, 1997, 2011) critically examines discourses and socioeconomic structures and processes, the interconnections among growing social inequality, inequitable power relations, global market forces, and the need for an ethos of social justice – though unfortunately not with significant reference to the environment.

9 Although due to different causes, the following scenario brought home to the author the necessity for EE to be real-world. In Australia, it is common for primary schools to develop wonderful vegetable gardens, which the students tend, and from which they eat; parents generally greatly appreciate this initiative, and schools can become well known for their gardens. One stand-out school had a vegetable garden, but the teacher in charge of EE at the school did not mention to students the problem that was going to arise over the long summer holiday: with temperatures in the 30s and even 40s (Celsius), the garden would have to be watered every day. The students returned from their long holiday to find their gardens a wasteland – they were dismayed. They had to learn the lesson the hard way.

References

Almeida, S., & Cutter-Mackenzie, A. (2011). The historical, present and future-ness of environmental education in India. *Australian Journal of Environmental Education, 27*(1), 122–133.

Apple, M. W. (2004). *Ideology and Curriculum* (3rd edn). New York and London: RoutledgeFalmer.

Bamberg, S., & Möser, G. (2007). Twenty years after Hines, Hungerford, and Tomera: A new meta-analysis of psycho-social determinants of pro-environmental behaviour. *Journal of Environmental Psychology, 27*(1), 14–25.

Barrett, M. J., Hart, P., Nolan, K., & Sammel, A. (2005). Challenges in implementing action oriented sustainability education. In W. L. Filho (Ed.), *Handbook of Sustainability Research* (Vol. 20, pp. 505–534). Frankfurt am Main: Peter Lang.

Beattie, H. (2012). Amplifying student voice: The missing link in school transformation. *Management in Education, 26*(3), 158–160. doi:10.1177/0892020612445700.

Berryman, T., & Sauvé, L. (2016). Ruling relationships in sustainable development and education for sustainable development *The Journal of Environmental Education, 47*(2), 104–117. doi:10.1080/00958964.2015.1092934.

Bjork, C. (2005). *Indonesian Education: Teachers, Schools, and Central Bureaucracy*. New York: Routledge.

Bjork, C. (2013). Teacher training, school norms and teacher effectiveness in Indonesia. In D. Suryadarma & G. W. Jones (Eds), *Education in Indonesia* (pp. 53–67). Singapore: ISEAS Publishing.

Bowler, D., Buyung-Ali, L., Knight, T., & Pullin, A. (2010). A systematic review of evidence for the added benefits to health of exposure to natural environments. *BMC Public Health, 10*, 456–466.

Busher, H. (2012). Students as expert witnesses of teaching and learning. *Management in Education, 26*(3), 113–119. doi:10.1177/0892020612445679.

Cook-Sather, A. (2002). Authorizing students' perspectives: Toward trust, dialogue, and change in education. *Educational Researcher, 31*(4), 3–14. doi:10.2307/3594363.

Cutter-Mackenzie, A. (2010). Australian Waste Wise schools program: Its past, present and future. *Journal of Environmental Education, 41*(3), 165–178.

Dewey, J. (1916 (1925 printing)). *Democracy and Education: An Introduction to the Philosophy of Education* New York: Macmillan.

Dyment, J., Hill, A., & Emery, S. (2015). Sustainability as a cross-curricular priority in the Australian Curriculum: A Tasmanian investigation. *Environmental Education Research, 21*(8), 1105–1126.

Ferguson, T. (2008). "Nature" and the "environment" in Jamaica's primary school curriculum guides. *Environmental Education Research, 14*(5), 559–577. doi:10.1080/135046 20802345966.

Ferreira, J., Ryan, L., & Tilbury, D. (2006). Whole-school approaches to sustainability: A review of models for professional development in pre-service teacher education. Retrieved from http://aries.mq.edu.au/projects/preservice/files/TeacherEduDec06.pdf.

Fien, J., & Corcoran, P. B. (1996). Learning for a sustainable environment: Professional development and teacher education in environmental education in the Asia-Pacific region. *Environmental Education Research, 2*(2), 227–236.

Freire, P. (1996). *Pedagogy of the oppressed* (M. B. Ramos, Trans. New rev. edn). London: Penguin.

Giroux, H. A. (1992). *Border Crossings: Cultural Workers and the Politics of Education.* New York, London: Routledge.

Giroux, H. A. (1997). *Pedagogy and the Politics of Hope: Theory, Culture, and Schooling: A Critical Reader.* Boulder, CO: Westview Press.

Giroux, H. A. (2011). *On Critical Pedagogy.* New York: Continuum.

Gough, A. (1999). Kids don't like wearing the same jeans as their Mums and Dads: So whose "life" should be in significant life experiences research? *Environmental Education Research, 5*(4), 383–394.

Gough, N. (2008 [2003]). Thinking globally in environmental education: Implications for internationalizing curriculum inquiry. In W. F. Pinar (Ed.), *International Handbook of Curriculum Research* (pp. 53–72). Mahwah, NJ: Laurence Erlbaum Associates.

Gruenewald, D. A. (2004). A Foucauldian analysis of environmental education: Toward the socioecological challenge of the Earth Charter. *Curriculum Inquiry, 34*(1), 71–107.

Hart, P. (2008). Ontological/epistemological pluralism within complex contested EE/ESD landscapes: Beyond politics and mirrors. In E. Gonzalez-Gaudiano & M. Peters (Eds), *Environmental Education: Identity, Politics and Citizenship* (pp. 25–38). Rotterdam: Sense Publishers.

Hines, J. M., Hungerford, H. R., & Tomera, A. N. (1986/87). Analysis and synthesis of research on responsible environmental behaviour: A meta-analysis. *Journal of Environmental Education, 18*, 1–8.

Huckle, J., & Wals, A. E. J. (2015). The UN decade of education for sustainable development: Business as usual in the end. *Environmental Education Research, 21*(3), 491–505. doi:http://dx.doi.org/10.1080/13504622.2015.1011084.

Hung, R. (2014). In search of ecopedagogy: Emplacing nature in the light of Proust and Thoreau. *Educational Philosophy and Theory, 46*, 1387–1401.

Jiang, K. (2004). Analysis of research findings on environmental education in secondary vocational schools in Shanghai. *Chinese Education & Society, 37*(4), 32–38.

Jickling, B. (1992). Why I don't want my children to be educated for sustainable development. *Journal of Environmental Education, 23*(4), 5–8.

Jickling, B., & Wals, A. E. J. (2008). Globalization and environmental education: Looking beyond sustainable development. *Journal of Curriculum Studies, 40*(1), 1–21.

Jóhannesson, I. A., Norðdahl, K., Óskarsdóttir, G., Pálsdóttir, A., & Pétursdóttir, B. (2011). Curriculum analysis and education for sustainable development in Iceland. *Environmental Education Research*, *17*(3), 375–391.

Jucker, R. (2002). "Sustainability? Never heard of it!" Some basics we shouldn't ignore when engaging in education for sustainability. *International Journal of Sustainability in Higher Education*, *3*(1), 8–18.

Kadji-Beltran, C. (2002). *Evaluation of Environmental Education Programmes as a Means for Policy Making and Implementation Support: The Case of Cyprus Primary Education*. PhD dissertation, University of Warwick, UK.

Kahn, R. (2010). *Critical Pedagogy, Ecoliteracy, and Planetary Crisis: The Ecopedagogy Movement*. New York: Peter Lang Publishers.

Klein, N. (2014). *This Changes Everything: Capitalism vs. The Climate*. London: Penguin Books.

Kollmuss, A., & Agyeman, J. (2002). Mind the Gap: Why do people act environmentally and what are the barriers to pro-environmental behavior? *Environmental Education Research*, *8*(3), 239–260.

Korfiatis, K. J. (2005). Environmental education and the science of ecology: Exploration of an uneasy relationship. *Environmental Education Research*, *11*(2), 235–248.

Li, D., & Chen, J. (2015). Significant life experiences on the formation of environmental action among Chinese college students. *Environmental Education Research*, *21*(4), 612–630. doi:10.1080/13504622.2014.927830.

Lucas, A. M. (1972). *Environment and Environmental Education: Conceptual Issues and Curriculum Implications*. PhD dissertation, The Ohio State University, USA.

Maulidya, F., Mudzakir, A., & Sanjaya, Y. (2014). Case study the environmental literacy of fast learner middle school students in Indonesia. *International Journal of Science and Research*, *3*(1), 193–197.

McIntyre, D., Pedder, D., & Rudduck, J. (2005). Pupil voice: Comfortable and uncomfortable learnings for teachers. *Research Papers in Education*, *20*(2), 149–168. doi:10.1080/02671520500077970.

Mead, R. (2017, December 11). Two schools of thought. *The New Yorker*, 34–41.

Mendikbud (Kementrian Pendidikan dan Kebudayaan) [Ministry of Education and Culture]. (2013). *Kurikulum 2013, Kompetensi Dasar, Sekolah Menengah Atas (SMA)/ Madrasah Aliyah (MA) [2013 Curriculum, Basic Competencies, Senior High School/Islamic Senior High School]*. Kementrian Pendidikan dan Kebudayaan [Ministry of Education and Culture].

Misiaszek, G. W. (2015). Ecopedagogy and citizenship in the age of globalisation: Connections between environmental and global citizenship education to save the planet. *European Journal of Education*, *50*(3), 280–292. doi:10.1111/ejed.12138.

Misiaszek, G. W. (2016). Ecopedagogy as an element of citizenship education: The dialectic of global/local spheres of citizenship and critical environmental pedagogies. *International Review of Education*, *62*, 587–607. doi:10.1007/s11159-016-9587-0.

Nagel, M. (2005). Constructing apathy: How environmentalism and environmental education may be fostering "learned hopelessness" in children. *Australian Journal of Environmental Education*, *21*, 71–80.

National Environmental Education Foundation. (2013). Benefits of environmental education. Retrieved from http://eeweek.org/resources/EE_benefits.

Nilan, P. (2018). Smoke gets in your eyes: Student environmentalism in the Palembang haze in Indonesia. *Indonesia and the Malay World*, *46*(136), 325–342. doi:10.1080/1363 9811.2018.1496624.

Nilan, P., Parker, L., Bennett, L., & Robinson, K. (2011). Indonesian youth looking towards the future. *Journal of Youth Studies, 14*(6), 709–728.

Nomura, K. (2009). A perspective on education for sustainable development: Historical development of environmental education in Indonesia. *International Journal of Educational Development, 29,* 621–627.

North American Association for Environmental Education. (2012). *Excellence in Environmental Education: Guidelines for Learning (K-12).* Retrieved from Washington, DC: http://eelinked.naaee.net/n/guidelines/posts/Download-Your-Copy-of-the-Guidelines.

O'Boyle, A. (2013). Valuing the talk of young people: Are we nearly there yet? *London Review of Education, 11*(2), 127–139. doi:10.1080/14748460.2013.799809.

Palmer, J. A., Suggate, J., Bajd, B., Hart, P., Ho, R. K. P., Ofwono-Orecho, J. K. W., … Van Staden, C. (1998). An overview of significant influences and formative experiences on the development of adults' environmental awareness in nine countries. *Environmental Education Research, 4*(4), 445–464. doi:10.1080/1350462980040408.

Parker, L. (2003). *From Subjects to Citizens: Balinese Villagers in the Indonesian Nation-state.* Copenhagen: Nordic Institute of Asian Studies Press.

Payne, P. (2001). Identity and environmental education. *Environmental Education Research, 7*(1), 67–88.

Payne, P. (2014). Vagabonding slowly: Ecopedagogy, metaphors, figurations, and nomadic ethics. *Canadian Journal of Environmental Education, 19,* 47–69.

Payne, P. (2015). Critical curriculum theory and slow ecopedagogical activism. *Australian Journal of Environmental Education, 31*(2), 165–193.

Payne, P. (2017). Ecopedagogy and radical pedagogy: Post-critical transgressions in environmental and geography education. *The Journal of Environmental Education, 48*(2), 130–138. doi:10.1080/00958964.2016.1237462.

Pinar, W. F., Reynolds, W. M., Slattery, P., & Taubman, P. M. (1995). Understanding curriculum as international text. In *Understanding Curriculum: An Introduction to the Study of Historical and Contemporary Curriculum Discourses* (pp. 792–843). New York: Peter Lang.

Powers, A. L. (2004). An evaluation of four place-based education programs. *Environmental Education Research, 35*(4), 17–32.

Reibelt, L. M., Richter, T., Rendigs, A., & Mantilla-Contreras, J. (2017). Malagasy conservationists and environmental educators: Life paths into conservation. *Sustainability, 9*(2), 227. doi:10.3390/su9020227.

Reid, A. D., & Payne, P. G. (2011). Producing knowledge and (de)constructing identities: A critical commentary on environmental education and its research. *British Journal of Sociology of Education, 32*(1), 155–165.

Reid, D. (1999). *Sustainable Development: An Introductory Guide.* London: Earthscan Publications.

Rudduck, J. (2002). The transformative potential of consulting young people about teaching, learning and schooling. *Scottish Educational Review, 34*(2), 123–137.

Rudduck, J., & Flutter, J. (2000). Pupil participation and pupil perspective: "Carving a new order of experience". *Cambridge Journal of Education, 30*(1), 75–89. doi:10.1080/03057640050005780.

Sachs, W. (1999). Sustainable development and the crisis of nature: On the political anatomy of an oxymoron. In F. Fischer & M. A. Hajer (Eds), *Living with Nature: Environmental Politics as Cultural Discourse* (pp. 23–41). Oxford: Oxford University Press.

Sauvé, L. (1996). Environmental education and sustainable development: A further appraisal. *Canadian Journal of Environmental Education, 1,* 7–34.

Scott, W. (2011). Sustainable schools and the exercising of responsible citizenship – a review essay. *Environmental Education Research, 17*(3), 409–423. doi:10.1080/135046 22.2010.535724.

Simon, H. A. (1973). The structure of ill structured problems. *Artificial Intelligence, 4*(3–4), 181–201. doi:http://dx.doi.org/10.1016/0004-3702(73)90011-8.

Steele, A. (2011). Beyond contradiction: Exploring the work of secondary science teachers as they embed environmental education in curricula. *International Journal of Environmental and Science Education, 6*(1), 1–22.

Stern, M. J., Powell, R. B., & Hill, D. (2013). Environmental education program evaluation in the new millennium: What do we measure and what have we learned? *Environmental Education Research,* 1–31. doi:10.1080/13504622.2013.838749.

Stevenson, R. B. (2007). Schooling and environmental education: Contradictions in purpose and practice. *Environmental Education Research, 13*(2), 139–153.

Stevenson, R. B., & Evans, N. S. (2011). The distinctive characteristics of environmental education research in Australia: An historical and comparative analysis. *Australian Journal of Environmental Education, 27*(1), 24–45. doi:10.1017/S0814062600 000057.

Stevenson, R. B., Wals, A. E., Dillon, J., & Brody, M. (2013). Introduction: An orientation to environmental education and the handbook. In R. B. Stevenson, A. E. Wals, J. Dillon, & M. Brody (Eds.), *International Handbook of Research on Environmental Education* (pp. 1–12). New York: Routledge.

Sward, L. L. (1999). Significant life experiences affecting the environmental sensitivity of El Salvadoran environmental professionals. *Environmental Education Research, 5*(2), 201–206. doi:10.1080/1350462990050206.

Tilbury, D., Coleman, V., & Garlick, D. (2005). A national review of environmental education and its contribution to sustainability in Australia: School education. Retrieved from http://aries.mq.edu.au/projects/national_review/.

UNESCO. (1977). *The Tbilisi Declaration – Intergovernmental Conference on Environmental Education Final Report, 14–26 October 1977.* Tbilisi (USSR).

UNESCO. (2006). *Framework for the UNDESD International Implementation Scheme.* Retrieved from http://unesdoc.unesco.org/images/0014/001486/148650e.pdf.

World Commission on Environment and Development. (1987). *Report of the World Commission on Environment and Development: Our Common Future.* Retrieved from www. un-documents.net/our-common-future.pdf.

Young, R. E. (1980). The controlling curriculum and the practical ideology of teachers. *Journal of Sociology, 16*(2), 62–70. doi:10.1177/144078338001600212.

4 Introducing Indonesia

[E]nvironmental values are not deeply embedded in society, leading to under-valuation of natural resources and environmental services.

(World Bank, 2014c)

This chapter assumes that the reader knows very little about Indonesia. It surveys topics such as the demography of Indonesia, the economy and the broad socioeconomic context in which our study is embedded, politics and government, religion and culture, and the environment. The second half of the chapter presents what is known about environmental awareness in Indonesia, underlining the observation that the populace and the government have very little knowledge and even awareness of the dire environmental problems that Indonesia faces – let alone what to do about them. Finally, the chapter examines environmental non-government organisations (ENGOs) and the role they have played thus far in plugging that hole and trying to spread environmental concern and enthusiasm among the populace.

Population, economy and socioeconomic situation

Indonesia is an important developing country, with a particular significance environmentally. Its diverse population of 258 million (UNDESA, 2015, p. 14) makes it the fourth largest country in the world.[1] Its total fertility rate of 2.36 per cent per annum (UNDESA, 2015, p. 39) is quite low, owing to a successful family planning program and increasing prosperity. Currently the median age of the population is 28.4 years; by 2050 it will be 36.5, and by 2100, 43.5 years (UNDESA, 2015, p. 34).[2] While Indonesia is often characterised as having a young, productive population, its median age is quite high for a developing country, due to its successful family planning program since the early 1970s. Demographic changes, notably rapid fertility decline and mortality improvements since the 1960s, have led to accelerated population ageing and the emergence of non-communicable diseases in many areas, coexistent with areas of persistent high fertility, infant and maternal mortality and infectious diseases. Indonesia now presents simultaneously with First and Third World health profiles. The average life expectancy at birth is now 69 years (UNDP HDI Report 2016, Table 1, p. 199). Although the proportion of elders is still modest, with people over 60 comprising 8.9 per cent of the population, by 2100 they will comprise 35.4 per cent (UNDESA, 2015, p. 28).

The demographic cohort of interest for this study of young people and education is the youth cohort: children in the age range 0–14 years comprise one-quarter of the total population of Indonesia (25.02%) and those in the 15–24 age range comprise 16.99% of the total population. The latter figure equates to 22.5 million males and 21.7 million females – a large number by any standard (CIA, 2018). By 2050, this cohort will be 50 years old, will all live in cities and towns and be the power-holders and decision-makers in society.

The most startling change in demographic distribution is the shift from rural to urban areas. In 1950, about 88 per cent of the population was rural; by 2010 it was 50 per cent; and by 2050 it will be only 30 per cent (UNDESA Population Division, 2014). These figures do not just represent large numbers of people physically moving to towns and cities, the rapid growth of towns and cities and concomitant loss of arable and other land. They also signify massive social changes, e.g. in employment, access to natural resources, social and community relations, levels of mobility and in/dependence; environmentally, implications include higher levels of energy consumption, longer supply chains, the need for more sophisticated infrastructure (notably sanitation and water), concentrated air and noise pollution, problems of food security and waste disposal, and of course loss of land and biodiversity. Indonesia is extremely vulnerable to natural "disasters", such as earthquakes and volcanic eruptions, and urban development magnifies the risk to humans of these occurrences, as well as hazards such as flash flooding (which in turn are often caused by overpopulation of hilly terrain leading to loss of forest cover).

The recent expansion of the middle classes is perhaps the demographic feature that has been most often noted – mainly because of its economic significance. While the actual size of the middle class is debatable – estimates range from 35 to 50 million (Bhaduri & Monroe, 2010; Robb, 2015; *The Economist*, 2011) – there is no doubt of the consumption trend. Much of Indonesia's economic growth has been fuelled by higher rates of domestic demand – for energy, industrial production, for more vehicles and roads, more infrastructure, more hotels, more fast food outlets, supermarkets and convenience stores, and more malls. Motorbikes dominate the roads, but the growth in the number of cars is staggering; air-conditioners are now common; just about everybody has a mobile phone. There is almost no discussion of the deleterious environmental impacts of untrammelled consumption in Indonesia.

Indonesia's economy has been developing steadily since the 1970s and is nowadays often described as "booming". Indonesia's economic performance through the so-called "New Order" period, 1966–1998, was strong; the Asian financial crisis of 1997–98 hit hard, and helped to bring about the fall of the Soeharto government. Since then, there has been sustained economic growth, even through the Global Financial Crisis. In the period 2007–2018, the annual growth in GDP has averaged at more than 5.5 per cent, and GDP per capita doubled (Indonesia Investments 2019).

Indonesia is now into the ranks of the middle-income countries, albeit the lower middle rank. On the Human Development Index it ranks 113 of 188 countries, and again belongs to the "medium" human development group (UNDP, 2016, p. 200). Poverty is steadily declining (World Bank, 2014b): "the poor" now make up around 11.3 per cent of the population, down from 18.1 per cent in 2002 (World Bank, 2016a, p. 8). Nevertheless, despite continuing economic growth, still nearly half the population clusters around the US$2-a-day poverty line and there is a large bulge in population just above the poverty line (World Bank, 2014a).

Economic development is geographically uneven, with eastern Indonesia consistently ranked as undeveloped and with eastern provinces such as the two in Papua and those in Nusa Tenggara consistently scoring lowest in the HDI rankings, highest in the maternal mortality statistics and lowest in levels of education. Indonesia's Gini index, which measures income inequality, now sits at a highly unequal 0.41 and is growing rapidly (World Bank, 2015, p. 42). Shockingly, the richest 1 per cent of the population controls 50.3 per cent of total national wealth, third in the world after Russia and Thailand (World Bank, 2016a, p. 18). Rural–urban disparities are large, with poverty concentrated in rural and remote areas.

Politics and government

For a seemingly interminable 32 years, Indonesia was under the so-called New Order government of ex-military General Suharto (1966–1998). Although he led a successful development program – introduced an effective family planning program, brought primary school enrolments to almost 100 per cent, and oversaw sustained economic growth and declining rates of poverty – his rule was authoritarian, corruption was rife and natural resources not managed sustainably. After his fall, Indonesia turned to democracy but simultaneously erupted into ethnic-religious violence. It patently failed to uphold its national motto of Unity in Diversity, as trouble spots flared in ugly, violent conflict. This mainly ethnic violence was over by 2004, although a separatist movement in Papua continues.

Under Suharto, government was highly centralised and stratified. Even young school children knew the top-down government structure of province, district, sub-district and village, with a large civil service and parallel military structure (Parker, 2003). The military was able to monitor civil administration by virtue of its "*dwi-fungsi*" role (dual function – military and socio-political control).

Post-Suharto, democratisation and decentralisation effectively strengthened ethnic and regional identities. Democratisation opened the way for expression of tension, and decentralisation meant that there was plenty to fight for: local elites played up tensions within communities in their own self-interest (Aspinall

& Mietzner, 2010; Turner *et al.*, 2003). Although there was "no clearly stated rationale for decentralization" in Indonesia (Turner *et al.*, 2003, p. iii), demo-cratisation entailed decentralisation. It was the notion of local community participation that articulated democratisation and decentralisation (Turner *et al.*, 2003, p. 6). Decentralisation connoted the devolution of finances, power and control of local affairs to local authorities, the accountability of local authorities to local constituencies, and the equitable distribution of each region's own wealth to its region (Aspinall & Fealy, 2003, p. 2).

Two ministries central to this study, the Ministry of the Environment and the Ministry of Education, were both decentralised. In environmental matters, decentralisation has been largely seen as detrimental to the environment, with more to fight for at the local level in terms of economic dividends, the opening of opportunities for local business people and politicians as well as larger corpo-rations to avoid national regulations that protect the environment, and lack of willpower and capacity in the bureaucracy and police to implement environ-mental protection policies (Setiawan & Hadi, 2007; World Bank, 2014c). However, Nomura (2007) argued that democratisation fostered the democrat-isation of environmental NGOs in Indonesia.

Education was to be decentralised as one important element of democrat-isation. The top-down, homogeneous system of education had to be made more responsive to local cultures and local needs; the management and curriculum of schools had to be devolved to lower levels of governance; and local educational institutions and ordinary people had to be given a voice, and empowered. Theoretically, the decentralisation of education would contribute to a "deepen-ing" of the culture of democracy (Parker & Raihani, 2011). The Indonesian education system is the subject of Chapter 5 and this issue will be discussed in more detail there.

Indonesia has now been democratic again for 20 years. There are regular free elections, a bicameral elected executive, a large number of political parties, and a free press, and the role of the military in politics and government has decreased significantly since the New Order. Interestingly, although Indonesia has become noticeably more Islamic since the 1980s, none of the Presidents since Suharto have come from Islamic parties.

Religion and culture

There are officially only six religions in Indonesia. According to the 2010 census, 87.5 per cent of the population follows Islam, 9.9 per cent Christianity (combining Protestants and Catholics), 1.7 per cent Hinduism, and smaller per-centages Buddhism and Confucianism (Ananta *et al.*, 2015, p. 257). An estim-ated 20 million persons practise animism and other types of traditional belief systems (Oslo Coalition, 2008). The first *sila* or principle in the Pancasila, which is the state ideology, is belief in one supreme God. Thus, Indonesia is neither a secular state nor an Islamic state, but it is a religious state. All citizens must have a religious identity, which is entered on to their ID card, is recorded in all official certificates such as birth and marriage certificates, and is recorded at all public offices, e.g. when children enrol at school. But more than this, "religion and faith are essential parts of national culture and daily life" in Indonesia (Azra, 2005, p. 1). Being religious is normal in Indonesia. In the public sphere, including schools, formal and informal events are opened with prayers; Indonesia has public holidays that mark important days in its six formally recognised religions; villages, schools and communities receive state funding for the building of mosques, temples and other houses of worship. In the sphere of education, religion is an inherent and very public element of schooling. Under all national governments since Independence, Religion has been a compulsory and examinable subject at all levels of schooling.

Since the 1980s, Islamisation has produced a more Islamic public space in Indonesia (Fealy & White, 2008). While Islam is very diverse in Indonesia, and there is a continuum from tolerant moderation to terrorism (Barton, 2005; van Bruinessen, 2002; van Bruinessen, 2013), fundamentalism has become an important force in Indonesian society. Although Islamist terrorism is the prism through which the Western media have mainly represented Indonesia since 2002, inside the country the trend has been towards more scrupulous observance of the five pillars of Islam and a much more public expression of piety. This has come to be labelled the "conservative turn" in Indonesian Islam (van Bruinessen, 2013). The religious nature of Indonesian society is not always realised by outside observers, and it will become clear through this book that this is an aspect of Indonesia that shapes understandings of the environment generally – i.e. that it is a creation of God – and therefore affects efforts to raise environmental awareness and promote pro-environment practice.

With more than 300 ethnicities and 700 living languages, Indonesia is one of the most culturally and linguistically diverse countries on earth.[3] There is no way to do justice to them in this potted introduction. In this book, because the field sites were in Java, the emphasis is on Javanese culture, but it is important to establish from the outset that Javanese culture does not represent Indonesia. There are stereotypes associated with each ethnic group – e.g. the Javanese are said to be obsessed with status differentiation, self-control and politeness; the Chinese are often considered arrogant and materialistic; the Minangkabau are clever with trade and speak their minds; and so on. Some ethnic groups map neatly onto a homeland, e.g. the Balinese live mainly in Bali; others are scattered, e.g. the Chinese and Malays; others have a homeland but are known for their diaspora, e.g. the Minangkabau and the Buginese. Ethnic conflicts, for instance after the fall of Suharto, have often been struggles for control of natural resources or economic interests. Some ethnic identities are difficult to define, e.g. the "Malays" (Fee, 2001; Long, 2013; Sakai, 2009). Sometimes an ethnic identity coincides with a distinct religion, e.g. the Balinese follow Balinese Hinduism; many ethnic groups are Muslim; and sometimes there are two ethnicities that follow the same religion and live intermingled, such that it can be difficult to determine who is which, e.g. the Muslim Makassarese and Buginese living in Makassar. Historically, Javanese culture has had a dominant impact across the archipelago, but at times the so-called "outer islands" have pushed back, feeling exploited, or neglected, and resentment has boiled over. Ethnic identity in Indonesia is complex.

The contemporary nation-state is basically the creation of the Netherlands in the late period of colonisation of the Netherlands East Indies, but it has been quite a feat of nation-building that Indonesia has survived and flourished to this day. The motto of Indonesia is "Unity in Diversity" and successive governments have worked hard, notably through the indoctrination of Pancasila, the state ideology, to unite the country and establish an "Indonesian culture". At times, such as under President Suharto, the interests of the different ethnic groups were downplayed (Acciaioli, 1985; Foulcher, 1990) but, throughout the 75 years of independence, the dominance of Java has prevailed. This pattern goes back to the time of the Hindu kingdoms – notably Majapahit, the greatest of the Javanese empires – and has provided Indonesia with a model of hierarchical government, terminology, style of leadership, and ideal culture (Anderson, 1990 [1972]; Cribb, 2001).

The island of Java is home to two main ethnicities: the Javanese, who dominate in central and eastern Java, and the Sundanese, in west Java. The majority religion in both is Islam. They have different languages but the salient feature of both is the importance of levels of language, with different vocabularies to denote relative rank and respect. The Javanese are far and away the largest ethnicity in Indonesia, comprising 40 per cent of the total number of Indonesian citizens. Javanese culture figures large in the ethnographic chapters of this book. In many spheres – particularly in the character of the bureaucracy, in school culture, leadership and family life – the strength of Javanese culture is evident, and must be taken into account as we look for ways to instantiate pro-environment sensibilities and conduct.

Environment

As a tropical, archipelagic nation, straddling the equator and with an active volcanic geology, Indonesia has an abundance of natural resources, including oil and gas and forests. In central and western Indonesia, such as in Java, a tropical, predictable climate and fertile soils enable agricultural production, traditionally supporting the populace mainly through sustainable wet rice agriculture, supplemented by dry rice and cash crops such as coffee, rubber, sugar and spices. The eastern half of Indonesia is much drier, less volcanic and less fertile, and is marked with much higher incidences of poverty, malnutrition and the like. Indonesia is strategically important, providing a narrow conduit between the Indian Ocean and the Pacific, linking Europe, the Middle East and South Asia to East Asia.

To even the most casual visitor to Indonesia, it is obvious that Indonesia is dirty: pedestrians and people in cars and buses mindlessly drop rubbish on the street – even when there are signs prohibiting the dumping of rubbish. National parks are full of litter – walking trails are lined with discarded plastic bottles.

Indonesia is the second worst plastics polluter in the world after China (Jambeck *et al.*, 2015). Rivers are obviously polluted. The air in cities such as Jakarta and even in non-industrial cities such as Yogyakarta, is so full of exhaust fumes and smoke that one quickly develops a cough and gets used to scrubbing one's face and cleaning out one's eyes every night. Evidence of low levels of environmental awareness abounds: people in cities turn on taps and let water flow continuously; televisions and lights are left on all day, even when no one is home.

Indonesia's environmental problems range from loss of species and reduction of biodiversity, destruction of habitat and massive deforestation through burning and logging, extensive mono-cultural planting (particularly palm oil), erosion, mining and over-exploitation of resources of all kinds, to industrial and vehicular air pollution, water pollution, uncontrolled urban development and excessive carbon emissions.

Natural resources are being rapidly depleted, with direct consequences for the people who rely on them. Climate change is reducing crop production, increasing risks of flooding and landslides, threatening coastal communities, and increasing the spread of vector-borne diseases. Poor sanitation, pollution and inadequate waste disposal will continue to have negative effects on the health of humans and the environment. Human vulnerability to environmental risks varies according to social position and identity (Phillips *et al.*, 2010).

There are so many environmental problems, and they're so interconnected, it's hard to know where to start if you're an environmentalist. This summary from the World Bank is useful, though it's important to note that it is focused on climate change:

> Indonesia is the third largest emitter of greenhouse gases in the developing world after China and India. These emissions stem largely from deforestation, peatland conversion, and associated fires, together with electricity generated by coal-fired power plants and the consumption of fossil fuels in the energy and transport sectors, also associated with high fuel subsidies and rapid urbanization. Composed of over 13,000 islands, Indonesia is also one of the most vulnerable countries to the rising adverse impacts of global climate change, including extreme weather events – tropical storms and droughts – and sea level rise, particularly on account of the concentration of much of its population in lowland areas.
>
> (World Bank, 2016b, p. xi)

More generally, available data suggest that Indonesia's GHG emissions have continued to rise in recent years, at least through 2012, due to persisting high rates of deforestation, peatland conversion, and fires, as well as growing fossil fuel-based energy consumption. Electricity subsidies were finally reduced somewhat as of late 2013 and geothermal energy investments increased in part with financial support from the World Bank and the Clean Technology Fund (CTF). However, the share of renewables in Indonesia's energy mix remains very low (around 3 per cent) and is expanding very slowly, as coal and oil continue to strongly predominate. Forest and land use management also persist as major challenges, while REDD+ implementation has advanced very slowly and had very limited results on the ground to date.

(World Bank, 2016b, p. xiii)

Environmental awareness in Indonesia

We know very little about environmental values and awareness in Indonesia. There are almost no statistical data on perceptions of the environment or of environmental awareness in Indonesia. Given the many environmental problems in Indonesia, this lacuna is quite serious.

In the Global North, researchers have addressed how, when and why individuals become environmentalists, but countries in the Global South are much less studied. Chawla (1998, 1999), studying life histories ("life paths") of individual environmental activists in the US and Norway, identified these influences: childhood experiences in nature; experiences of environmental destruction; environmental values held by the family; environmental organisations; role models (friends or teachers); and education (Chawla, 1998, 1999). Her study is often too cavalierly encapsulated in the acronym SLE (Significant Life Experiences); it was controversial and raised many questions for further research (Chawla, 1998, 2001; Gough, 1999). To our knowledge, these influences have not been identified for environmentalists in Indonesia.[4]

However, our goal in this book is not to seek out individual environmentalists to find their motivations. Rather, we are interested in seeing how formal education can create practising pro-environment citizens, who share an environmentalist subjectivity. Chawla noted that she believes that besides individual environmental activists who variously demonstrate, petition, lobby and participate in direct action to save wilderness, stop fracking, etc., environmentalism requires

> a large population of citizens who support the protection of the environment in other ways as well: through their voting records on state and local referenda, through holding politicians accountable for their environmental positions, through recycling, reducing consumption and other day-to-day behaviors.
>
> (Chawla, 2001, p. 455)

The processes of creating new environmental subjectivities and practice *en masse*, and thus a new social movement for environmental sustainability, are complex, but all point to the value of education, whether formal or informal. Many hundreds of studies have been done in Western countries, using different disciplines and various theoretical frameworks. For most countries of the Global North, we have small- and large-scale survey data on attitudes towards the environment, e.g. many using the NEP (New Environmental Paradigm)

(Dunlap, 2008; Dunlap & Van Liere, 2008 [1978]; Dunlap *et al.*, 2000). Kollmuss and Agyeman (2002) usefully surveyed this literature, noting the ubiquity of the "gap" between environmental sensibilities and environmental action. Work has begun in a few non-Western contexts – including a Japanese study of Hong Kong, Japan, Thailand, and Vietnam (Nickum & Rambo, 2003; Rambo *et al.*, 2003); another of China and Thailand (Nisihira *et al.*, 1997); there are a few specialised studies (e.g. Montana & Mlambo, 2019); and a few large comparative (quantitative) studies (e.g. Davino *et al.*, 2018). However, as far as we are aware, no one has conducted a large-scale survey of attitudes towards the environment in Indonesia,[5] nor has there been other research about how education can contribute to creating an environmental subjectivity among peoples in Indonesia.

What we have are a few statements, like the one at the beginning of the chapter from the World Bank, and acknowledgements from the Indonesia government, such as the 2004 statement above, that are probably accurate but are not based on survey evidence, and a few small-scale studies with particular interests, e.g. a study of the attitudes to environmental issues of teachers in high schools in Jakarta (Hadisuwarno, 1997); and a study of the knowledge of students on the sustainable management of natural resources at the pre-eminent university in Indonesia in the field, Institut Pertanian Bogor (Koch *et al.*, 2013). There are two related surveys done by Pew on perceptions of climate change. The 2006 survey, conducted in both developed and developing countries, asked "Have you ever heard of the environmental problem of global warming?" and Indonesia was one of the most *climate-unaware* countries in the world, with a considerably majority of the sample answering "No" (cited in Leiserowitz, 2007, p. 4). By 2015, in answer to a different question, 41 per cent of a different Indonesia sample said "Global climate change is a very serious problem" (Pew Research Center, 2015, p. 13).

Ma'ruf, Surya, and Apriliany (2016) surveyed the relationships among environmental knowledge, attitudes and intended behaviour of students at three different universities in Padang, Jakarta and Denpasar, Indonesia. While their results showed that students had good knowledge about the environment, there were only two questions on this in their survey. Their study confirmed findings elsewhere, that knowledge affects attitudes, and that attitudes affect environmental behaviour intention; but their study revealed that only 32.3 per cent of environmental behaviour intention was affected by environmental knowledge and attitudes.

The study by Kusmawan *et al.* (2009) was of Chemistry students in senior high school in Indonesia. Kusmawan was interested to find out how different teaching styles affected students' beliefs and attitudes towards the environment. He found that "positive changes in students' beliefs and attitudes were somewhat greater among the students taking a more active approach to environmental learning than those in the group who were taught only in the usual classroom manner" and that "More active learning approaches seemed to promote cohesion between beliefs, attitudes and intentions, with participation in community issues having a greater impact on student ecological affinity than field research projects." (Kusmawan *et al.*, 2009, p. 257).

Bohensky, Smajgl, and Brewer (2103) surveyed 6310 households in rural, peri-urban and urban areas in East Kalimantan in 2007 and Central Java in 2008, to investigate people's engagement with climate change. They were interested in the steps and sequences of engagement: observation (experience) of climate change in their lifetime, risk perception, reactive action and proactive action. A surprisingly strong majority of respondents had observed climate change (81.9 per cent); 70.7 per cent considered climate change a risk; 38.9 per cent were taking reactive measures and 28.2 per cent were taking proactive action. The authors were surprised at the high level of awareness, given the study by Leiserowitz (2007), but dismayed at the low level of conversion into action – though many studies (summarised in Kollmuss & Agyeman, 2002), have observed this "gap". One must question the validity of people's reported observations of climate change, as scientists generally require a minimum period of 30 years as constituting a unit of time in which genuine climate change could be considered. Anything less than this is just "weather".[6]

Van der Laarse (2016) argued that there is a growing environmental movement and awareness in Indonesia, and we would not challenge that, but van der Laarse does acknowledge a string of earlier scholars who "have been quite negative about the existence of an environmental awareness in Indonesia" (van der Laarse, 2016, p. 9), for example:

> In reality, neither the global population nor the Indonesian population as a whole feel responsible for their present citizens, let alone for future generations.
>
> > (Colombijn, 1998, p. 306)

> Indonesian consumers' environment consciousness is still weak.
>
> > (Sudiyanti, 2009, p. 2)

> Among the Indonesian public the conservationist constituency is very small, consisting mainly of a few university faculty and natural scientists. For the great majority of Indonesians, the environmentalists' concern with complex interdependencies and long-term consequences appear irrelevant to their lives.
>
> > (Aden, 1975, p. 988)

It is rather difficult to make any generalised conclusions from the small isolated studies identified above, but this World Bank assessment seems accurate:

> Public awareness is an essential part of the effort to address Indonesia's environmental problems, from disaster risks to biodiversity conservation. Informed and aware citizens can take action to address environmental issues, and can form constituencies for improved efforts at the political and

local government level. At a broader level, however, environmental values are not deeply embedded in society, leading to undervaluation of natural resources and environmental services. Participation and voice in decision making is an essential element of good governance. Recent environmental disasters (floods, mud, fires, erosion) have stimulated greater environmental concern, but further analysis of knowledge, attitudes and practices would be needed to determine how far or deep this understanding goes outside of urban centers, and what tools can best be used to build on this basic awareness.

(World Bank, 2014c)

Our own impressions from long-term fieldwork in Bali, Yogyakarta, Surabaya and West Sumatra and for shorter periods elsewhere in Indonesia, support the statements of the World Bank and the government (below). One of our team members, Pam Nilan, surveyed 804 undergraduate university students in four different cities: Palembang in Sumatra and Jakarta, Bandung and Yogyakarta in Java, and found that more university students said they were not an environmentalist (52.2 per cent) than said there were (47.8 per cent). Another way to measure environmental sensibilities (or the lack of) in democratic Indonesia is to look at the platforms of political parties. It has to be said that not only is there not a "green" party in Indonesia, but also few political parties ever mention the environment. Research by Indonesia's foremost and nationwide environmental organisation, WALHI, found that in the first free elections after the fall of President Suharto in 1998, only four of the 48 political parties to contest the election had placed the environment on their main agenda (WALHI, 2016).

Government attitudes

In Indonesia, the establishment of environmental responsibility at the level of national government is credited to Emil Salim, Indonesia's first Minister for the Environment. He was one of the 23 members of the Brundtland Commission, which came up with the concept and standard definition of Sustainable Development: "Sustainable development is development that meets the needs of the present without compromising the ability of future generations to meet their own needs" (World Commission on Environment and Development, 1987, p. 41). While the concept has been much criticised by environmentalists, the term (*pembangunan berkelanjutan* in Indonesian) has some purchase among government officials in Indonesia and is used in its rhetoric in international fora.

Indonesian researchers including Mohamad Soerjani (1993) and the Japanese researcher, Ko Nomura, began the task of researching EE in Indonesia (Nomura, 2007; Nomura & Abe, 2005; Nomura & Hendarti, 2005). Soerjani's work is among the earliest. He was the Director of the Centre for Research of Human Resources and the Environment, and the Chair of Postgraduate Study in Environmental Science at Indonesia's pre-eminent university, Universitas

Indonesia. We quote from his work at length because it reflects the attitudes towards the environment and how to deal with it, during the New Order (1966–1998):

> Indonesia is a country rich in natural resources, but lacks sufficient qualified personnel to manage these resources in a sustainable way.... While the country is still facing many challenges to the achievement of prosperity and equality among all its people, its rich ... natural resources are gradually being managed and developed in a more sustainable way.... The islands have an abundance of natural resources and in 1988–89 a population of over 176 million people ... a unique tropical climate and fertile land covered by thick tropical rainforests and replenished by volcanic eruptions, but also a rich supply of resources and minerals with significant economic importance.... Indonesia's human resources are also abundant, but at present a serious lack of education and technical skills, as well as uneven population distribution, hinder their optimal contribution to national development.... The Government of Indonesia's present environmental concern is largely due to population pressure resulting from the high rate of population growth and the inadequate level of education.
>
> (Soerjani, 1993, pp. 146, 147, 149–150)

The main tropes in this government discourse were: the environment comprises rich natural resources available for exploitation; these resources need to be managed by educated people; Indonesia has too many people and too many poor and uneducated people, and this causes environmental destruction. However, it must also be acknowledged that Soerjani understood that centralised economic growth, without consideration of its environmental impact, would be deleterious to the environment, and he supported the introduction of legislation for environmental impact assessments of development projects.

This sense of abundant natural resources is ubiquitous in government discourse on the environment in Indonesia and is a feature of school curricula (discussed in Chapter 6). While not inaccurate, the constant iteration of abundance, without qualification, has the discursive effect of enabling unlimited exploitation while neglecting the need for careful stewardship and sustainable consumption.

Under Soeharto, Indonesia participated in the two landmark United Nations Environment conferences, the UN Conference on the Human Environment, held in Stockholm in 1972, and the UN Conference on Environment and Development (UNCED, the "Earth Summit") held in Rio de Janeiro in 1992.

> There was a ten-member Indonesian NGO delegation at Rio. The top priority for the Indonesian and Malaysian governments at UNCED was to avoid any restrictions on the freedom to exploit their forests. They denied that forests are a global common, and together these two states were strong enough to block the formulation of a new international regime. Developed

countries regarded the final text on forestry as worse than no declaration at all, and Friends of the Earth and Greenpeace condemned the forestry principles of Agenda 21 as a "chain-saw decision". In the wake of the Earth Summit, President Soeharto proclaimed 1993 as the Year of the Environment for Indonesia (Inside Indonesia June 1993; McCoy and McCully 1994: 115; Porter and Brown 1996: 117, 124–126). The billboards that brought notice of the Year of the Environment to the general public displayed popular threatened species (rhinoceros, birds of paradise, rafflesia) but avoided logging, traffic congestion, industrial pollution and other socio-economic issues.

(Colombijn, 1998, p. 315)

The Government has acknowledged that more needs to be done. The Indonesian Minister for the Environment wrote: "We must acknowledge that our attention to and consciousness of environmental problems is still very low. This is caused by the fact that most people in Indonesian society have not yet awoken to a real perception of the environment" (Kantor Menteri, 2004).

Indonesia participated in the UN's Decade of Education for Sustainable Development (2005–2014) and there have been various government commitments to providing environmental education in Indonesia, e.g. a joint agreement between the Minister of State for the Environment and the Minister of National Education, 3 June 2005, KEP-07/MENLH/06/2005 and 05/VI/KB/2005 on the Development of Environmental Education, and Law No. 32 of 2009 on Environmental Protection and Management.

Under President Susilo Bambang Yudhoyono, Indonesia presented globally as a "good environmental citizen" – notably in intervening actively in the 2007 Bali Climate Change Conference to make that a success. Unfortunately, his successor, President Joko Widodo, has turned back some initiatives and not even presented well internationally on the environment front (McLellan, 2015). Indonesia still lacks the political will and the bureaucratic expertise to implement environmental regulations, and there is a serious lack of leadership and interest in environmental matters from politicians.

Government capacity

The public service in Indonesia is a large and sluggish institution, consisting of some 4.7 million civil servants, plus the police and military and a host of auxiliary, honorary and other workers (Buehler, 2011, p. 66). It is marked by a high level of underemployment, low levels of education and capacity, systemic corruption and a low level of inter-agency coordination.

The bureaucracy is highly inefficient and ineffective in delivering public services at both the national (McLeod 2005) and sub-national level (Von Luebke 2009: 225) and, according to USAID (2009: 47), the quality of service delivery has stagnated over the last 10 years.

(Buehler, 2011, p. 66)

As noted above, the (now) Ministry of the Environment and Forestry is one of the decentralised ministries. However, when it comes to care and management of the environment, both in Jakarta and in the regions, the general traits of the civil service are very obvious:

> First, despite the substantial investment in environment and natural resources policy and staff development, actual implementation of rules and procedures has been poor and slow due to weak commitment by sector agencies, low awareness in local departments and capacity challenges at all levels. Also, awareness about the expected negative environmental impacts of sustained economic growth and the mechanisms for stakeholders to hold government agencies accountable for their performance are weak. Second, there is little integration of environmental considerations at the planning and programmatic levels, especially in the public investment planning process and in regional plans for land and resource use.
>
> (World Bank, 2014c)

We will see in the ethnographic chapters (Chapters 7–10) how important this lack of capacity with regards to the environment is for EE in schools.

NGOs, universities and environmental awareness

There is not yet an environmentalist ethos prevalent in society. It is true that there has long been an environmental movement in Indonesia, and this has been studied elsewhere (e.g. Aditjondro, 1990, 1998; Colombijn, 1998; Cribb, 1988; Eldridge, 1995; van der Laarse, 2016). Environmentalism has long been associated with universities and environmental NGOs in Indonesia – attempting to fill the void of government activity in this space. This was at least understandable during the New Order, when rampant exploitation of natural resources was permitted because of crony capitalism. It was not in the interest of government to regulate natural resource management or foster environmentalism (Gordon, 1998; Setiawan & Hadi, 2007). Further, although NGOs had to work closely with government during authoritarian rule, protest about environmental destruction was a way of raising class and development issues "under the radar".

> "The only way to be anti-Suharto is through the environmental movement," one NGO activist went so far as to say. "There is no other way to talk about Suharto or human rights."
>
> ("Hands off! Fighting for the Environment in Asia," Newsweek (International Edition) (22 April 1991), 45, cited in Gordon, 1998, p. 2)

Under democracy, the reasons for the lack of purchase of ENGOs among the broader society are more elusive. It remains to be seen if the burgeoning middle classes will take up the environment cause as they have in the Global North.

The environmental movement really began in the universities. In the 1970s, five Environmental Study Centres (ESCs) were established at universities; by 1994 there were 53, raising "environmental awareness among academics in Indonesia" (Setiawan & Hadi, 2007, p. 75). Collaborating with ENGOs, ESC members were active in training, research, environmental monitoring and community outreach. Aditjondro (1998, p. 45) reported that a "radical strand of environmentalism" emerged from the convergence of campus and non-campus-based activists in the 1980s, as student groups helped NGO activists to organise dispossessed peasants in land sovereignty claims (Peluso, Afiff, & Rachman, 2008). Although some agrarian studies departments showed a radical environmental trend in the 1990s (White, 2005, p. 130), humanities and social science disciplines, rather than physical and biological sciences, now address pressing environmental issues for Indonesia, often as studies of the environment in student theses (White, 2005, p. 131). This disciplinary location is interesting, given that in schools, it is usually presumed that the natural science subjects and teachers are the repository of environmental knowledge and awareness. Student groups have more recently focused on the preservation of endangered animals and challenged the destruction of forests and peatlands for palm oil cultivation. In the new millennium, the popularity of campus environmental and nature conservation groups continues, with the growing involvement of Muslim student groups.

Nevertheless, the heart of the environmental movement is the environmental NGOs (ENGOs). We are using this term to cover a wide variety of organisations: they range from the nationwide umbrella organisation, WALHI, and local branches of the large international ENGOs such as Greenpeace and The Nature Conservancy (TNC), to small, ephemeral, single-issue groups for which the word "organisation" is not really appropriate. Despite their activism and advocacy, we would argue, contra Nomura (Nomura, 2009; Nomura & Abe, 2005), that ENGOs in Indonesia have not been active in EE. This assessment comes from our quest on the ground in many places in Indonesia to find ENGOs that are actively "doing" EE. In Bogor, RMI (the Indonesian Institute for Forest and Environment) is very active and innovative (http://rmibogor.id/; Tillah & Rahman, 2017); and the occasional single-issue ENGO, such as COP (The Centre for Orangutan Protection) is doing a good job. Many of the ENGOs putatively based in Yogyakarta only have an online presence, and most have no expertise in *education*. Admittedly there is a grey area between advocacy and education, where both occur, and there are occasional events, such as Coral Day or Earth Day, at which a day of fun activities sometimes incorporates an educational element. In our experience this might consist of a video showing or posters conveying facts – but no one reads them. There is an EE Network, Jaringin Pendidikan Lingkungan (JPL),[7] which occasionally organises "professional development" events for activists from ENGOs. A JPL representative identified "the evaluation of the effectiveness of our EE efforts" as the most pressing issue for JPL (Interview 9 June 2011).

Although most activists in ENGOs in Indonesia can be assumed to be of the educated middle classes (Hadiwinata, 2003), it is small farmers, the landless, and lower (often rural) classes that generally engage in much of the activism and protest against development projects (e.g. Aditjondro, 1998; Warren, 1998).

> In Indonesia there is no mass movement like the green lobby in the North. The group of people that defend the environment breaks down into the direct victims of ecological degradation and a mixed group of professionals in the government, consultancies, environmental study centres of universities and NGOs.
>
> (Colombijn, 1998, p. 312)

In the context of a developing country, it is important to recognise that many local environmental issues are also issues of social justice, and when reported in the media, it is the social conflict, rather than the environmental damage, that is reported, e.g. when large dams, highways or "ecotourism" resorts displace local farmers in the name of Development (Colombijn, 1998). It is worth quoting Colombijn at length to convey the connection between what Western outsiders might see as an environmental problem and what locals see as a social justice issue:

> Local groups perceive ecological deterioration as a social problem only when their livelihood is immediately threatened. At the root of almost every environmental problem in Indonesia, once it has been defined as such, is a social conflict: citizens against the government, the poor against big entrepreneurs, local people against government protégées. The inherent social conflict makes even minor changes that could reduce local environmental degradation difficult to achieve.
>
> (Colombijn, 1998, p. 329)

> Sawah cultivators see their rice wither when golf links consume too much water, and their yields may decline by a third. Laundresses get itchy skin from washing clothes in the river when a new factory upstream discharges toxic effluent. Shifting cultivators run the risk of overexploiting a diminished resource base when a logging concessionaire closes forest land to which the cultivators have traditional but unregistered rights. Women in urban kampung have increasing difficulties drawing water when a nearby factory uses an electric pump. Urban dwellers may also suffer from industrial noise, dust, or smoke. These people are not concerned about erosion, loss of biodiversity, global warming, or marine pollution. They face immediate loss of income or health.
>
> They feel that they are victims, and they ask themselves: "why in my back yard?" (wimby).
>
> (Colombijn, 1998, p. 323)

Thus, once a felt "environmental problem" is identified, it becomes a social issue, and not infrequently a social conflict. This may be one of the explanations for why ENGOs have not been successful in raising a widespread environmental consciousness in Indonesia: not unreasonably, social inequalities "hijack" debates and surpass the environmental issue as the problem.

Conclusion

This chapter has introduced Indonesia to the uninitiated. The principal arguments have been that while Indonesia has reduced the proportion of the population who are poor, it is still a developing country with a very large and diverse population, a difficult geography, and severe environmental problems. Population growth and rampant exploitation of Indonesia's rich natural resources are major causes of environmental concern. A rapidly growing, and consuming, middle class and urbanisation are two salients of Indonesia's demography, and both contribute to ecological destruction. Some of the successes of Indonesia in the last 20 years are the return of democratic government, improved educational access, and the beginnings of a social welfare system. In the face of government ignorance and ineffectiveness in managing its rich natural resources, particularly under conditions of decentralised government, ENGOs have stepped into the space. However, they have not been effective in educating the populace about the environment. Overall, we underline that the prevailing ignorance and lack of awareness around the environment is one of Indonesia's major problems. An effective, non-corrupt democratic government would have to respond to nationwide pressure to address Indonesia's dire environmental problems. Unfortunately, there is no such upward pressure in Indonesia.

The next chapter introduces the formal education system in Indonesia.

Notes

1 The UN estimates it will be 313 million by 2100 (UNDESA, 2015, p. 20) but 322 million by 2050, with a medium variant projection (UNDESA, 2015, p. 20).
2 Europe's median age currently is 42 years; the median age for the least developed countries as a whole is 20 years in 2015 (UNDESA, 2015, p. 9).
3 The anthropological study of ethnicities has shifted from the study of "ethnic groups" to the study of "ethnic identities" and "ethnicities", in line with the postmodern turn. While fascinating, it is not necessary for the purposes of this book to explore the ramifications of this shift for environmental education in Indonesia. A solid statistical study of Indonesia's ethnicity, using the most recent census data of 2010 is Ananta *et al.* (2015). They describe some of the difficulties of classifying ethnicities in Indonesia, and note that, in any case, people should be able to self-identify their ethnicity.
4 The exception is Nilan and Wibawanto (2015).
5 In this book we are not talking about ethno-environmental attitudes and knowledge, i.e. local knowledge and "traditional" understanding of local ecologies. While we recognise that these are important in the context of natural resource management at the local level (Laumonier, Bourgeois, & Pfund, 2008), such understandings are outside the scope of this book.

6 The World Meteorological Organisation, Inter-governmental Panel on Climate Change and other eminent organisations use a 30-year period as the standard reference period with regard to climate change. The period 1961 to 1990 is the last standard reference period and 1991–2020 is the next standard reference period.
7 See Nomura (2009); Nomura and Abe (2005); Nomura and Hendarti (2005).

References

Acciaioli, G. (1985). Culture as art: From practice to spectacle in Indonesia. *Canberra Anthropology*, 8(1 and 2), 148–174.

Aden, J. B. (1975). The relevance of environmental protection in Indonesia. *Ecology Law Quarterly*, 4(4), 987–1006.

Aditjondro, G. J. (1990). *The Emerging Environmental Movement in Indonesia.* Salatiga: Universitas Kristen Satya Wacana.

Aditjondro, G. J. (1998). Large dam victims and their defenders: The emergence of an anti-dam movement in Indonesia. In P. Hirsch & C. Warren (Eds), *The Politics of Environment in Southeast Asia: Resources and Resistance* (pp. 29–54). London: Routledge.

Ananta, A., Arifin, E. N., Hasbullah, M. S., Handayani, N. S., & Pramono, A. (2015). *Demography of Indonesia's Ethnicity.* Singapore: Institute of Southeast Asian Studies.

Anderson, B. R. O. G. (1990 [1972]). The idea of power in Javanese culture. In *Language and Power: Exploring Political Cultures in Indonesia* (pp. 17–77). Ithaca, NY: Cornell University Press.

Aspinall, E., & Fealy, G. (Eds). (2003). *Local Power and Politics in Indonesia: Decentralisation & Democratisation.* Singapore: Institute of Southeast Asian Studies.

Aspinall, E., & Mietzner, M. (Eds). (2010). *Problems of Democratisation in Indonesia: Elections, Institutions and Society.* Singapore: Institute of Southeast Asian Studies.

Azra, A. (2005). *Teaching Tolerance through Education in Indonesia.* Paper presented at the Symposium on Cultivating Wisdom, Harvesting Peace. Educating for a Culture of Peace through Values, Virtues, and Spirituality of Diverse Cultures, Faiths, and Civilizations, Griffith University.

Barton, G. (2005). *Jemaah Islamiyah: Radical Islamism in Indonesia.* Singapore: Ridge Books, an imprint of Singapore University Press.

Bhaduri, R., & Monroe, L. (2010). Indonesia's growth led by its private sector. Retrieved from http://english.alrroya.com/content/indonesia'sgrowth-led-its-private-sector.

Bohensky, E. L., Smajgl, A., & Brewer, T. (2013). Patterns in household-level engagement with climate change in Indonesia. *Nature Climate Change*, 3(April), 348–351. doi:10.1038/NCLIMATE1762.

Buehler, M. (2011). Indonesia's law on public services: Changing state-society relations or continuing politics as usual? *Bulletin of Indonesian Economic Studies*, 47(1), 65–86.

Chawla, L. (1998). Research methods to investigate significant life experiences: Review and recommendations. *Environmental Education Research*, 4(4), 383–397.

Chawla, L. (1999). Life paths into effective environmental action. *Journal of Environmental Education*, 31(1), 15–26.

Chawla, L. (2001). Significant life experiences revisited once again: Response to vol 5(4) "Five critical commentaries on significant life experience research in environmental education". *Environmental Education Research*, 7(4), 451–461.

CIA (Central Intelligence Agency). (2018). *The World Factbook.* Retrieved from www.cia.gov/library/publications/the-world-factbook/geos/id.html.

Colombijn, F. (1998). Global and local perspectives on Indonesia's environmental problems and the role of NGOs. *Bijdragen tot de Taal-, Land en Volkenkunde, 154*(2),
305–334.

Cribb, R. (1988). *The Politics of Environmental Protection in Indonesia*. Clayton, Vic.: The
Centre of Southeast Asian Studies, Monash University.

Cribb, R. (2001). Independence for Java? New national projects for an old empire. In G.
Lloyd & S. Smith (Eds), *Indonesia Today: Challenges of History* (pp. 298–310). Singapore: Institute of Southeast Asian Studies.

Davino, C., Vinzi, V. E., Santacreu-Vasut, E., & Vranceanu, R. (2018). An attitude
model of environmental action: Evidence from developing and developed countries.
Social Indicators Research, August, 1–28. doi:10.1007/s11205-018-1983-3.

Dunlap, R. E. (2008). The new environmental paradigm scale: From marginality to
worldwide use. *The Journal of Environmental Education, 40*(1), 3–18.

Dunlap, R. E., & Van Liere, K. D. (2008 [1978]). The "New Environmental Paradigm":
A proposed measuring instrument and preliminary results [Reprint of original 1978
article]. *The Journal of Environmental Education, 40*(1), 19–28.

Dunlap, R. E., Van Liere, K. D., Mertig, A. G., & Jones, R. E. (2000). Measuring
endorsement of the new ecological paradigm: A revised NEP scale. *Journal of Social
Issues, 56*(3), 425–442.

Eldridge, P. J. (1995). *Non-Government Organisations and Democratic Participation in Indonesia*. Kuala Lumpur: Oxford University Press.

Fealy, G., & White, S. (Eds). (2008). *Expressing Islam: Religious Life and Politics in Indonesia*. Singapore: Institute of Southeast Asian Studies.

Fee, L. K. (2001). The construction of Malay identity across nations Malaysia, Singapore, and Indonesia. *Bijdragen tot de Taal-, Land en Volkenkunde, 157*(4), 861–879.

Foulcher, K. (1990). The construction of an Indonesian national culture: Patterns of
hegemony and resistance. In A. Budiman (Ed.), *State and Civil Society in Indonesia*
(pp. 301–320). Clayton, Vic.: Centre of Southeast Asian Studies, Monash University.

Gordon, J. (1998). NGOs, the environment and political pluralism in new order Indonesia. *Explorations: A Graduate Student Journal of Southeast Asian Studies, 2*(2), 47–68.

Gough, S. (1999). Significant life experiences (SLE) research: A view from somewhere.
Environmental Education Research, 5(4), 353–363.

Hadisuwarno, H. (1997). *High School Teachers' Knowledge of and Attitude Toward Environmental Issues in Jakarta, Indonesia*. PhD Thesis, Florida State University.

Hadiwinata, B. (2003). *The Politics of NGOs in Indonesia: Developing Democracy and Managing a Movement*. London: RoutledgeCurzon.

Indonesia Investments. (2019). *Gross Domestic Product of Indonesia*. Accessed 12
June 2019. www.indonesia-investments.com/finance/macroeconomic-indicators/gross-
domestic-product-of-indonesia/item253.

Jambeck, J. R., Geyer, R., Wilcox, C., Siegler, T. R., Perryman, M., Andrady, A., …
Law, K. L. (2015). Plastic waste inputs from land into the ocean. *Science, 347*(6223),
768–771. doi:10.1126/science.1260352.

Kantor Menteri Negara Lingkungan Hidup (Office of the State Ministry for the Environment). (2004). *Kebijakan Pendidikan Lingkungan Hidup*. Jakarta: Kementerian
Lingkungan Hidup (Ministry of the Environment).

Koch, S., Barkmann, J., Strack, M., Sundawati, L., & Bögeholz, S. (2013). Knowledge of
Indonesian university students on the sustainable management of natural resources.
Sustainability, 5(4), 1443–1460. doi:10.3390/su5041443.

Kollmuss, A., & Agyeman, J. (2002). Mind the gap: Why do people act environmentally and what are the barriers to pro-environmental behavior? *Environmental Education Research*, 8(3), 239–260.

Kusmawan, U., O'Toole, J. M., Reynolds, R., & Bourke, S. (2009). Beliefs, attitudes, intentions and locality: The impact of different teaching approaches on the ecological affinity of Indonesian secondary school students. *International Research in Geographical and Environmental Education*, 18(3), 157–169.

Laumonier, Y., Bourgeois, R., & Pfund, J.-L. (2008). Accounting for the ecological dimension in participatory research and development: Lessons learned from Indonesia and Madagascar. *Ecology and Society*, 13(1), 15.

Leiserowitz, A. (2007). *International Public Opinion, Perception, and Understanding of Global Climate Change*. UNDP (United Nations Development Program) Occasional Paper. Retrieved from https://core.ac.uk/download/pdf/6248846.pdf.

Long, N. (2013). *Being Malay in Indonesia: Histories, Hopes and Citizenship in the Riau Archipelago*. Singapore: Asian Studies Association of Australia in association with NUS Press and NIAS Press.

Ma'ruf, Surya, S., & Apriliany, P. D. (2016). Knowledge, attitudes and behavior of university students towards environmental issues in Indonesia. *Sains Humanika*, 8(1–2), 81–88.

McLellan, S. (2015). Climate policy under SBY and Jokowi: Making progress or going backwards? *Australian Outlook*. Retrieved from http://www.internationalaffairs.org.au/australianoutlook/climate-policy-under-sby-and-jokowi-making-progress-or-going-backwards/.

Montana, M., & Mlambo, D. (2019). Environmental awareness and biodiversity conservation among resettled communal farmers in Gwayi Valley Conservation Area, Zimbabwe. *International Journal of Sustainable Development & World Ecology*, 26(3), 242–250. doi:10.1080/13504509.2018.1544946.

Nickum, J. E., & Rambo, A. T. (2003). Methodology and major findings of a comparative research project on environmental consciousness in Hong Kong (China), Japan, Thailand, and Vietnam. *Southeast Asian Studies*, 41(1), 5–14.

Nilan, P., & Wibawanto, G. R. (2015). "Becoming" an environmentalist in Indonesia. *Geoforum*, 62(2), 61–69.

Nisihira, S., Kojima, R., Okamoto, H., & Fujisaki, S. (Eds). (1997). *Environmental Awareness in Developing Countries: The Cases of China and Thailand*. Tokyo: Institute of Developing Economies.

Nomura, K. (2007). Democratisation and environmental non-governmental organisations in Indonesia. *Journal of Contemporary Asia*, 37(4), 495–517.

Nomura, K. (2009). A perspective on education for sustainable development: Historical development of environmental education in Indonesia. *International Journal of Educational Development*, 29, 621–627.

Nomura, K., & Abe, O. (2005). The environmental education network in Indonesia. In K. Nomura & L. Hendarti (Eds), *Environmental Education and NGOs in Indonesia* (pp. 125–137). Jakarta: Yayasan Obor Indonesia.

Nomura, K., & Hendarti, L. (Eds). (2005). *Environmental Education and NGOs in Indonesia*. Jakarta: Yayasan Obor Indonesia.

Oslo Coalition. (2008). *Indonesia Report*. Retrieved from www.oslocoalition.org/resources_upr_reviews.php.

Parker, L. (2003). *From Subjects to Citizens: Balinese Villagers in the Indonesian Nation-state*. Copenhagen: Nordic Institute of Asian Studies Press.

Parker, L., & Raihani. (2011). Democratizing Indonesia through education? Community participation in Islamic schooling. *Education Management, Administration and Leadership, 39*(6), 712–732.

Peluso, N. L., Afiff, S., & Rachman, N. F. (2008). Claiming the grounds for reform: Agrarian and environmental movements in Indonesia. *Journal of Agrarian Change, 8*(2 and 3), 377–407.

Pew Research Center. (2015). *Global Concern about Climate Change; Broad Support for Limiting Emission*. Retrieved from www.pewglobal.org/2015/11/05/global-concern-about-climate-change-broad-support-for-limiting-emissions/.

Phillips, B. D., Thomas, D. S. K. T., Fothergill, A., & Blinn-Pike, L. (Eds). (2010). *Social Vulnerability to Disasters*. Boca Raton, London, New York: CRC Press, Taylor & Francis Group.

Rambo, A. T., Midori, A.-U., Lee, Y. F., Nickum, J. E. and Takashi, O. (2003). Environmental consciousness in Southeast and East Asia: Comparative studies of public perceptions of environmental problems in Hong Kong (China), Japan, Thailand and Vietnam (Preface). *Southeast Asian Studies, 41*(1), 3–4.

Robb, A. (2015, 1 December). Growing Australian businesses with Indonesia's middle class: Trade. *The Australian Financial Review*. Retrieved from http://search.proquest.com.ezproxy.library.wa.edu.au/docview/1737481454?accountid=14681&rfr_id= uinfo%3Axri%2Fsid&3Aprimo.

Sakai, M. (2009). Creating a new centre in the periphery of Indonesia: Sumatran Malay identity politics. In J. H. Walker, G. Banks, & M. Sakai (Eds), *The Politics of the Periphery in Indonesia: Social and Geographical Perspectives* (pp. 62–83). Singapore: NUS Press.

Setiawan, B. B., & Hadi, S. P. (2007). Regional autonomy and local resource management in Indonesia. *Asia Pacific Viewpoint, 48*(1), 72–84.

Soerjani, M. (1993). Ecological concepts as a basis for environmental education in Indonesia. In M. Hale (Ed.), *Ecology in Education* (pp. 145–160). Cambridge: Cambridge University Press.

Sudiyanti, S. (2009). *Predicting Women Purchase Intention for Green Food Products in Indonesia*. Master's thesis, Universitas Gadjah Mada.

The Economist. (2011). Indonesia's middle class – Missing BRIC in the wall. A consumer boom masks familiar problems in South-East Asia's biggest economy *The Economist*, 21 July 2011.

Tillah, M. & Rahman, F. (2017). Fighting for existence. *Inside Indonesia*, 127. Retrieved from www.insideindonesia.org/fighting-for-existence?highlight=WyJ0aWxsYWgiXQ%3D%3D.

Turner, M., Podger, O., Sumardjono, M., & Tirthayasa, W. K. (Eds). (2003). *Decentralisation in Indonesia: Redesigning the State*. Canberra: Asia Pacific Press.

UNDESA Population Division. (2014). World Urbanisation Prospects, the 2014 revision. Country Profile: Indonesia. Retrieved from https://esa.un.org/unpd/wup/Country-Profiles/.

UNDESA Population Division. (2015). *World Population Prospects: The 2015 Revision, Key Findings and Advance Tables* (ESA/P/WP.241). Retrieved from https://esa.un.org/unpd/wpp/publications/files/key_findings_wpp_2015.pdf.

UNDP. (2016). *Human Development Report 2016: Human Development for Everyone*. Retrieved from http://hdr.undp.org/sites/default/files/2016_human_development_report.pdf.

van Bruinessen, M. (2002). Genealogies of Islamic radicalism in post–Suharto Indonesia. *South East Asia Research, 10*(2), 117–154.

van Bruinessen, M. (Ed.) (2013). *Contemporary Developments in Indonesian Islam: Explaining the "Conservative Turn".* Singapore: Institute of Southeast Asian Studies.

van der Laarse, M. C. (2016). *Environmentalism in Indonesia Today.* Masters in Asian Studies, Leiden University.

WALHI. (2016). Sejarah. Retrieved from www.walhi.or.id/sejarah/.

Warren, C. (1998). Tanah Lot: The cultural and environmental politics of resort development in Bali. In P. Hirsch & C. Warren (Eds), *The Politics of Environment in Southeast Asia: Resources & Resistance* (pp. 229–261). London: Routledge.

White, B. (2005). Between apologia and critical discourse: Agrarian traditions and scholarly engagement in Indonesia. In V. Hadiz & D. Dhakidae (Eds), *Social Science and Power in Indonesia* (pp. 107–142). Jakarta: Equinox.

World Bank. (2014a). Indonesia overview. Retrieved from www.worldbank.org/en/country/indonesia/overview.

World Bank. (2014b). Poverty & equity: Indonesia. Retrieved from http://povertydata.worldbank.org/poverty/country/IDN.

World Bank. (2014c). World Bank and environment in Indonesia. Retrieved from www.worldbank.org/en/country/indonesia/brief/world-bank-and-environment-in-indonesia.

World Bank. (2015). Reforming amid uncertainty. *Indonesia Economic Quarterly.* Retrieved from http://pubdocs.worldbank.org/en/844171450085661051/IEQ-DEC-2015-ENG.pdf.

World Bank. (2016a). *Indonesia's Rising Divide.* The World Bank, Australian Aid.

World Bank. (2016b). *Project Performance Assessment Report. Indonesia. Indonesia Climate Change Development Policy Loan (IBRD-71950).* Retrieved from http://ieg.worldbank.org/Data/reports/PPAR_Indonesia.Clmt_.ChgDPL.pdf.

World Commission on Environment and Development. (1987). *Report of the World Commission on Environment and Development: Our Common Future.* Retrieved from www.un-documents.net/our-common-future.pdf.

5 Education in Indonesia

This book studies, first, how schools teach environmental education in Indonesia, then, based on our findings, suggests a way forward for schools in order to better educate young people for a future where there will be serious environmental problems. The main focus is on formal education, in schools, because schools have the most reach, and on students in senior high school, aged 16–18 years. These students are ideal because they are articulate; they have long been exposed to the values inculcated by schools, families and communities; they have some awareness of community values and political issues; and they are the cohort that best embodies the values, knowledge and skills that the Indonesian education system aims to imbue in its citizens. This chapter describes the education system. A chapter is devoted to this, because (a) it is our contention that many of the shortcomings with EE are shortcomings with the education system in general, and (b) "the way forward" must take into account the situation on the ground: what is possible and realistic.

Indonesia is an education success story – particularly with regard to the education of women. At the end of the colonial era, only about 2 per cent of Indonesian women were literate, and nearly 11 per cent of men. Now, in the early twenty-first century, there is almost universal literacy for those under 25 years of age, an adult literacy rate (percentage of those 15 years and above) of 95.12 per cent (UNESCO, 2017) and almost universal completion of primary school. Nine years of education became compulsory in 1994. Gender differences in educational attainment in the first nine years of schooling have largely been dissolved.

The main concern with education these days is with the quality of the education that these impressive numbers of young people receive. Education bureaucrats realise the problem, and indeed often cite Indonesia's woeful performance in international tests such as the OECD's PISA. The 2013 tests placed Indonesia the second lowest of all 72 countries tested (at 71). The good news is that the 2015 tests showed a comparative improvement, with Indonesia placed 62 of 72, and that girls achieved slightly higher scores than boys across the board (OECD, 2016b).

The poor quality of the education that students receive is a critical reason for the lack of environmental knowledge and understanding that we have found among Indonesian school students, so this chapter takes some time to explain the system. First, it examines the structure of the system, which is conventional except for the parallel stream of Islamic education; then it explores issues of access to schooling. It notes the significant improvements in enrolments over the last 50 years and the gender equality revealed in the statistics, but also remaining inequalities by socioeconomic status, province and remoteness. It then outlines the main objectives of education in Indonesia, highlighting the continuing emphasis on nationalism and the objective to create loyal, and these days pious, citizens, as well as the abiding need to produce effective workers. A key section examines the notoriously poor quality of education in Indonesia, exploring the causes and the wide-ranging efforts to improve the quality of teaching and the management of schools over the last two decades. The important and related issues of curriculum and textbooks are examined in Chapter 6.

Structure of the education system

The basic education system consists of three levels: primary school (six years, children aged 6–12 years), junior high school (three years, ages 13–15 years) and senior high school (three years, ages 16–18). There are also pre-school play-groups for children aged 2–4 years and kindergartens for children aged 4–6. The vast majority of these are run by private organisations. Of course there is also tertiary education. As one would expect in the fourth most populous nation in the world, the education system is immense, and quite diverse.

An examination of enrolment figures in all levels of schooling reveals, not unexpectedly, that the Indonesian education system is like a pyramid, with primary school at the bottom. Virtually all children attend primary school and most primary schools are state schools. State schools dominate the junior high school level too. At senior high school level there are two basic types of school: general (SMA, Sekolah Menengah Atas) and vocational (SMK, Sekolah Menengah Kejuruan). These schools and all private religious schools other than *madrasah* and *pesantren* (see below) are under the aegis of the Ministry of Education and Culture (hereafter MOEC). There is a parallel Islamic system of education run by the Ministry of Religion (MoR), with primary schools, junior high, and Madrasah Aliyah (MA) at the senior high school level. These schools use the same curricula and have the same national examinations as non-religious schools for secular subjects, but the secular subjects make up only 70 per cent of the curriculum: these schools offer extra subjects (and longer hours) at school, making up the remaining 30 per cent of their curriculum. There are private and state schools in all of the categories. The trend has been for the two systems of education to grow closer together – for instance, a graduate of a MOEC primary school may now attend an Islamic junior high school, and a graduate of an MA may attend a non-religious state university (Jackson & Parker, 2008).

If we divide the education system into public and private sectors and examine the private sector, the pyramid is inverted for private schools. Only 7 per cent of primary schools are private; 56 per cent of junior secondary schools are private; and 67 per cent of senior secondary schools (World Bank, 2014). State schools dominate the education system, though private religious and vocational schools are very important.

Access to schooling

The size of the education system is an important factor, not only when we consider access to schools but also when we consider the quality of education. Table 5.1 shows the large number of schools, teachers and students in the system. There are more primary school students in Indonesia than the total population of Australia.

In 2014, the Gross Enrolment Ratio (GER – the number of pupils enrolled in a given level of education, regardless of age, expressed as a percentage of the population in the theoretical age group for the same level of education) for primary school was 105.74 per cent, with a Gender Parity Index (GPI – the ratio of the number of female students enrolled in a particular level of education to the number of male students in that level) of 0.98 (signifying almost gender parity). For lower secondary, in 2014, the GER was 90.68 per cent, with a GPI of 1.06; and for upper secondary the GER was 73.9 per cent, with a GPI of 0.92 (UNESCO, 2017).[1] These enrolment figures have been increasing steadily over the decades, particularly since 1970, and represent one of the principle achievements of the Suharto era (1966–1998).

Our ethnographic study was conducted in two cities: Yogyakarta and Surabaya. It is important to note that, by comparison with other regions of Indonesia, these are high-achieving regions in terms of access to school and quality of education. The government acknowledges that there are significant rural–urban differences at all levels of schooling (BPS 2011). There are also significant inter-provincial and intra-provincial differences. For instance, in Papua, the junior high school GER was 60.05 for 2010 and for senior high, 48.20 (BPS 2017); and the World Bank noted that "Only 55 percent of children from low-income families are enrolled in junior secondary schools" (World Bank, 2014).

Table 5.1 Size of the education system in Indonesia, 2015/2016

Level of schooling	No. of schools	No. of teachers	No. of students
Kindergarten	113,498	366,635	4,495,432
Primary school	172,096	1,795,613	25,885,053
Junior secondary	53,957	681,422	10,040,277
Senior secondary	33,191	569,265	8,647,394

Source: MOEC (2016).

The poor's likelihood of enrollment varies by region, even within the same income quintile. The poor in Papua have low net enrollment rates even at primary school level (80 percent). In fact the regional differences dominate conditions to such an extent that the richest quintile in Papua still has lower enrollment rates (92 percent) than the poorest quintile in Sumatra (World Bank 2006). At the junior secondary school level, the level of access varies even more widely across provinces. Indonesia has almost universal enrollment at the primary level across provinces. However, vast differences in enrollment rates emerge for children between 13–15 years.

(World Bank, 2007: p. 22)

Kristiansen and Pratikno, who researched the impact of decentralisation in Indonesia on access to and quality of primary and secondary education, concluded that "huge social and geographical disparities exist" (Kristiansen & Pratikno, 2006, p. 513). They studied four districts chosen for their range of incomes and variety of level of urbanisation as well as centrality. The gender balance in their overall sample was 1.13, made up of 53 per cent boys and 47 per cent girls, and with great gender inequality at higher levels of schooling (1.33 at senior high school level). They also note "substantial regional differences" in gender in/equality (Kristiansen & Pratikno, 2006, p. 526).

An elaboration of geographical differences in education can be made by introducing the category of "remote" schools. An examination of teacher supply, for example, shows that "Urban and rural areas schools have substantial over-supplies [of teachers] (with 68 percent and 52 percent of the schools having an oversupply, respectively), while remote schools have serious teacher shortages, with 66 percent of the schools being undersupplied" (World Bank, 2007, p. 28). We can expect that in disadvantaged regions, and particularly in remote areas, gender inequality in access to schooling is greater than elsewhere.

At high school, the different types of school map class, gender and other differences (Parker, 2009). State high schools are the province and the source of the middle classes. Civil servants and teachers typically send their sons and daughters to SMP and SMA. If children from lower socioeconomic groups make it to senior high school, they typically attend vocational schools. Vocational schools aim to equip their students to find work immediately after (or before completing) school. Parents and students usually choose vocational schools on the basis of hopes for future occupations, which are strongly gendered. Technical high schools cater for boys who want to learn about automotive mechanics, design and technology; business/administration schools are dominated by girls who want to learn office procedures and computer skills; tourism schools are more gender-equal. In the vocational schools, the disadvantages of class and gender inequalities are freely reproduced: "Attendance at vocational schools leads to significantly lower academic achievement as measured by national test score[s]" (Chen, 2009, p. 22).

The government had a goal of dramatically increasing the percentage of students attending vocational schools, but has gone silent on this over the last few years. The aim was to have 70 per cent of students at vocational schools and 30 per cent at general senior high schools (Chen, 2009, p. 5). The shift was intended to address employment problems and the serious mismatch between student job expectations and qualifications on the one hand and the needs and opportunities presented by the job market on the other. Enrolments in vocational schools are actually declining: in 1999, 1.6 million students were enrolled, but by 2006 there were only 1.2 million (Newhouse & Suryadarma, 2011, p. 301).

The religious school sector is dominated by Islamic schools, and this sector is growing rapidly.[2] Islamic schools are traditionally the province of poor rural people, and have long been considered to be both poor quality and poorly resourced. However, the articulation of the state *madrasah* (Islamic day schools) with the state Islamic universities (UIN), improved teacher education and the regulation that allows *madrasah* graduates to enter conventional (non-Islamic) state universities have meant an improvement in their academic reputations and appeal. Cutting across this traditional identification of religious schools with the rural poor are the reputations of some religious schools as high-quality, even prestigious schools. Christian schools have long enjoyed such a reputation in Indonesia, and some of the top schools in the country are Christian schools. It must also be acknowledged that there are some Islamic schools that have produced national leaders and boast an international reputation for academic performance.

Objectives of schooling

In Indonesia, there have been several national curricula since Independence in 1945, but the twin objectives of creating national unity and good citizens have been constant for the education system (Fearnley-Sander & Yulaelawati, 2008; Raihani, 2007). However, different regimes at different times have had their different emphases. Under the New Order there was great emphasis on the indoctrination of the state ideology of Pancasila, as a vehicle to unite and stabilise the country and create a nation of loyal citizens, and Development, which was the regime's main legitimising policy agenda.[3] For instance, the 1975 Curriculum states that:

> The purpose of National Education is to form a Pancasila- and Development-minded humanity…. The aims of Primary School General Education are that the graduates
>
> a.   Have good basic qualities as citizens;
> b.   Are healthy in body and mind;
> c.   Have knowledge, skills and basic attitudes that are needed for:
>
> 1.   Continuing studies;
> 2.   Working in society;
> 3.   Developing themselves for life in accordance with the principles of education.
>
> (Departemen Pendidikan, 1975, p. x)

Thus, being a good person, being an educated person with knowledge and skills and being a good citizen were conflated in the creation of a good subject, and this was the express purpose of national education. The New Order state did not assume that a child would be a good citizen simply by dint of birth within the borders of its territory. Children had to be taught the nature of the good citizen, become imbued with the values of the state, have considerable knowledge of its version of the history, geography and government of the nation-state, and eventually come to accept, know and identify with the Indonesian nation-state.

A quarter of a century later, after democratisation in 1998, there was a clear need to change the education system and its objectives. The Education Act No. 20/2003 stipulated the function and aim of education:

> National education functions to develop the capability, character, and civilization of the nation by enhancing its intellectual capacity, and is aimed at developing learners' potential so that they become persons imbued with human values who are faithful and pious to the one and only God; who possess good morals and noble character, who are healthy, knowledgeable, competent, creative, independent; and as citizens, are democratic and responsible.
>
> (Fearnley-Sander & Yulaelawati, 2008, pp. 111–112)

The emphasis on developing the individual's potential was new, as was the explicit aim of developing piety. By 2013, the year of the most recent Curriculum, again there is no mention of Pancasila, but piety and morality and character education are well and truly embedded, as we will see in the next chapter. The government sees the new Curriculum as an instrument for directing participants in education to become:

1   a humankind that is capable and proactive in answering the challenges of the era, which is always changing; and
2   an educated humankind that is faithful to God and pious, of honourable ethics, healthy, knowledgeable, capable, creative, and self-sufficient; and
3   citizens who are democratic and responsible (Mendikbud, 2012, p. 2).[4]

There is no explicit mention of the environment or of an objective to develop students' environmental consciousness or knowledge at this level of policy, but the shift to learning how to adapt to the challenges of the era, and the importance of democratic and responsible citizenship, are features of both the core documents that provide entry points for EE.

The quality of schooling

The big challenge for Indonesia now is the quality of the education available in schools. In this section we will first show some international test results to document this poor quality, then we will analyse some of the causes of the poor quality of teaching and learning.

As noted in the introduction to this chapter, Indonesia scores poorly in international tests, such as the PISA tests. The PISA tests assess 15-year-old students in the three core subjects of science, mathematics and reading. However, they do not assess the students' knowledge in these subjects. Rather, they assess the extent to which students who are at the end of their compulsory education "have acquired key knowledge and skills that are essential for full participation in modern societies" (Pellini, 2016). The assumption is that high quality education produces successful students who have the ability to apply their knowledge to real-world problems. For instance, the highest achieving students in PISA science tests (e.g. 25 per cent of students in Singapore) "… are sufficiently skilled in and knowledgeable about science to creatively and autonomously apply their knowledge and skills to a wide variety of situations, including unfamiliar ones" (OECD, 2016c).

Indonesia's performance is one of the lowest among the 69 PISA-participating countries in all three subject areas, for both girls and boys (OECD, 2016a). Interestingly, in all three subject areas, the score difference between the 10 per cent of students with the highest scores and the 10 per cent of students with the lowest scores is one of the smallest among PISA-participating countries (OECD, 2016a). In science, PISA expects that all students should have attained Level 2 by the time they leave compulsory education (OECD, 2016c). At Level 2, students can draw on their knowledge of basic science content and procedures to identify an appropriate explanation, interpret data, and identify the question being addressed in a simple experiment. About 20 per cent of students in OECD countries do not attain this baseline Level 2. The lowest acceptable score for test results at Level 2 in science is 410 and, in Indonesia, the mean score for girls was 405 and for boys 401 (OECD, 2016b). Overall, girls score a little higher than boys in all three subjects. Also interesting are some of the answers to attitudinal and aspirational questions. For instance, "The percentage of students [in Indonesia] who report feeling confident about their ability to complete tasks requiring competence in science" is the lowest among PISA-participating countries and economies, and very few students in Indonesia expect to pursue a career in science (OECD, 2016b).

This poor performance is concerning, not least because Indonesia now spends an incredible 20 per cent of the national budget on education and has been doing so for several years.[5] The quality of teaching was early identified as a problem, and this included teachers' low wages, the common problems of moonlighting and absenteeism, low level of qualifications in subject areas as well as in pedagogy, the mismatch of qualifications with teaching duties, and so on. Other serious problems included the over-supply of teachers in urban areas, such as in our field sites, Yogyakarta and Surabaya, and a serious shortage of teachers in rural and remote areas. A remote area allowance has been introduced to encourage teachers into remote or difficult sites. The government has invested heavily in significant pay rises, and upgrading of teachers' qualifications. However, in turn, the quality of the upgrading and professional development courses has been called into question (Bjork, 2013, p. 61).

Of course, being such a large education system, and working from teachers' very low knowledge base, improvements will take time. However, time will not fix another problem: the fact that most teachers are public servants (Bjork, 2013). The problem is that most teachers currently in the system became teachers not because they wanted to teach, but because they wanted to become civil servants. In earlier times, and still today in more backwoods areas, civil service was the most desired occupation because it meant employment and income security and an old-age pension (Branson & Miller, 1984). In a country with no social welfare system, a white-collar job and a guaranteed income for life were no mean considerations. Working as a civil servant is a matter of years of service, gradually rising in the ranks, if one is obedient and loyal. The expectation is of a long and boring life of service and obedience. It contrasts rather strongly with the ethos of teaching: a commitment to "make a difference", to inspire young people, a desire to teach and to improve student learning. Bjork's conclusion in an earlier study is still applicable:

> That stress on the teachers' duties as civil servants produced a culture of teaching that values obedience above all other behaviours. Educators are not recognised for their instructional excellence or commitment to their craft. Instead, they derive rewards from dutifully following the orders of their superiors. Teachers candidly told me that they considered the role of educator to be secondary to their civil servant identity.
>
> (Bjork, 2004, p. 252)

This role as a public servant instantiates what we call a "public servant mindset", which pervades government offices and schools. In the 1990s, Bjork studied the introduction of a local content module in curricula (LCC). The idea was that schools should be more responsive to local needs.

> When offered control of the LCC, they [teachers] demurred and continued to wait for their superiors to instruct them how to carry out their work. The mismatch between central expectations and local realities produced a state of paralysis at all levels of the education system. Central education officials assumed that teachers had assumed leadership over the LCC. In actuality, local educators continued to wait for direction from the capital.
>
> (Bjork, 2004, p. 251)

There is a sort of "double whammy" with this problem. Since democratisation after 1998, Ministry of Education policy has been to move away from the usual chalk-and-talk pedagogy and switch to a focus on

> learning as an active action by students to build meaning and understanding, while teaching is the responsibility of teachers to create situations

supportive [of] students' creativity, motivation, and responsibility for life-long education.

(Raihani, 2007, p. 178)

This was an explicit turnaround from previous curricula in that education was no longer considered to be a process of knowledge transfer from teacher to student but rather was to be a student-centred process that involved learning by experiencing, interaction and communication. The two Curricula since the introduction of this shift to a needs-based approach to student learning provide lists of "competencies" that students should achieve, but the expectation is that teachers will develop their syllabi and lesson plans themselves in such a way as to develop these competencies. One concomitant problem is that this is a lot of "extra" work for teachers. The perception, especially among older teachers who have the "public service mindset", is that this is an unreasonable demand, so many take the easy route and simply teach from the textbook. During fieldwork in Denpasar for another project, the first author was diligent and persistent in asking teachers for their syllabi and in two SMA, only two teachers could, or would, produce their syllabus.

The other, more fundamental problem is that most teachers have no idea how to do "active learning", or to create those situations that enable students to explore, be curious, discuss, problem solve, etc. Absent of any training in student-centred, open-ended learning, and lacking confidence in their own knowledge base, most teachers revert to the safety of the textbook and rote learning. Thus, the quality of the teaching and learning does not improve. We will say more about this in the following section on Pedagogy.

There is a lot more that could be said about the causes of the low quality of education that students receive. Here we are only focusing on those issues that we identified during fieldwork in schools in Yogyakarta and Surabaya, which are directly relevant to our study of EE in schools. Here we do not address issues of infrastructure or financing of schools. The management of schools is, however, relevant.

Under Soeharto, the education system was extremely centralised, homogeneous and top-down, as one would expect with an authoritarian regime. The fall of the Soeharto regime in 1998 and the ushering-in of a new era of democracy brought many changes to the education system. Some of these (such as legislation to make nine years of schooling compulsory (1994) and moves to decentralise and to introduce more local content) had begun during the 1990s (Bjork, 2003, p. 184), but there was a feeling that the sea-change in politics and government had to cleanse the education system and induce significant change in education too. Education was to be an important part of democratisation: it was commonly felt that "the people" had to be educated in the ways of democracy: that the homogeneous, top-down system of education had to be made more responsive to local cultures and local needs; that the management and curriculum of schools had to be devolved to lower levels of governance; and that local educational institutions and ordinary people had to be given a voice, and empowered. The decentralisation of education would contribute to a "deepening" of the culture of democracy (Parker & Raihani, 2011, p. 713). Responsibility, decision-making and authority were to be devolved to lower levels of government, and local authorities were to be more accountable to the local populace.

The major plank of decentralisation in education was the change from 2002 to School-Based Management (SBM). Schools were to have much more autonomy: the power to hire-and-fire staff, to have stronger control of their budgets, and to develop syllabi. The major new institution under SBM was to be the School Committee, which was to better integrate schools with local communities, and with parents (hitherto rarely considered as stake-holders in school education), and for community leaders to be active agents of change. The shift to SBM was to be part of the decentralisation package:

> The reforms include among other things the implementation of school-based management, a school-level curriculum, school-based teacher professional development, teacher certification, international benchmarking, and national examinations. The reforms are expected to synergically cause an on-going continuous school restructuring to become more autonomous in making local decisions, strong community participations, and effective in delivering quality education services.
>
> (Firman & Tola, 2008, p. 71)

Firman and Tola argue that this latter SBM directive has proved particularly problematic for schools, leading to confusion and a continued reliance on the old forms of curriculum – and we would add pedagogy. Fieldwork in various parts of the country and at various types of schools has generally revealed that the School Committees are not working as envisaged and that the level of parental and community interaction in schools is low. For instance, research in *madrasah* in West Sumatra and Yogyakarta, revealed the perception

> among parents and madrasah committee members, that parents should not engage in the academic life of the madrasah, whether through offering input on curricula development, pedagogical techniques, teaching resources, or supporting teachers in the classroom. Most madrasah committees and parents understood their proper role as confined to providing financial support.
>
> (Parker & Raihani, 2011, p. 727)

The OECD/ADB concluded that "Neither parents nor school committees are currently actively involved in school decision making and activities" (OECD/Asian Development Bank, 2015, p. 104).

It appears that the shift to SBM, in concert with some other policies (such as those regarding "international standard schools" – see Raihani, 2014, pp. 185–187), is having a "marketisation effect" on schools. In any one area, schools compete for the best students, and try to "sell" themselves with their *imej* (image) as academically strong, or old and eminent, as trying hard to improve with lots of young energetic teachers, or, as we will see, as having some specialty such as an environmental education program. Although there is supposed to be universal free basic education, School Committees can raise funds from parents, and they therefore compete for the "catchment" of parents with a strong socioeconomic status. Many schools appreciate greater autonomy in managing their budgets now, though it is also clear that a school's relationship with the District Office and *bupati* (district head) is key (see Rosser & Sulistiyanto, 2013). Financial accountability is still a problem. In *madrasah*, for instance, senior teachers were sometimes allowed to play a part in financial decision-making, but there was an

> apparently deliberate failure to provide a structural mechanism that allows openness and scrutiny in budgetary matters [and this] discourages parents and the community from questioning the use of funds received by the madrasah.
>
> (Parker & Raihani, 2011, p. 727)

However clear it is that SBM and the decentralisation of education in Indonesia were Indonesian responses to broad social and political movements within Indonesia, it is also important to note that the decentralisation and marketisation of education is a global neoliberal movement, supported by major international educational funding bodies such as the World Bank. Proponents of decentralisation argued that it would "lead to one or more of the following outcomes: a redistribution of power, increased efficiency, or greater sensitivity to local culture" (Bjork, 2003, p. 185). The World Bank, a major funder of education in Indonesia, strongly supported the move and funds a large governance evaluation program in education. It is easy to see decentralisation as part of global neoliberalism and the retreat of the state from the provision of educational services, but it is also necessary to weigh up decentralisation against the fact that recent legislation in Indonesia requires the government to spend 20 per cent of the state budget on education.

Finally, it is worth noting that it is difficult to generalise about the quality of the different types of schools. The private school category, for example, which is growing very quickly, includes some of the best schools in the country and many of the worst schools. Most private schools do not enjoy the same facilities as those in state schools (e.g. Bangay, 2005). Teacher qualifications in private schools are notoriously low; most teachers work part-time or on a casual basis; and a range of other factors combine so that, in general, the quality of education offered in private schools is inferior to that in state schools. The vast majority of private schools are Islamic schools. The students' results in the national examinations in private schools, be they madrasah or general schools, are lower than those in state schools (Bangay, 2005, p. 172). State general schools are usually regarded as academically respectable, though the range in quality can be great.

At the senior high level (SMA), however, private schools have mushroomed to take advantage of strong demand and a shortfall in government provision of SMA.

Pedagogy

It is probably clear by now that under the New Order, the primary pedagogy was rote learning:

> ... school teachers stuck to textbooks with leech-like enthusiasm, devoting the greater part of teaching time to reading the textbook: usually reading out loud rather than silent reading, reading en masse, reading by turn around the class, or individuals reading as appointed by the teacher, either standing in front of the class or sitting at their desks. Teachers in high schools seemed to be better trained and more confident of leaving the textbook to one side while they talked or asked questions.
>
> The bulk of a typical lesson consisted of this reading, though often, more recently, the repetition of the passage was interspersed with simple comprehension questions, usually provided at the end of the passage in the textbook or devised by the more active, generally younger, teachers.
>
> (Parker, 2003, p. 226)

So students do learn to read and supposedly to comprehend what they read. However, their understanding is open to question. Students are tested on their knowledge, usually with multiple choice tests (except in Maths):

> There was supposed to be only one right answer to these questions. I had considerable personal difficulty answering many of them. For example, the sixth question in a "Summative Test" for 5th grade in the subject called Moral Pancasila Education in 1991/92 was:
> Our state is friendly with other states which
>
> a. are large and rich   b. are close to our state
> c. love peace   d. are modern and progressive
>
> In fact, most of these questions were not difficult for students because they only required the students to have rote-learned sentences in textbooks which told them these "facts".
>
> (Parker, 2003, p. 227)

Students are assessed as having acquired knowledge when they can replicate the teacher's (or the textbook's) knowledge. The student is then *ipso facto* capable and clever. A child is said to be "still stupid" (*masih bodoh*) if s/he does not know something: people tend not to distinguish ignorance from stupidity. There is a huge belief that as long as students are exposed to some teaching, they will accept it and it will make them knowledgeable and even virtuous. For instance, there is currently a moral panic about young people's socialising practices and the prevalence of pre-marital sex (Parker, 2014). The antidote is to give them more training in Quranic recitation. No one thinks of the possibility that the result might be youth rebellion.

The facts or knowledge are not questioned or analysed: they are accepted as true and valid by virtue of the fact that they are being taught in school. We have sat through a Mathematics class where a teacher, who was lacking confidence, relied on the answers in the back of the textbook. The teacher had asked a student to carry out a long division sum on the blackboard; the student did so, and came up with the correct answer. However, it differed from the textbook answer. The teacher crossed out the student's correct answer and wrote the incorrect textbook answer beside it, without comment. The teacher felt he could not override the authority of the textbook, even though we are pretty sure he knew it was wrong. This is a uni-directional and hierarchical system of downwards knowledge transmission. Pushing back is not possible.

Although theoretically the pedagogy has shifted to active, student-centred learning, for the reasons mentioned above, the old pedagogies are still dominant. Bjork, for instance, reckoned that only 5 per cent of the lessons he witnessed included student discussion (Bjork, 2013, p. 60). Weston wrote, "teaching in most classrooms remains traditional, dominated by rote learning and intended changes in the curriculum have not been implemented at school level" (Weston, 2008, p. 21). Utomo wrote, "Teachers claimed to know what CBC [Competency Based Curriculum] is, but in actual classroom implementation of CBC, these teachers were lost, returning instead to the former curriculum, which they were more comfortable teaching" (Utomo 2005, p. v). In short, "Teachers have not adopted the role of the autonomous educator" (Bjork, 2004, p. 260).

The Ministry of Education has provided documentary support to teachers, with online lists of teaching and learning processes that could be used – reversed meaning of learning, student-centredness, learning by experience, developing social, cognitive, and emotional skills, developing curiosity, creative imagination, and the quality of believing in God, etc. – but this does not mean the teachers know how to do these things (Raihani, 2007, pp. 178–179).

Those in the Ministry of Education assume that because the policy documents have changed, practice has changed, but this is not the case. As the OECD/ADB report noted,

> Although the curriculum is developed and disseminated centrally, teachers are trained, monitored and supported at the district level. This creates a challenge in ensuring the national curriculum is well understood and used by teachers and that they also understand the extent to which they can adapt the curriculum to ensure it is relevant to their local context. The review team noted that the teachers and principals the team met were often aware of the organisational change in terms of students' choice of subjects

and orientation, but that they rarely talked about the core changes that the government intended to introduce, such as the teaching of critical thinking and creativity.

(OECD/Asian Development Bank, 2015, p. 113)

Conclusion

This chapter has provided an overview of the education system in Indonesia. There has been scant mention of environmental education, simply because this chapter aimed to present the main, relevant features of the system: it's a success story of providing the vast majority of young people with an education, but a low-quality education. Most criticism of the education system is economics-based: that high school graduates do not have the skills and knowledge required by employers; and young people are not finding jobs for which they are qualified – this is particularly the case for university graduates. Our criticisms are rather different: years of schooling have not taught young people to think critically or constructively about their environment. Their understanding of complex systems such as natural ecosystems or supply chains is inadequate; they accept the often-dubious knowledge that is transmitted to them as truth; they have no idea how to use their science or social science knowledge of "facts" to solve real-world problems. This is a serious indictment of an education system and it will have dire effects in the decades to come.

Notes

1 These two measures are those used by UNESCO. It should be noted that statistics in Indonesia tend to vary and are not particularly reliable. For instance, Statistics Indonesia published a primary school GER of 111.68 for 2010; a junior high GER of 80.59; and a senior high GER of 62.85 (BPS 2017).
2 In the period 2000–2005, the number of enrolments in religious schools at junior high school level grew by nearly 13 per cent, compared with 1.29 per cent for all junior high schools; at senior high school level enrolments grew by nearly 23 per cent (Diknas 2006) compared with 15.2 per cent for all senior high schools. There has also been a proliferation of types of Islamic school, e.g. Salafi (fundamentalist) schools, Sekolah Islam Terpadu (Integrated Islamic Schools), and so on.
3 See Parker (1992a, 2002, 2003); Leigh (1991, 1992).
4 All translations of the Curriculum and associated documents are the author's.
5 This percentage was mandated in the Education Law of 2003 but it was 2009 before Indonesia actually managed to comply with this Law (Suryadarma & Jones, 2013, p. 6).

References

Badan Pusat Statistik (BPS). (2017). Angka Partisipasi Kasar (APK) menurut Provinsi, 2003–2010 (Gross Enrolment Rates according to Province, 2003–2010). Retrieved from www.bps.go.id/linkTableDinamis/view/id/1049.

Bangay, C. (2005). Private education: Relevant or redundant? Private education, decentralization and national provision in Indonesia. *Compare, 35*(2), 167–179.

Bjork, C. (2003). Local responses to decentralization policy in Indonesia. *Comparative Education*, 47(2), 184–216.

Bjork, C. (2004). Decentralisation in education, institutional culture and teacher autonomy in Indonesia. *International Review of Education*, 50, 245–262.

Bjork, C. (2013). Teacher training, school norms and teacher effectiveness in Indonesia. In D. Suryadarma & G. W. Jones (Eds), *Education in Indonesia* (pp. 53–67). Singapore: ISEAS Publishing.

BPS (Badan Pusat Statistik) (Central Bureau of Statistics). (2011). Perkembangan Beberapa Indikator Utama Sosial-Ekonomi Indonesia (Trends in Several Socio-Economic Indicators for Indonesia). Jakarta.

Branson, J. & Miller, D. B. (1984). Education and the reproduction of sexual inequalities. In R. Burns & B. Sheehan (Eds), *Women And Education* (pp. 261–281). Bundoora, Vic.: La Trobe University.

Chen, D. (2009). *Vocational Schooling, Labor Market Outcomes, and College Entry*. Retrieved from Policy Research Working Paper 4814, The World Bank. Retrieved from https://elibrary.worldbank.org/doi/abs/10.1596/1813-9450-4814.

Departemen Pendidikan dan Kebudayaan (Department of Education and Culture). (1975). *Kurikulum Sekolah Dasar 1975, Garis-Garis Besar Program Pengajaran, Buku II.A.4, Bidang Studi Agama Hindu' (Primary School Curriculum 1975, Outlines of Teaching Program, Book II.A.4, The Hindu Religion Field of Study)*. Jakarta.

Diknas (Departemen Pendidikan Nasional, Department of National Education). (2006) Retrieved from www.depdiknas.go.id/statistik/thn0405/buku%20saku2004_files/sheet 010.htm (accessed 8 August 2006, no longer available).

Fearnley-Sander, M., & Yulaelawati, E. (2008). Citizenship discourse in the context of decentralisation: The case of Indonesia. In D. L. Grossman, W. O. Lee, & K. J. Kennedy (Eds), *Citizenship Curriculum in Asia and the Pacific* (pp. 111–126). Hong Kong: Comparative Education Research Centre, The University of Hong Kong.

Firman, H., & Tola, B. (2008). The future of schooling in Indonesia. *Journal of International Cooperation in Education*, 11(1), 71–84.

Jackson, E., & Parker, L. (2008). "Enriched with knowledge." Modernisation, Islamisation and the future of Islamic education in Indonesia. *Review of Indonesian and Malaysian Affairs*, 42(1), 21–54.

Kristiansen, S., & Pratikno. (2006). Decentralising education in Indonesia. *International Journal of Educational Development*, 26, 513–531. doi:10.1016/j.ijedudev.2005.12.003.

Leigh, B. (1991). Making the Indonesian state: The role of school texts. *Review of Indonesian and Malaysian Affairs*, 25(1), 17–43.

Leigh, B. (1992). The growth of the education system in the making of the state: A case study in Aceh, Indonesia. PhD thesis, University of Sydney.

Mendikbud (Kementrian Pendidikan dan Kebudayaan) [Ministry of Education and Culture]. (2012). Dokumen Kurikulum 2013 (Document of 2013 Curriculum). Jakarta.

MOEC (Ministry of Education and Culture) Centre for Educational Data and Statistics and Culture. (2016). Indonesia. Educational statistics in brief 2015/2016. Retrieved from http://publikasi.data.kemdikbud.go.id/uploadDir/isi_AA46E7FA-90A3-46D9-BDE6-CA6111248E94_.pdf.

Newhouse, D., & Suryadarma, D. (2011). The value of vocational education: High school type and labor market outcomes in Indonesia. *The World Bank Economic Review*, 25(2), 296–322. doi:10.1093/wber/lhr010.

OECD. (2016a). *Education GPS. Indonesia – Student Performance (PISA 2015).* Retrieved from http://gpseducation.oecd.org/CountryProfile?primaryCountry=IDN&treshold=10&topic=PI.

OECD. (2016b). *Program for International Student Assessment (PISA) Results from 2015: Indonesia.* Retrieved from www.oecd.org/education/pisa-2015-results-volume-i-9789264266490-en.htm.

OECD. (2016c). *Summary Description of the Seven Levels of Proficiency in Science in PISA 2015.* Retrieved from www.oecd.org/pisa/test/summary-description-seven-levels-of-proficiency-science-pisa-2015.htm.

OECD/Asian Development Bank. (2015). *Education in Indonesia: Rising to the Challenge.* Paris: OECD.

Parker, L. (1992a). The quality of schooling in a Balinese village. *Indonesia, 54,* 95–116.

Parker, L. (1992b). The creation of Indonesian citizens in Balinese primary schools. *Review of Indonesian and Malaysian Affairs, 26,* 42–70.

Parker, L. (2003). *From Subjects to Citizens: Balinese Villagers in the Indonesian Nation-state.* Copenhagen: Nordic Institute of Asian Studies Press.

Parker, L. (2009). Religion, class and schooled sexuality among Minangkabau teenage girls. *Bijdragen tot de Taal-, Land-en Volkenkunde, 165*(1), 62–94.

Parker, L. (2014). The moral panic about the socializing of young people in Minangkabau. *Wacana: Journal of the Humanities in Indonesia, 15*(1), 19–40.

Parker, L., & Raihani. (2011). Democratizing Indonesia through education? Community Participation in Islamic schooling. *Education Management, Administration and Leadership, 39*(6), 712–732.

Pellini, A. (2016). Indonesia's PISA results show need to use education resources more efficiently. *The Conversation,* 16 December.

Raihani. (2007). Education reforms in Indonesia in the twenty-first century. *International Education Journal, 8*(1), 172–183.

Raihani. (2014). *Creating Multicultural Citizens: A Portrayal of Contemporary Indonesian Education.* London and New York: Routledge.

Rosser, A., & Sulistiyanto, P. (2013). The politics of universal free basic education in decentralized Indonesia: Insights from Yogyakarta. *Pacific Affairs, 86*(3), 539–560.

Suryadarma, D., & Jones, G. W. (2013). Meeting the education challenge. In D. Suryadarma & G. W. Jones (Eds), *Education in Indonesia* (pp. 1–14). Singapore: ISEAS (Institute of Southeast Asian Studies).

UNESCO. (2017). Education: Gross enrolment ratio by level of education. Retrieved from http://data.uis.unesco.org/?queryid=142.

Utomo, E. (2005). Challenges of curriculum reform in the context of decentralization: The response of teachers to a competence-based curriculum (CBC) and its implementation in schools. PhD thesis, University of Pittsburgh, Pittsburgh.

Weston, S. (2008). *A Study of Junior Secondary Education in Indonesia: A Review of the Implementation of Nine Years Universal Basic Education.* United States Agency for International Development.

World Bank (2007). Investing in Indonesia's education: Allocation, equity, and efficiency of public expenditures. Retrieved from http://siteresources.worldbank.org/INTINDONESIA/Resources/Publication/280016-1152870963030/InvestEducationIndo.pdf.

World Bank (2014). World Bank and education in Indonesia. Retrieved from www.worldbank.org/en/country/indonesia/brief/world-bank-and-education-in-indonesia.

6 Religious environmental education?[1]

Introduction

In this chapter, we begin our examination of EE in schools in Indonesia. First we outline some early efforts at EE by NGOs and in schools, and the dismal results of these early efforts in schools. Then we survey the role of UNESCO, as the lead agency for EE in Indonesia. Disturbingly, we find that although UNESCO was instrumental in introducing EE to schools through the Adiwiyata Programme, nowadays it does not mention the environment or even education for sustainable development in its advocacy. We argue that although there is some rhetoric, and there are even publications about EE in schools in Indonesia, there is no substance.

The main body of the chapter presents the way the current school Curriculum of 2013 deals with human–environment interaction and environmental sustainability – in other words, the EE in the Curriculum. It evaluates the relative importance of the environment in the Curriculum and analyses the way the Curriculum constructs human–environment interaction. The main findings are that the new Curriculum neglects the interrelationships of economic development and environmental sustainability. It fails to discuss the causes of environmental problems and to address questions of agency and responsibility for environmental problems. It frames the environment within a creationist, religious worldview. The next section is a discussion of what we call "religious environmental education", arguing that in a religious country such as Indonesia, with such urgent environmental issues, such an approach has potential, but also problems. The final section, on textbooks, presents an example of how textbook writers miss opportunities to discuss the complex issues that EE demands.

The introduction of EE in Indonesia

NGOs

The earliest efforts of EE in Indonesia were made by the World Wide Fund for Nature (WWF) in 1974 (Nomura & Abe, 2005, p. 129). More recently, transnational NGOS such as WWF and TNC (The Nature Conservancy) have designed environmental education modules as part of their efforts to promote national park conservation (Acciaioli & Afiff, 2018). Environmental NGOs (ENGOs) have played a more important role in EE than state institutions, but they tend to be ephemeral and local, often built around single issues. WALHI, the peak nationwide ENGO, has occasionally made some efforts at the national level – for instance, the publication of the book, *Becoming an Environmentalist – It's Easy!*, discussed below, and a book series on living with natural disasters for small children – but has not been active in EE at the national level since those publications.[2] However, provincial WALHI offices have considerable autonomy and follow very different courses in different provinces, with very different mixes of social activism and environmental concerns.

The book called *Menjadi Environmentalis Itu Gampang: Sebuah Panduan bagi Pemula* (*Becoming an Environmentalist – It's Easy! A Guide for Beginners*), promised much (Munggoro & Armansyah, 2008). However, as it is a tome of 356 pages, WALHI admits that it was not an effective educational tool.[3] It is not targeted at any particular age-group of children, and was not written by teachers or others who know about pedagogy and learning. It is a very attractive and eye-catching book, with many photos, especially of street demonstrations, cartoons (see Figure 6.1), inset boxes of inspirational quotes from famous people (Mahatma Gandhi, Kartini, Sukarno), including "Communists" such as Tan Malaka and radical (formerly banned) author Pramoedya Ananta Toer, and uses the language of radical ideology (anarchy, capitalism, etc.). It is one of the few

Figure 6.1 A cartoon from *Becoming an Environmentalist – It's Easy! A Guide for Beginners*.

Source: Munggoro and Armansyah (2008, p. 26).

educational materials produced in Indonesia that presents critical and holistic environmental education. In our reading, it has been put together by recently graduated university students for university students.

The Environmental Education Network, Jaringan Pendidikan Lingkungan (hereafter JPL) was established in 1996 and is regarded as a model of best practice in NGO networking: it mobilises resources and facilitates information exchange among member organisations engaged in environmental education nationwide (Nomura & Abe, 2005).[4] Like many NGOs in Indonesia, it has been successful partly because it has attracted international patronage. This is obvious from its web page, where The Nature Conservancy figures large. However, JPL acknowledges that it needs expertise, and it needs to evaluate such programmes as are running.

Schools

In 2005, a book called *Environmental Education and NGOs in Indonesia* was published (Nomura & Hendarti, 2005). The Foreword was by Emil Salim, the first Minister for the Environment and to this day known as Indonesia's foremost environmentalist. Salim and the editors highlighted that NGOs had been very active in EE in Indonesia: "they have been a major actor in promoting environmental education and raising environmental awareness in Indonesia for a long time" (Nomura & Hendarti, 2005, p. x). Several chapters were upbeat about the work of ENGOs in EE. The chapter on schools was a sad contrast.

Written by an official in the then Department of National Education, Parus, it described the introduction of the "National Project for the Programme on Population" in 1976, introduced in primary and secondary schools in 1978 and renamed the "Project for Education of Population and Environment" (Parus, 2005, p. 66). There is almost no information about the content of these projects, apart from the tradition of "Clean Friday" and voluntary labour service. In 2004, the project was renamed the Environmental Education Programme (PLH, Pendidikan Lingkungan Hidup) and its activities were supposedly: information dissemination in the form of bulletins/magazines and a website; teacher training at national and local levels; provision of books to school libraries; competitions for academic papers, painting/drawing, and songs for teachers and students; the establishment of model schools for environmental culture; seminars/workshops; the procurement of consultants (Parus, 2005, p. 67).

However, it has not been successful. Parus openly and sadly presented a list of the problems associated with PLH:

- The present system of environmental education training is not effective.
- The teaching methods are dominated by lectures.
- Curriculum are [*sic*] tight and time for environmental education is limited; besides, the materials are difficult to be incorporated into the curriculum.

- There is no clear target for the implementation of environmental education.
- The involvement of other institutions in the implementation of environmental education is still low.

(Parus, 2005, p. 67)

Then he identified the "root causes" of the problems:

- School management has not bought in to the basic concept and strategy of environmental education at primary and secondary schools. This can be interpreted [sic] in the lack of middle and long term plans.
- The implementation of [the] environmental education project has not incorporated a "project cycle" method (plan, implement, evaluate, replan, etc.).
- Project management is mostly carried out by a project team that works independently with little involvement by Directorate General of Primary and Secondary Schools or other institutions in the PKLH
- The promotion of environmental education by and within the Directorate General of Primary and Secondary Schools is still low, so that the project lacks attention and support in its implementation at field level.
- The project is not accompanied by sufficient human resources.

(Parus, 2005, p. 68)

The absence of interest and commitment by the then Department of Education is obvious.

UNESCO

UNESCO has been the leading agency in shaping the direction of EE globally through the Tbilisi Declaration of 1978 (resulting from the 1977 UNESCO-UNEP Intergovernmental Conference on Environmental Education), Agenda 21 (the global action plan from the United Nations Conference on Environment and Development, 1992), the UNESCO Decade for Education for Sustainable Development (DESD, 2005–2014), The Bonn Declaration of 2009 and various other charters, publications and programmes.[5]

UNESCO is also the lead agency for EE in Indonesia and was instrumental in the introduction of the Adiwiyata schools EE program in 2006 as part of the DESD (Seta & Mochtar, 2014).[6] Despite this, the UNESCO Education Program in Jakarta currently does not use the terms EE or ESD when pushing its agenda to local organisations and the Ministry of Education and Culture. Instead it focuses on the Sustainable Development Goals, in particular, Sustainable Development Goal 4.7.[7] The reason behind this is an observed lack of interest and support for the "environment" by decision makers.[8]

The Adiwiyata Programme is an environmental education programme that was created by the (then) national Ministry of Environment and introduced to 10 pilot schools in 2006. The programme has grown to include 7278 schools from across Indonesia (Kementerian Lingkungan Hidup dan Kehutanan, 2017).[9] Schools choose to participate in Adiwiyata, and we will examine this choice in more detail in Chapter 9. There are about 260,000 schools in Indonesia (see Table 5.1), so the coverage of the Adiwiyata Programme can be described as very limited. The Adiwiyata Programme "aims to develop students who take responsibility in efforts to protect and manage the natural environment through school governance which supports sustainable development" (Kementerian Lingkungan Hidup Indonesia, 2014) (our translation). Its two main principles reported in official documents are: *participation* – school communities are involved in school management, which includes the whole process of planning, implementation and evaluation in accordance with role and responsibilities; and *sustainability* – all activities must be done comprehensively in a planned and continuous manner (Kementerian Lingkungan Hidup Indonesia, 2014, p. 3).

Rather than building capacity for change (or the empowerment of learners to bring about change), the aims of the Adiwiyata Programme focus on building the capacity of the school, and on the different levels of management. Adiwiyata schools have the opportunity to be assessed annually in order to move up the Adiwiyata rankings. The ranking levels are: District/City (*Kabupaten/Kota*), Province (*Propinsi*), National (*Nasional*) and Independent (*Mandiri*).[10] The Adiwiyata Programme is highly prescriptive and schools are scored (between 0 and 5 marks) for actions related to areas of policy, curriculum, participation in activities and environmental management. Although the Programme rhetoric says that it aims to produce children who take responsibility for the environment, it actually focuses on school management of the environment above all else, with a reliance on figures and numbers to document the achievement of standards. As described in Chapter 8, the reliance on numbers has resulted in forced participation in activities with few learning outcomes.

Thus, while some NGOs have been active in EE, and we will look at one in particular in Chapters 9 and 10, the formal system of education suffers from a lack of interest in EE by the Ministry of Education and Culture. There is some complacency in the Ministry because it can point to the Adiwiyata Programme as "doing environmental education". The Programme continues to be the responsibility of the Ministry of the Environment and Forests, and the two ministries are like siloes. Our visits to the Ministry of Education in Jakarta revealed that there is no "desk" or position or person in the Ministry who is responsible for Adiwiyata or for EE in general. In reality, although there are a few reports and books, and even if the Adiwiyata Programme were a high quality programme, its limited reach would still mean that the Ministry of Education is actually not "off the hook". No one is taking responsibility for environmental education in schools today.

The 2013 Curriculum

National curricula are the product of the state, and can be read as statements of what the state wants its citizens to know and how they want their citizens to develop. Most obviously, there is great potential for the state (and its government) to disseminate its ideology. National school curricula typically present a selection of knowledge, fixed and stable values and universalising truths, which the state has deemed essential for its child citizens to learn. Unsurprisingly, they usually aim to create ideal citizens, who are loyal and patriotic and embody the aspirations of the nation, and Indonesia shares this aim with most countries. This is often, but not always, a process of indoctrination. Indoctrination is at its most powerful when it seems most natural. Students who are not being taught to be critical will usually not realise that they are being indoctrinated; nevertheless, despite the odds, education is a site of conflict and negotiation (Apple, 2004; for Indonesia, Parker, 2002).

Sociologists of education have often pointed to the way curricula assume certain types of knowledge, which can privilege the already-powerful and wealthy and reproduce class inequalities (e.g. Bourdieu, 1973, 1974; Bourdieu & Passeron, 1990). Scholars of critical pedagogy such as Apple (2004), Freire (1974, 1996) and Giroux (1992, 1997, 2011) have also examined how curricula and other aspects of schooling benefit dominant groups and disadvantage the subordinate.

As noted in Chapter 5, since the resignation of military-man President Suharto and the introduction of both democracy and decentralisation, the education system has been quite strenuously reformed. Among the reforms was the implementation of a new Curriculum, known as KTSP (School-Based Curriculum) 2006, based on the Competency-Based Curriculum which was introduced on a pilot basis through 2004 and had a legal basis in Law 20/2003. KTSP was an outcomes-based, competency model that incorporated a new democratic, student-centred ethos (Fearnley-Sander & Yulaelawati, 2008; Raihani, 2007). The new 2013 Curriculum is a revised competency-based model.

The government has referred to the 2013 Curriculum as a "perfecting" of the curriculum (Mendikbud, 2012, p. 2), saying that the rapidly developing economy required "a young generation which has an entrepreneurial spirit, which is strong, creative, tenacious, honest and independent" (Mendikbud, 2012, pp. 7–8). Concern with Indonesia's poor performance in international tests such as PISA (Program for International Student Assessment) is frequently mentioned in government documents about the 2013 Curriculum (e.g. Kementrian, 2013, slides 53–58; Mendikbud, 2012, p. 9). The PISA test assesses students in Reading, Maths and Science, but strangely the new Curriculum does not directly address these weaknesses. Instead, there is a new emphasis on religious and character education. This means less time for more "academic" subjects, such as English, e.g. formerly, in Grade X, the first grade of senior high school, there were four English classes per week and now there are only two classes per week. This perceived hollowing-out of academic subjects has attracted the ire of many educationists in Indonesia and seems to contradict the stated concern with Indonesia's international ranking.

In justifying the introduction of a new curriculum, the 2013 Curriculum documents cite the "threat of disintegration" in relation to "geography, ethnicity, economic potential, and diversity in development progress from region to region" and the need to "form an Indonesian 'humankind' which is capable of balancing the needs of the individual and society to progress their identity as part of the Indonesian nation and the need to integrate as one entity, the nation of Indonesia" (Mendikbud, 2012, pp. 7–8). The creation of a united nation-state is a long-standing concern for Indonesia.

It is hard to overstate the importance of curricula and textbooks in the Indonesian school context. Although the rhetoric in the two most recent Curricula advocates a learner-centred pedagogy, there is considerable inertia in this huge education system, and the stultifying effects of decades of a homogeneous, authoritarian education system, with its principal pedagogy of rote learning for national exams, are still much in evidence in classrooms. Borrowing from a study of education in India by Kumar (1988), Leigh (1991) accurately described school culture under the New Order regime as a "textbook culture", and even now, teachers rely heavily on textbooks. The textbooks for most subjects are produced by independent authors, based on the Curriculum. Those approved by the Ministry of Education and Culture are distributed to schools and are free for students.

The implementation of the 2013 Curriculum was extremely problematic. After being introduced, teacher backlash was extremely vocal, as teachers claimed to be struggling with it; there was a rash of media reports; the 2013 Curriculum was recalled; and then it was announced that schools could decide for themselves whether or not to continue with it. Some schools decided mid-term to revert to the 2006 Curriculum; others continued with the new one. The 2013 Curriculum has now become the only Curriculum.

The objectives of the Curriculum, presented in Chapter 5,[11] say nothing specific about the environment but theoretically allow space for students to learn

1   how to respond to environmental/sustainability challenges;
2   that a religious person should be a responsible environmental steward; to have some sense of environmental ethics; to enjoy health borne of a healthy environment; to be knowledgeable about human–environment interdependence; to be capable, creative and self-sufficient in using natural resources; and
3   to become environmentally responsible citizens.

In the 2013 Curriculum there are four core target competencies across all levels of schooling and all subjects, which refer to (1) religious attitudes; (2) social attitudes; (3) knowledge; and (4) the application of knowledge. Thus, according to the Curriculum, Maths and Dance teachers are also responsible for teaching religious and social attitudes. For senior high school there are four core competencies that are the basic objectives throughout the three grades. Then there are minimum basic competencies for each grade and each subject in each grade. These competencies are the standard by which the central government designs the national examinations, which are held at the end of senior high school.[12] Individual teachers are supposed to independently develop their subject syllabi, including formulating learning objectives, selecting content and teaching strategies, and developing learning evaluations. Teachers have expressed concern with the new Curriculum, for a number of reasons, including the difficulty of assessing the religious core competencies (e.g. Wahyuni 2013).

The four core competencies for Grades X–XII (all subjects) are:

1   Religious competencies: To live and practise the teachings of the religion of the student.

2   Social competencies: To live and practise behaviour that is honest, disciplined, responsible, caring (helping one another, being cooperative, tolerant and peaceful), polite, responsive and pro-active; showing an attitude of being a part of the solution to various problems in interaction in an effective way in the social and *natural environments*; as well as reflecting the nation in socialising in the world.

3   Knowledge: To understand, apply and analyse[13] factual, conceptual, procedural and metacognitive knowledge based on curiosity about knowledge, technology, the arts, culture and humanities from the viewpoints of humanity, nationality, matters pertaining to the state, and civilisation, related to the causes of phenomena and events, as well as applying procedural knowledge to the specific field of studies, in keeping with talent, and an interest in overcoming problems.

4   Application of knowledge: To process, allow and present,[14] both abstractly and concretely, the development of what is studied in school, in an independent as well as effective and creative way, and to be able to use methods that suit the conventions of knowledge.

(Mendikbud, 2013, p. 7, emphasis added)

In the main Curriculum document for all subjects and grades, these four core competencies are repeated for each subject in each grade; so it is only the basic competencies that change according to subject and grade. We can see that a pro-environment element only appears explicitly in core competency 2, where students are to learn to show "an attitude of being a part of the solution to various problems in interaction in an effective way in the social and natural environments." However, as we will see, this core competency is translated into basic competencies to an insignificant extent.

Tables 6.1 and 6.2 set out the subjects and the number of classes each week in each subject and grade. Table 6.1 shows the compulsory subjects that all students must take; Table 6.2 sets out the specialised subjects that students are offered when they choose one of three major disciplinary streams in Grade X:

Table 6.1 Curriculum 2013: compulsory subjects

Subject	Number of classes per week*		
	Grade X	Grade XI	Grade XII
Group A (compulsory)			
Religious and Character Education	3	3	3
Pancasila and Citizenship Education	2	2	2
Indonesian	4	4	4
Maths	4	4	4
Indonesian History	2	2	2
English	2	2	2
Group B (compulsory)			
Cultural Arts**	2	2	2
Physical Education, Sport & Health	3	3	3
Craft & Entrepreneurship	2	2	2
Total Groups A and B	24	24	24
Group C (Choice) (see Table 6.2)			
Academic specialisation	18	20	20
Total that must be taken each week	42	44	44

Source: Mednikbud (2013, p. 3).

Notes
* One "hour" of lessons is actually 45 minutes.
** Can be Regional Language.

Maths and Natural Science, Social Science, and Languages and Culture. The text after each table identifies only those competencies that are *for or pro-environment* (not those that are *about* the environment).

Examination of the curricula for compulsory subjects for all students shows that there is little attention to the environment and that "care" is the main value taught with respect to the environment. However, there is room for teaching about environmental issues, e.g. if relevant readings in Indonesian and English were studied for Grade XII. Unfortunately the subject Pancasila and Citizenship Education makes no mention of environmental citizenship. Perhaps strangely, given the religious emphasis in the Science subjects, discussed below, the subject Religion and Character Education is devoid of environmental references. The exception is the Buddhism subject, which is only studied by Buddhist students. They should "develop environment-friendly behaviour and … be responsible, as a form of caring about the environment". The subject Craft has some emphasis on putting local natural resources to good use.

All students must choose one of three subject specialisations: Maths and Natural Science, Social Science or Languages and Culture, as in Table 6.2.

Table 6.2 Curriculum 2013: subjects of choice

Subject	Number of classes per week*		
	Grade X	Grade XI	Grade XII
Group C Specialisation			
I Maths & Natural Science			
Maths	3	4	4
Biology	3	4	4
Physics	3	4	4
Chemistry	3	4	4
II Social Science			
Geography	3	4	4
History	3	4	4
Sociology	3	4	4
Economics	3	4	4
III Languages and Culture			
Indonesian Language and Literature	3	4	4
English Language and Literature	3	4	4
Other Foreign Languages and Literature	3	4	4
Anthropology	3	4	4
Subject of Choice and Depth			
Cross-interest and/or Deep Interest	6	4	4
Total Group C	18	20	20
Total Available Hours per week[1]	66	76	76

Source: Mendikbud (2013, p. 4).

Note
1 The difference between the number of hours that must be taken each week, i.e., 42, 44 and 44 for
   Grades X, XI and XII, and the total available each week, 66, 76 and 76, is accounted for by the
   difference between non-Islamic (SMA and SMK) and Islamic schools (MA). In the latter, Islamic
   subjects account for 30 per cent of the total curriculum.

I Maths and Natural Science

In the Biology curriculum for Grade X, the religious competency objective is
(1.1) to admire, guard and conserve the order and complexity of God's creation.
Students must be able to (3.8) describe the biodiversity of Indonesia, and preservation efforts as well as the utilisation of natural resources. They must be able
to (3.9) identify types of waste and recycling as well as make recycled products.
The practical competency, like the knowledge competency, includes activities
and skills such as being able to (4.10) look for data about the threats to preservation for various animals and plants unique to Indonesia, and arrange the
results in the form of a report; (4.16) make recycled products that can be utilised for life. In Grade XII, we come across evolution for the first time: students
must be able to (3.11) describe the theory, principles and mechanism of biological evolution from the study of the literature, old theory as well as new trends in
evolution theory; and to differentiate Darwin's evolution theory from other

theories [unidentified]. One further item of interest is that students should be able to present the results of a discussion about the process of cloning from the points of view of science and religion.

In Physics Grade X, the religious competencies are that students should (1.1) increase their faith through realising the connection between the order and complexity of nature and the greater universe (on the one hand) and the greatness of God who created it (on the other); and realise the greatness of God who created water as the principle element of life with the characteristic that it makes it possible for living creatures to grow and develop. In Grade XI, the religious competency includes that students should (1.2) realise the greatness of God who arranged the characteristics of the sun and earth so that, having the force of gravity, the ability to orbit, and the right temperature, it is suitable for humankind to live on the surface of the earth. Students have to analyse the movements of the planets and sun based on the laws of Newton (3.2). In Grade XII, students are to (1.2) realise the greatness of God who created the balance of change in the electric and magnetic fields which are mutually tied together thus enabling humankind to develop technology to make it easier to live.

In Chemistry Grade X, the religious competencies to be achieved by students are (1.1) to realise the order and complexity of electron configuration in the atom as a form of the greatness of The One Great God; and to be thankful for the wealth of nature in Indonesia, in the form of oil, coal and gas as well as various other mining materials as a blessing of The One Great God that can be used for the prosperity of the people of Indonesia. In Grade XI, these religious competencies are repeated. The social competency objective is for students to (2.4) behave so as to guard the environment and be frugal in utilising natural resources. Many of the knowledge competencies in this grade have to do with mining processes, including negative and positive impacts (4.10); and (4.11) students are to be able to present an analysis of the impact of the burning of hydrocarbons on climate change (the increase in the earth's temperature).

Summary: Those students who take the Maths and Natural Science specialisation are not only taught the old-style "naturalist, apolitical and scientific" knowledge *about* the environment (Jóhannesson *et al.*, 2011, p. 377); they are also required to understand some of the natural cycles and the impacts of humans; endangered species; the need for recycling and frugality with using resources; the negative impacts of mining and of burning of hydrocarbons for the earth's climate. Alongside these pro-environment lessons there are also messages about how the earth's resources are there for humans to use and the wealth of natural resources in Indonesia. However, the salient feature of the Science curriculum for the Western reader is the integration of religious messages into Science. Students are to realise and be thankful that God created the universe in such a way that it is suitable for humans to live in, that it is to some extent knowable by humans (e.g. through science), and that so much has been provided for our utilisation. The rash of religious competencies that suddenly appear in the Natural Science curriculum – when they were absent from the compulsory subjects – suggests that curriculum writers were particularly interested in framing Science within Religion. The same framing occurs in Geography.

II Social Science

In Geography, Grade X, the religious competences to be achieved are: (1.1) to appreciate the natural situation of the universe as well as its contents as the creation of the One Great God; (1.2) to be thankful for the creation of the earth as a place to live as the gift of God the Compassionate One; (1.3) to be thankful for our own existence as citizens of Indonesia with thought patterns and actions that show piety towards the One Great God. One of the social competencies is (2.2) to show behaviour that is responsible as a creature that is a part of universal nature. In Grade XI Geography, students are to (2.1) show caring behaviour towards environmental problems in Indonesia and the world; (2.2) show responsive attitudes in avoiding and overcoming environmental problems; and (2.3) show responsible attitudes in guarding the sustainability of the surrounding environment. Knowledge competencies are also pro-environment: (3.4) to show wisdom in the utilisation of natural resources in agricultural, mining, industrial and service activities; and (3.5) evaluate actions that are appropriate for environmental sustainability in connection to sustainable development. In Grade XII, the religious competencies are: (1.1) to appreciate the differences in the potential of different areas as a blessing from God Almighty; and (1.2) be thankful for the blessing of God the Merciful for cooperation between the regions in fulfilling humankind's needs.

In Economics, Grade X students learn to (1.1) be thankful for resources as a blessing from God in fulfilling needs. They should be able to (3.2) analyse scarcity (the connection between resources and humankind's needs) and strategies for overcoming the problem of scarce resources.

Summary: Students who take the Social Sciences specialisation are mainly exposed to EE in Geography. Again, the salient message is the religious one, that students should be grateful to God for creating the earth in such a way that humans can live on it. There are some awkward insertions: e.g. that students should be thankful that God created the different regions of Indonesia with different resources, and the possibility of regional cooperation in order to fulfil humankind's needs – a nod to the nation-state as God-given, and a reference to the regionally uneven distribution of resources in this decentralised government. Second to the religious messages in Geography are messages about caring for the environment, using resources wisely and responsibly, and we get the first and only mention of environmental sustainability and sustainable development (Mendikbud, 2013, p. 134). In the Social Sciences, a third theme is the ecological basis of society, and the notion that resources are scarce, but, as ever, this is linked to the need for students to be thankful for resources, which are a blessing from God.

III Languages and Culture

The subjects in the Languages and Culture specialisation have the potential for teachers to insert "environment-friendly" knowledge and attitudes, but this is not mandated or elaborated, and Anthropology – seemingly an ideal host subject for EE – is devoid of competencies that refer to the environment.

The relative importance of the environment in the curriculum

First, it is important to note that in contrast to many other countries (see Dyment *et al.*, 2015), the Indonesian Ministry of Education and Culture did not announce any particular policy with regard to the embedding of EE in the new Curriculum as part of the UNDESD.

The above presentation of the new Curriculum in Indonesia could be misleading in that it only presents the occurrences of EE, so there is little sense of the weight or importance of the environment in the Curriculum relative to other topics. The Curriculum document for senior high schools is 204 pages long. The Curriculum for the subject Religion and Character Education is the longest at 26 pages; within that, the Curriculum for Islam is the longest, at seven pages, Christianity takes up three pages, and Catholicism two; for the sake of comparison, compulsory Maths comes in "second" with 11 pages; and the two most pro-environment subjects, Biology and Geography, take nine and six pages respectively. To judge the relative importance of the different subjects, the amount of space devoted to the different subjects must be balanced against the number of hours taught in each subject as indicated in Tables 6.1 and 6.2, and remember the three subject area specialisations. Geography is the most pro-environment subject in its content, but is only taught for three or four classes a week, and only to those students who chose the Social Science specialisation. It must also be remembered that the religious core competencies appear for every subject in every grade. For these reasons, the second point to note is that the salient feature of this Curriculum is the teaching of Religion and Character Education. This accords with the stated goals of the Curriculum (above) and core competency one, the religious competency.

In order to judge the relative importance of EE in the Curriculum, one can also count the frequency with which key environment terms occur. For example, the term *berkelanjutan* (sustainable) only occurs once, in Geography; the term *lestari*, meaning eternal, continuing or permanent, used a lot in environmentalist discourse in Indonesia, only occurs five times in different verbal and nominal forms (to make something everlasting, conservation) in Biology and Geography; the word *hemat* (careful, prudent, e.g. with reference to use of natural resources) only occurs once, in Chemistry; the word *punah* (extinct, e.g. of species) does not occur in the Curriculum. However, this is not a very reliable method, as some words are often used in a general sense, which may or may not be taken to refer to the environment, e.g. the values of *peduli* (care) and *bertanggung jawab* (to be responsible).

A third conclusion is that true EE carries very little weight in the new Curriculum. Those students who take the Languages and Culture specialisation will not touch EE; those who take the Social Science stream will be cognisant of interconnections between economic development and environmental sustainability, mainly through the study of Geography; and those who take the most prestigious specialisation, Maths and Natural Science, will understand that the environment was provided for humans by God to utilise as best they can, through science.

Religious environmental education?

For our purposes, the striking features of this Curriculum, in comparison to previous curricula in Indonesia and to curricula elsewhere,[15] are the dominance of religious competencies and the relationship between the religion priority and the representation of the environment.

The EE literature on formal religion is rather meagre. While the environmental movement in the West (insofar as one can generalise) has often been positive about spirituality and selected aspects of Eastern religions, particularly Buddhism, the academic literature on EE in schools is basically a Western, secular, science-based discourse (N. Gough, 2008 [2003]). EE has often been institutionalised within science education (Gruenewald, 2004, pp. 72–73). Engagement with the "great religions" of the world is quite scant. Local and Indigenous knowledge is taken seriously up to a point – sometimes that point is the reality of national examinations; Outdoor Education and place-based pedagogy also offer something different, often based on the experience of being "in nature".[16]

Some scholars have observed that "nature" has disappeared from conventional EE. Bonnett (2007) argues that the instrumentalism of contemporary Western society, in concert with the seductions (or perhaps compromising ease) of sustainable development, has led to the ignoring of "nature" in EE. Others, such as Stevenson, argue that what began as "nature study", to learn about, understand and appreciate the natural environment, and the conservation movement, has gradually morphed into EE in schools in such a way that the original goals have been inevitably compromised by the organisational culture of schooling (Stevenson, 2007). Others, such as the special issue of the *Canadian Journal of Environmental Education* on Religion and Environmental Education (vol. 11, 2006), ask, "Where is the place for religion in environmental education?" In answering that question, Beringer suggests that EE

> bear[s] responsibility to re-introduce, on a cultural level and global scale, lost dimensions of a religious–spiritual knowledge of nature. This includes reclaiming environmental ethics embedded in timeless metaphysical, epistemological, and ontological understandings of the cosmos, and validating non-scientific ways of knowing.
>
> (Beringer, 2006, p. 26)

In that same special issue, Hitzhusen (2006, p. 12) advocates religious EE but privileges a version of religion which is that of the liberal multiculturalist. He says that, in order to avoid controversy when introducing religion into EE, environmental educators can "describe without advancing particular religious or ethical teachings". He advocates searching for commonalities among religions, mobilising common "ecotheology" resources. While that approach might well work well in Canada, or other broadly secular countries, I suggest it would not be acceptable in Indonesia. For one thing, the question in a religious country is much more likely to be, "Where is the place for the environment in Religion?".

Religious education can take different forms and have different aims. A common typology for religious education is one that is striking in its resemblance to Lucas' (1972) characterisation of EE in its "prepositional" framing (Reid in Stevenson & Evans, 2011, p. 25). It was first suggested by Grimmitt (1987): there is teaching *about* religion, teaching *from* religion, and teaching *into* religion. Teaching *about* religion refers to the religious studies approach, which involves teaching about the world religions in a neutral and objective fashion. Teaching *from* religion places students' learning at the centre, and gives them the opportunity to consider major social and moral issues from a religious perspective. Teaching *into* religion, also known as the confessional approach, is the teaching of a single religious tradition, by insiders, with the intention of nurturing the faith of student believers. This has been the most common approach to religious education traditionally, worldwide, and continues until today in Indonesia. The idea of Religious Education in Indonesia is to enable students to become "better Muslims" (or Christians, Buddhists, etc.). Indeed the Education Law of 2003 stipulates that students can only be taught their own registered religion and only by someone of that faith. With this model, Hitzhusen's advocacy of religious EE and hopeful inclusiveness is rather put to the test. However, he does make the point that one can use religious belief in EE "by empowering students to develop their environmental values within whatever pre-existing value system they already occupy" (Hitzhusen, 2006, p. 13).

There is no doubt that religious and spiritual cosmologies have the potential to offer something different, often complementary, to science-based EE. With that in mind, we now consider the relationship between Religion and the Environment in the new Curriculum in Indonesia, examining first creationism and divine and human agency in the Curriculum and the Qur'an;[17] second, the desired affect and values that the Curriculum identifies that follow from the basic belief that God created the universe; third, religious resources that are neglected in the Curriculum; and finally some problems with religious EE.

Creationism

The new Curriculum presents the idea that everything is God's creation as fact. Guessoum, who has written extensively on the relation between Islam and science, says,

[O]ne of the most beautiful and most often quoted verses from the Qur'an is "*Those who reflect on the creation of the heavens and the earth (and say): Our Lord! Thou hast not created this in vain! Glory be to Thee*" (Q 3:191). [I]n this verse already we begin to see the place of natural theology as presented in the Holy Book: it is an approach for confirming – not establishing – God's existence. Indeed, the Qur'an takes the belief in the Creator as rather self-evident.

(Guessoum, 2012, p. 376)

The Qur'an is thick with similar messages.[18] There is frequent repetition of the fact that God created the world in the Curriculum but not, as Guessoum notes, to establish this as fact – it is given. For instance, in the Physics curriculum for Grade X, students are "to increase their faith through realising the connection between the order and complexity of nature and the greater universe [on the one hand] and the greatness of God who created it [on the other]". The main subject in these sentences is not that God created the world – it is just a subordinate clause ("who created it"). This "fact" is made into "taken-for-granted background knowledge" (Fairclough, 2013, p. 31) and is subsumed into the main point, that students need to realise the greatness of God. This subsuming has the twin effects of naturalising and neutralising the "fact", making it, and the concomitant (that students need to realise the greatness of God) seem orderly, non-ideological and as "common sense" (Fairclough, 2013, p. 31).

The framing of the environment as God's creation also constructs a religious epistemology or way of knowing that is different to scientific epistemology, insofar as one can reduce these to unitary categories. The Curriculum posits that we know that God is great because we can see his creation. Science, on the other hand, requires the deployment of our cognitive faculties, particularly our capacity for rationality, in a particular, disciplined, systematic (some might say reductionist) way to discover knowledge. It boils down to the clash of faith versus reason.

Another aspect of creationism in the Curriculum is its particularism. By this we mean that the Curriculum mirrors the Qur'an in its itemisation of the particular nature (*fitra*) of the different components of creation (Chishti, 2003). There are many awkward insertions of religious messages in the science curricula (especially in Physics and Chemistry) and elsewhere (Geography, Economics), where students must thank God for the particular way He created the earth and the universe, i.e. to enable the existence of humankind. The text of the Curriculum sometimes mirrors the Qur'an:

It is God who raised up the heavens with no visible supports and then established Himself on the throne; He has subjected the sun and the moon each to pursue its course for an appointed time; He regulates all things....

The Quran, 2005, 13: 2)

In the Curriculum for Physics Grade XI, students should

> realise the greatness of God who arranged the characteristics of the sun and earth so that, having the force of gravity, the ability to orbit, and the right temperature, it is suitable for humankind to live on the surface of earth.

Instrumentalism

In the Curriculum, as in the Qur'an, God's creation is constantly linked to His provisioning of human beings. In this way, humans are constructed as separate from the environment: the environment was created *for humans*. In the Curriculum, the environment is frequently depicted as a resource, or as containing resources, available for human exploitation, e.g. in Economics, Grade X students learn to "be thankful for resources as a blessing from God in fulfilling needs"; and in Craft and Entrepreneurship, students learn to appreciate the diversity of materials in their local area as the gift of God. This way of presenting the environment – as resource for human exploitation – could be seen to facilitate understanding that humans have the right to treat nature however they please, one of the root causes of ecological destruction (Stibbe, 2004, p. 246). There are verses in the Qur'an, e.g. "We established you [people] on the earth and provided you with a means of livelihood there – small thanks you give!" (*The Quran*, 2005, 7: 10) and in Muslim scholarly writings that set up humans as having dominion over the earth, e.g. "Humans are at the top of the pyramid of creatures of God because they have spiritual connection with the provider of their privilege, God." (M. I. Dien, 2000, p. 76 citing Qutb, S., Muqawwimat, Cairo, 1986). Of course, this anthropocentric instrumentalism is not unique to Islam.

Nationalism

Nationalism is a strong theme in the Curriculum, following a long-standing tradition in Indonesian education. Both "Indonesia" and "the environment" are absorbed into the scope of God's creation, but again for instrumentalist and utilitarian purposes. For instance, in Chemistry Grade X, students must understand the wealth of nature in Indonesia, particularly in the form of oil, coal and gas, as God's blessing that can be used to make Indonesians prosperous. In Geography, students are to be thankful to God for their own existence as citizens of Indonesia; and, in a rather prosaic reference to regional resource wealth inequalities in Indonesia, they are to appreciate the differences between different regions as a blessing from God Almighty while being thankful to God the Merciful for the gift of regional cooperation to fulfil humankind's needs. Overall, "resource nationalism" is on display: Indonesia has an abundance of natural resources, and is proud of it; it is a sign of God's favour.

Agency and responsibility

In the Curriculum, there is more explicit agency ascribed to God than there is to humans with respect to the environment. In the Qur'an, as we have seen, there are several plain verses that tell that everything is God's creation, that the laws of nature were laid down by God, and that "He knows what is needed to run this universe" (Masri, 1992, p. 5). There is considerable elaboration of the one-way agency of God with respect to both humans and the environment; the return flow is one of submission, positive affect – appreciation, thanks and gratitude – and consciousness – realisation of the greatness of God. There is little elaboration of triangular interrelations among God, humans and the environment and even less about two-way interactions between humans and the environment.

There seems little attention to causation in the Curriculum: e.g. if we can attribute the good things in creation to God (e.g. the air that is "just right" for humans to breathe), how can we account for the bad things – for disease and earthquakes? Fairclough focuses on the constructive function of language (following Foucault on power as both a force that says no as well as a constructive force): particular and repetitive language can have the discursive effect of naturalising what might otherwise be seen as arbitrary or artificial (Fairclough, 2013, p. 31ff.), and enables the avoidance of causation, responsibility or agency. For instance, the constant refrain in the Curriculum that students appreciate the wonder of God and His creation has the effect of making it appear orderly, natural and as "common sense".

In the Curriculum, human agency with respect to environmental damage and exploitation is limited. In fact, considering the scale of environmental destruction in Indonesia,[19] there is very little in the Curriculum about that destruction, about responsibility for that destruction and about the possibilities for human conservation, recycling, consumption or other ways to responsibly and sustainably use these "resources". Given the rapid economic development occurring in Indonesia, it would seem the time is ripe for such discussion.

In the few instances where the Curriculum does mention environmental problems, the language used is distant, agent-less but authoritative (Stibbe, 2004, pp. 244–245). For instance, in Chemistry Grade XI, the focus is on mining, "including positive and negative impacts" and "the impact of the burning of hydrocarbons on climate change". Here there is no agent who does the burning or negatively impacts the environment through mining. There is a singular lack of attention to actual causes of environmental damage; there is what Fairclough calls "a 'logic of appearances' rather than 'explanatory logic'" (Fairclough, 2003, p. 95). "This 'mystification and obfuscation ... of agency and responsibility' (Fairclough, 2003, p. 13) obscures the economic, political and cultural causes of ecological destruction ..." (Stibbe, 2004, p. 245). If humans are not identified as being responsible for ecological destruction we arrive at that very difficult question of who is responsible, which can be a politico-economic question about power and the condition of the world under late capitalism, or a philosophical question about responsibility for evil in the world.

On the few occasions when sensitive issues such as negative impacts of the actions of humans on the environment are presented, they are presented at a distance, as an "authoritative expression of fact" (Stibbe, 2004, p. 245). There is great distance and objectivity – no affect, no controversy, no agency, no conflict, no responsibility. This is what Apple calls a "valuative consensus", and such common values are part of the "hidden curriculum" of the Curriculum (Apple, 2004, p. 59 and Chapter 55).

Values and affect

Appreciation and gratitude: the Curriculum often couples thanks to God for His creation with the objective of realising the greatness of God. Within an Islamic context, giving thanks to God for his beneficence and being conscious of His greatness are aspects of worship and religious duty (*ibadah*), reflecting that the meaning of Islam is submission to God (M. I. Dien, 2000, p. 76; Frisk, 2009). Many of the religious competencies in the Curriculum require that students thank God for the blessing of the phenomenon that they are learning about.

A cynical observer might see these insertions of gratitude as an easy way for teachers to accommodate the requirements of religious Curriculum planners. Giving thanks to God is quick and easy, and stops there; it does not lead to existential questioning or theological contestations and therefore to pedagogical complications. Of course, one could think of some associated problems, such as the occurrence of natural disasters (very common in Indonesia) – should students be thanking God for them? Clearly, students are to be grateful that their country was provided with oil and gas, but the negative impacts of the utilisation of such resources and their development are not presented as being related to God, nor are students to thank God for these negative impacts. In fact the three-way interaction of God, humans and environment is only positive, and on the few occasions that negative impacts of development are mentioned, it is not in association with God. One could add that the emphasis on gratitude and appreciation obviates the need for taking personal responsibility for the environment.

The value that appears most frequently in the Curriculum is "honesty", but this is never linked to the environment. Other values that are quite dominant are that students should be "caring" and "responsible", though again, not explicitly linked to the environment. In several subjects, particularly in Geography, students are to show "environment-friendly" behaviours.

Missing religious resources

Remembering Bonnett's (2007) and Beringer's (2006) identification of the anthropocentrism and spiritual impoverishment of current models of EE in the West, it is important to note that there are ethical "resources" or values within Islam that could be mobilised to teach students a different approach to natural resources. The values of *rahmah* (mercy, kindness, compassion), justice (*adl*) and *mizan* (balance, equilibrium, harmony) in the universe are key themes in Islamic discourse characterising the unifying principle of *tawhid* (Setia, 2007). Several scholars have written about the values and principles explicated in the Qur'an and the Hadith that emphasise humans' responsibility, as God's stewards appointed to look after the earth, to conserve resources, to value water and even to hold population growth in check (Al-Jayyousi, 2011; M. I. Dien, 2000; M. Y. I. Dien, 1992; Khalid, 1992; Masri, 1992; Rice, 2006; Saniotis, 2012); e.g.

> 54 Your Lord is God, who created the heavens and earth in six Days, then established Himself on the throne; He makes the night cover the day in swift pursuit; He created the sun, moon, and stars to be subservient to His command; all creation and command belong to Him. Exalted be God, Lord of all the worlds! 55 Call on your Lord humbly and privately – He does not like those who transgress His bounds: 56 do not corrupt the earth after it has been set right.[20]
>
> (*The Quran*, 2005, 7: 54–56)

These environmentalist Islamic scholars conventionally describe the role of humans as stewards, with the duty of looking after the earth (Saniotis, 2012, p. 157); and humans will be called to account for their stewardship of the earth when they enter the hereafter (*akhirat*). The Qur'an is said to have appointed humans as the *khalifah*, or "vice-regent" of Allah on earth. However, this role is not particularly clear in the Qur'an and the translation of the term *khalifah* is arguable.[21] Other scholars have pointed to the continuing anthropocentrism of the human stewardship approach (e.g. Bonnett, 2007, p. 710; Hitzhusen, 2006, p. 13) and the perceived need for human management of "resources".

Another qualification is that most of the Islamic literature on the environment is of an apologetic nature (Mangunjaya, 2011, p. 40), and there is now an effort to play "catch-up", seeking Hadith and pronouncements in *fiqh* (jurisprudence) that are relevant to these environmentally sensitive times. Some scholars are honest about this: for instance, KH Husein Muhammad, the head of the large Islamic boarding school, Pondok Pesantren Dar al-Tauhid, in Cirebon, northwest Java said:

> "In point of fact the issue of natural and environmental conservation has not been discussed explicitly [i.e. in those terms] in the classical books of Islam."
>
> Accordingly, efforts to research into and respond to environmental issues are made through what he called "marginal *fiqh*", meaning that the ideas of different *ulema* and the material found in dispersed Hadiths need to be gathered and studied systematically.
>
> (Mangunjaya, 2011, p. 42)[22]

Potential problems with religious EE

Secular readers might say that there is creationism on the one hand and hard science and evolution on the other, i.e. a clash of cosmologies, as well as of epistemologies. Some readers could well reject the compatibility of Science and Religion, and hence the possibility of religious EE, out of hand. In the Curriculum, the emphasis on the world as the creation of God is not presented as a contest of cosmologies: rather, science and evolution exist within the overall cosmology of God's creation. This is in line with Islamic views that there are two types of science: one that is atheistic and outside Islam, because it is based on nothing but human observation, experimentation and thought, and the science that is within Islam, which starts with the creation of the universe by God from the free will of God to create (Negus, 1992, pp. 39–40). In the Curriculum, it is this second type of science that is presented. Islamic creationism is presented as a "valuative consensus" (Apple, 2004, p. 59) – it is not open for discussion. The case could be made that the Curriculum particularly mentions religion in the Science specialisation – certainly a rash of religious competency objectives regarding the environment suddenly appeared in Biology, when in other subjects the environment was mainly mentioned in association with care for the environment and environment-friendly behaviour.

A recent comparative study of Muslim high school students in Islamic schools in Australia and Malaysia showed that Australian Muslim students perceived significant epistemological dissonance between the theory of evolution and their religion (Robottom & Norhaidah, 2008, pp. 155–156), suggesting that Islamic understandings of science are diverse. The same study showed that students could distinguish between evolution as knowledge and Islam as belief, and they could happily learn about evolution without believing in it. The new Curriculum in Indonesia certainly suggests that this would be an interesting place for further research.

Textbooks

While the curriculum plays an important role in setting the standards for education in Indonesia, and textbook writers follow the curriculum, most teachers teach directly from textbooks and rely on these texts and examinations as guides for what is taught in the classroom (Parker, 2002). As noted above, Leigh's (1991) designation of this as a "textbook culture" is still pertinent. During fieldwork (described below), we found that teachers were not very familiar with the new curriculum requirements, but knew the textbooks and examination content well. Some textbooks are produced by the Ministry of Education and Culture and are distributed to schools. Other textbooks are produced by various publishers. These books must first be approved by the Ministry of Education and Culture to be used in schools. If schools use unapproved textbooks, they do not receive funding to cover the cost of the books by the government. Some private schools choose their own text books (and ask parents to pay for them), and it is not uncommon for private primary schools to produce their own textbooks. The

Principal from an Islamic primary school in Surabaya explained that by doing this they could integrate the messages that they felt were most important, integrate local content and in some cases, use names and photos of students in the book to make it relatable and interesting for the students.

There is plenty of opportunity to include environmental themes in textbooks of every subject, but, disappointingly, this barely occurs. In instances where environmental themes are included, they tend to be presented in a factual manner, for example, reporting an environmental event such as a landslide yet making no comment on what might cause a landslide. Failing to recognise humankind's contribution to such "natural disasters" promotes simplistic, isolationist thinking. One such example is evident in the Year 10 English textbook (Kementerian Pendidikan dan Kebudayaan Indonesia, 2014). This government-produced textbook has a chapter on "Visiting Ecotourism Destination" [*sic*]. One of the tasks is for students to read a passage and answer some questions. The text (Figure 6.2) is about the Tanjung Puting National Park. The article makes passing mention that the rehabilitation centre is a place for ex-captive orangutan and a preservation site. It makes no comment on the fact that orangutan are a protected and "critically endangered" species; that they are only found in Indonesia; that it is illegal to have them as pets; or that they are under threat from the illegal wildlife trade, palm oil expansion and habitat loss. It fails to mention that many of the orangutan at the centre are babies whose mothers have been killed (and are not ex-captive). It also fails to examine the basics of eco-tourism, the impacts (positive and negative) of eco-tourism or what makes a destination appealing to eco-tourists. The passage jumps to asking students what they would do if they met an orangutan in the jungle. It does not mention that with dwindling numbers of orangutan it is extremely unlikely that they would ever come across one; that tourists should not approach orangutan; or that an eco-tourist has a responsibility to help protect and conserve these animals. It also assumes that students live in cities and that cities have parks.

There was ample opportunity to ask questions of the students that would encourage them to think more deeply about the situation. For example – Why should we preserve the rainforest? Why do we need a preservation site? What role does tourism play in conservation? Where do you think these ex-captive orangutan come from? Why is it illegal to keep orangutan as pets? The final and most obvious place to raise these issues is in the "Points to Ponder" section. Instead of focusing on the eco-tourism or environmental elements most pertinent to the issue (conservation of rainforest and an endangered species), students are asked about rubbish and what tourists should do with it (referring to a simple "green and clean" message, Figure 6.3).

This passage and related questions are a perfect example of the silences around human behaviour related to environmental problems in Indonesian school textbooks. There is obviously no interest in environmentalism in the Ministry of Education and Culture. With a few simple changes, this activity could have been an excellent example of how to integrate EE across subjects, but instead it is an example of missed opportunities in Indonesian textbooks.

D READING

Task 1:
Now, read the text carefully. What do you think about the place described below?

TANJUNG PUTING NATIONAL PARK

One of the internationally famous *ecotourism destinations* in Indonesia is Tanjung Puting National Park in the southwest of Central Kalimantan peninsula. Tanjung Puting National Park offers *impressive* experience to its visitors. This is called a park, but *unlike* any park that you have seen in your city, this is a jungle! It is a real jungle, which is home to the most incredible animals in the world: orang utans and proboscis monkeys! The male proboscis monkeys are interesting because they have *enormous* snout. So, imagine yourself to be in the jungle and meet these special animals in their original habitat. What will you do when you meet them?

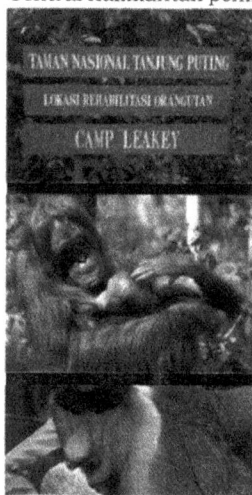

Source: http://oranguatanexplore.com
Picture 6.2

To see orang utans we should go to Camp Leakey. Camp Leaky is located in the *heart* of Tanjung Puting National Park. This is a rehabilitation place for *ex-captive* orang utans and also a *preservation* site. This camp was established by Birute Galdikas, an important scientist who has studied orang utans since 1971.

To reach the place, we should take a boat down Sekonyer river. The boat is popularly called *perahu klotok* which is a *boathouse* that can accommodate four people. The trip by the boat to Camp Leaky takes three days and two nights.

The traveling in the boat offers another *unforgettable* experience. You sleep, cook, and eat in that klotok, night and day during your journey into the jungle. In daylight, on your way to Camp Leaky, you can see trees filled with proboscis monkeys. At night, you can enjoy the clear sky and the amazingly bright stars as the only lights for the night.

Text sources: 1. www.lonelyplanet.com. 2. www.Indonesian.travel.com. 3. www. Exploguide.com

70 Kelas X SMA/MA/SMK/MAK Semester I

Figure 6.2 Reading passage from Year 10 English language textbook.
Source: Kementerian Pendidikan dan Kebudayaan Indonesia (2014).

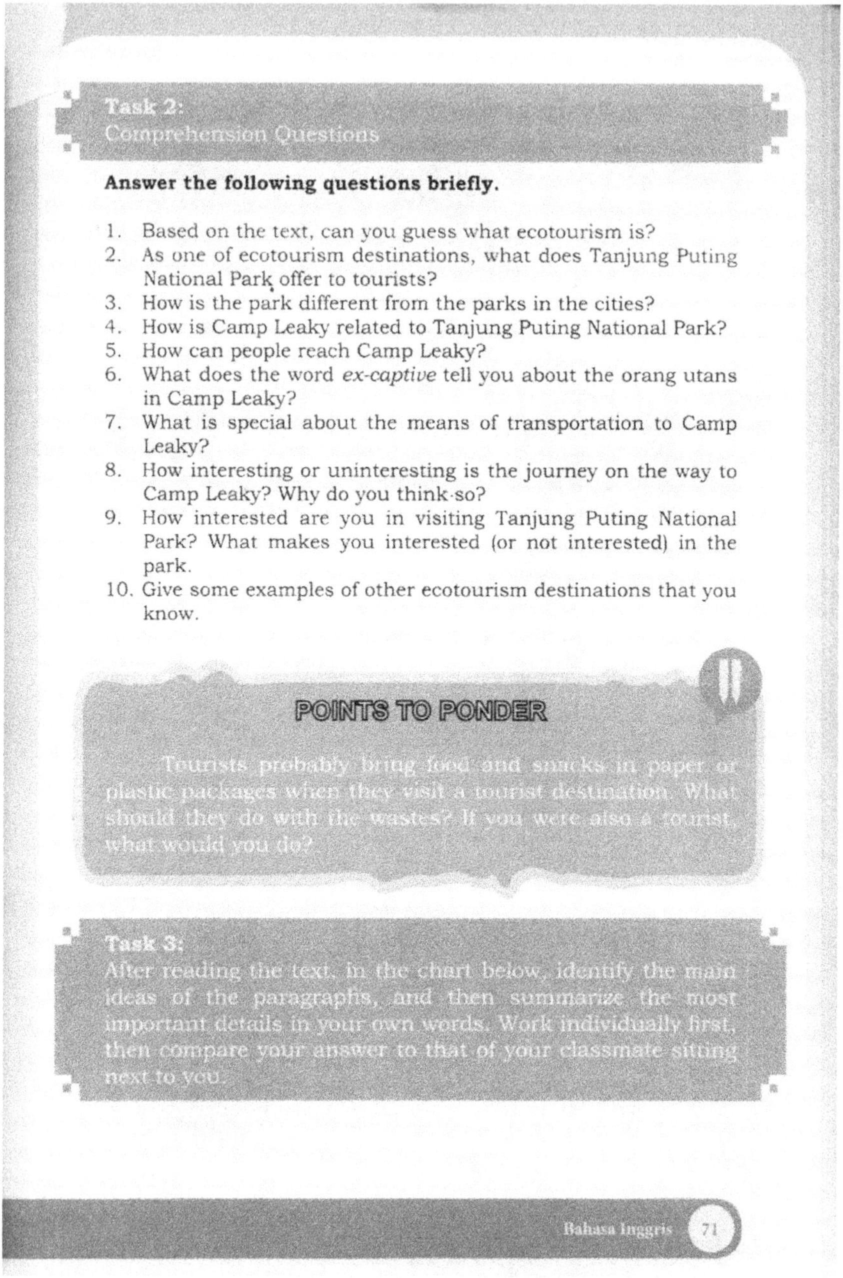

Task 2:
Comprehension Questions

Answer the following questions briefly.

1. Based on the text, can you guess what ecotourism is?
2. As one of ecotourism destinations, what does Tanjung Puting National Park offer to tourists?
3. How is the park different from the parks in the cities?
4. How is Camp Leaky related to Tanjung Puting National Park?
5. How can people reach Camp Leaky?
6. What does the word *ex-captive* tell you about the orang utans in Camp Leaky?
7. What is special about the means of transportation to Camp Leaky?
8. How interesting or uninteresting is the journey on the way to Camp Leaky? Why do you think so?
9. How interested are you in visiting Tanjung Puting National Park? What makes you interested (or not interested) in the park.
10. Give some examples of other ecotourism destinations that you know.

POINTS TO PONDER

Tourists probably bring food and snacks in paper or plastic packages when they visit a tourist destination. What should they do with the wastes? If you were also a tourist, what would you do?

Task 3:
After reading the text, in the chart below, identify the main ideas of the paragraphs, and then summarize the most important details in your own words. Work individually first, then compare your answer to that of your classmate sitting next to you.

Bahasa Inggris 71

Figure 6.3 Comprehension questions in Year 10 English language textbook.
Source: Kementerian Pendidikan dan Kebudayaan Indonesia (2014).

Conclusion

The 2013 Curriculum represents a missed opportunity for environmental education in Indonesia. In the new Curriculum, there is mention of pro-environment behaviour as a competency objective and identification of values such as caring for the environment and being responsible for the environment in several subjects. Critical analysis of the Curriculum shows us that the environment most often appears as the creation of God, for which students are to be grateful. Particularly in the optional Natural Science subjects, the expression of thanks to God for creating an environment suitable for humankind is prioritised, and usually linked to appreciation of the greatness of God. Human agency for the development of natural resources is present in the Curriculum, but responsibility for the destruction of the natural world is neglected. The only subject that deals with the relationship between environmental sustainability and economic development in a thorough-going way is Geography, an optional Social Science subject. Given the rapidly developing economy and the justification for the new Curriculum, the neglect of this relationship is lamentable.

Secular scholars might have epistemological difficulties with the presentation of the environment in religious terms. Given the urgency of the environmental problems, and the low level of environmental awareness in Indonesia, it might behove us not to be too precious about religious EE. In the context of Indonesia, where worldviews are generally religious, the religious identity of individuals is not only assumed but also required, and public space is increasingly religious, the framing of the environment within a religious cosmology makes sense. Such a framing has the potential to minimise cognitive dissonance in school learning and enable the development of religious environmental ethics and practice, using values from within religious teachings. There are also problems with such an approach, not least for students who want to continue with Science at university (Robottom & Norhaidah, 2008). Certainly this is an area for further research and consideration.

Remembering Indonesian government commitments to provide education for sustainable development and environmental sustainability, we can conclude that what is missing from the 2013 Curriculum is education that alerts students to human responsibility for the destruction of the natural world; that informs them about the socio-econo-political systems that drive this destruction; that develops their understanding of their own place in these systems; and that equips them with hope and the capacity for pro-environment social action.

Notes

1 A slightly different version of the body of this chapter appears as Parker (2016).
2 Interview with WALHI representative, 8 June 2010.
3 Interview with WALHI representative, 8 June 2010.
4 Interview with representative of JPL, 9 June 2011.
5 See A. Gough (2013) for a "history" of the field.

6 This book paints a positive picture of efforts at EE, based on the UNESCO/Ministry of Education partnership.

7 The SDG Target 4.7 does include ESD, but the Officer explicitly said his Office does not talk about education for the environment because they could not gain traction. Target 4.7 includes the following:

> by 2030 ensure all learners acquire knowledge and skills needed to promote sustainable development, including among others through education for sustainable development and sustainable lifestyles, human rights, gender equality, promotion of a culture of peace and non-violence, global citizenship, and appreciation of cultural diversity and of culture's contribution to sustainable development.
>
> (UNESCO, n.d.)

8 Interview with National Program Officer for Education, Jakarta, 31 January 2017.

9 While Adiwiyata is not a compulsory programme, some schools are told that they must participate in order to improve the school's reputation or to give it a distinctive image in order to attract "better" students. This number reflects schools that are participating in the program at any level.

10 The two Yogyakarta schools in this book were National level.

11 The government sees the new Curriculum as

> … an instrument for directing participants in education to become:
>
> 1 a humankind which is capable and proactive in answering the challenges of the era, which is always changing; and
> 2 an educated humankind which is faithful to God and pious, of honourable ethics, healthy, knowledgeable, capable, creative, and self-sufficient; and
> 3 citizens who are democratic and responsible.
>
> (Mendikbud, 2012, p. 2)

12 These exams have a standardising and homogenising effect, flowing down through the whole system. Many teachers and students consider that the national examinations contradict the principles of student-learning centredness and local content of the competency-based curriculum. After much discussion, the current government is continuing with the external exams, but high school graduation is no longer determined by exam marks.

13 For Grade XII, the words "and evaluate" are added.

14 For Grade XII, the words "and create" are added.

15 See, for example, Dyment *et al.* (2015); Jóhannesson *et al.* (2011).

16 There are many examples, see, for example the special issue of *Environmental Education Research*, 14(3), 2008.

17 From this point, for the sake of simplicity and clarity, we are talking about Islam, rather than the other five religions acknowledged in the Curriculum.

18 The Qur'an does not have a chronologically-ordered beginning, like the Bible. Instead, creation declarations are scattered throughout. For instance,

> It was He who created all that is on the earth for you, then turned to the sky and made the seven heavens; it is He who has knowledge of all things.
>
> (*The Quran*, 2005, 2: 29)

> 3 [I]t is He who spread out the earth, placed firm mountains and rivers on it, and made two of every kind of fruit; He draws the veil of night over the day. There truly are signs in this for people who reflect. 4 There are, in the land, neighbouring plots, gardens of vineyards, cornfields, palm trees in clusters or otherwise, all watered with the same water, yet We make some of them taste better than others: there truly are signs in this for people who reason.
>
> (*The Quran*, 2005, 13: 3–4)

1 It is the Lord of Mercy 2 who taught the Qur'an. 3 He created man 4 and taught him to communicate. 5 The sun and the moon follow their calculated courses; 6 the plants and the trees submit to His designs; 7 He has raised up the sky. He has set the balance 8 so that you may not exceed in the balance: 9 weigh with justice and do not fall short in the balance. 10 He set down the Earth for His creatures, 11 with its fruits, its palm trees with sheathed clusters, 12 its husked grain, its fragrant plants.

(The Quran, 2005, 55: 1–12)

19 The forest fires of 2015 have been reported as "almost certainly the greatest environmental disaster of the 21st century – so far." (Monbiot, 2015).
20 Saniotis translates 7: 56 as "Do no mischief on the earth after it has been created" (Saniotis, 2012, p. 157).
21 The verse that is customarily cited to show this appointment is:

[Prophet], when your Lord told the angels, "I am putting a successor [*khalifa*] on earth," they said, "How can You put someone there who will cause damage and bloodshed, when we celebrate Your praise and proclaim Your holiness?" but He said, "I know things you do not."

(The Quran, 2005, 2: 30)

The translator Haleem notes that "The term *khalifa* is normally translated as 'viceregent' or 'deputy'. While this is one meaning of the term, its basic meaning is 'successor'" (*The Quran*, 2005, 4).
22 He also said that he had never found the Arabic word, *al-bia*, now used for "environment", used in that sense in the books of *fiqh* of the classical period (Mangunjaya, 2011, p. 42).

References

Acciaioli, G., & Afiff, S. (2018). Neoliberal conservation in central Kalimantan, Indonesia. *Indonesia and the Malay World*, 46(136), 241–262. doi:10.1080/13639811.2018.1531968.

Al-Jayyousi, O. R. (2011). *Islam and Sustainable Development: New Worldviews*. Burlington, VT: Gower Pub.

Apple, M. W. (2004). *Ideology and Curriculum* (3rd ed.). New York and London: RoutledgeFalmer.

Beringer, A. (2006). Reclaiming a sacred cosmology: Seyyed Hossein Nasr, the perennial philosophy, and sustainability education. *Canadian Journal of Environmental Education*, 11(1), 26–42.

Bonnett, M. (2007). Environmental education and the issue of nature. *Journal of Curriculum Studies*, 39(6), 707–721.

Bourdieu, P. (1973). Cultural reproduction and social reproduction. In R. Brown (Ed.), *Knowledge, Education and Cultural Change: Papers in the Sociology of Education* (pp. 71–112). London: Tavistock.

Bourdieu, P. (1974). The school as a conservative force: Scholastic and cultural inequalities. In J. Eggleston (Ed.), *Contemporary Research in the Sociology of Education* (pp. 32–46). London: Methuen.

Bourdieu, P., & Passeron, J. C. (1990). *Reproduction in Education, Society and Culture* (R. Nice, Trans.). London: Sage, with Theory, Culture and Society.

Chishti, S. K. K. (2003). Fitra: An Islamic model for humans and the environment. In R. C. Foltz, F. M. Denny, & A. Baharuddin (Eds), *Islam and Ecology* (pp. 67–84). Cambridge, MA: Harvard University Press.

Dien, M. I. (2000). *The Environmental Dimensions of Islam*. Cambridge, UK: Lutterworth Press.

Dien, M. Y. I. (1992). Islamic ethics and the environment. In F. M. Khalid & J. O'Brien (Eds), *Islam and Ecology* (pp. 25–36). London: Cassell.

Dyment, J., Hill, A., & Emery, S. (2015). Sustainability as a cross-curricular priority in the Australian Curriculum: A Tasmanian investigation. *Environmental Education Research, 21*(8), 1105–1126.

Fairclough, N. (2003). *Analysing Discourse: Textual Analysis for Social Research*. London and New York: Routledge.

Fairclough, N. (2013). *Critical Discourse Analysis: The Critical Study of Language*. Retrieved from http://UWA.eblib.com.au/patron/FullRecord.aspx?p=1397484.

Fearnley-Sander, M., & Yulaelawati, E. (2008). Citizenship discourse in the context of decentralisation: The case of Indonesia. In D. L. Grossman, W. O. Lee, & K. J. Kennedy (Eds), *Citizenship Curriculum in Asia and the Pacific* (pp. 111–126). Hong Kong: Comparative Education Research Centre, The University of Hong Kong.

Freire, P. (1974). *Education for Critical Consciousness* (M. B. Ramos, L. Bigwood, & M. Marshall, Trans.). London: Sheed and Ward.

Freire, P. (1996). *Pedagogy of the Oppressed* (M. B. Ramos, Trans. New rev. edn). London: Penguin.

Frisk, S. (2009). *Submitting to God: Women and Islam in Urban Malaysia*. Copenhagen: NIAS Press.

Giroux, H. A. (1992). *Border Crossings: Cultural Workers and the Politics of Education*. New York; London Routledge.

Giroux, H. A. (1997). *Pedagogy and the Politics of Hope: Theory, Culture, and Schooling: A Critical Reader*. Boulder, CO: Westview Press.

Giroux, H. A. (2011). *On Critical Pedagogy*. New York: Continuum.

Gough, A. (2013). The emergence of environmental education research. In R. B. Stevenson, A. E. Wals, J. Dillon, & M. Brody (Eds), *International Handbook of Research on Environmental Education* (pp. 13–22). New York: Routledge.

Gough, N. (2008 [2003]). Thinking globally in environmental education: Implications for internationalizing curriculum inquiry. In W. F. Pinar (Ed.), *International Handbook of Curriculum Research* (pp. 53–72). Mahwah, NJ: Laurence Erlbaum Associates.

Grimmitt, M. H. (1987). *Religious Education and Human Development: The Relationship Between Studying Religions and Personal, Social and Moral Education*. Great Wakering: McCrimmons.

Gruenewald, D. A. (2004). A Foucauldian analysis of environmental education: Toward the socioecological challenge of the Earth Charter. *Curriculum Inquiry, 34*(1), 71–107.

Guessoum, N. (2012). Nidhal Guessoum's *Reconciliation of Islam and Science*: Issues and agendas of Islam and Science. *Zygon, 47*(2), 367–387.

Hitzhusen, G. E. (2006). Religion and environmental education: Building on common ground. *Canadian Journal of Environmental Education, 11*(1), 9–25.

Jóhannesson, I. A., Norðdahl, K., Óskarsdóttir, G., Pálsdóttir, A., & Pétursdóttir, B. (2011). Curriculum analysis and education for sustainable development in Iceland. *Environmental Education Research, 17*(3), 375–391.

Kementerian Lingkungan Hidup dan Kehutanan. (2017, 2 August 2017). Rayakan Hari Lingkungan Hidup 2017, KLHK Berikan 65 Penghargaan untuk Masyarakat, Sekolah dan Pemerintah Daerah. Retrieved from http://ppid.menlhk.go.id/siaran_pers/browse/678.

Kementerian Lingkungan Hidup Indonesia. (2014). Informasi Mengenai Adiwiyata. Retrieved from www.menlh.go.id/informasi-mengenai-adiwiyata/.

Kementerian Pendidikan dan Kebudayaan Indonesia. (2014). *Bahasa Inggris* (Vol. Semster 1). Jakarta: Kementerian Pendidikan dan Kebudayaan Indonesia.

Kementrian Pendidikan dan Kebudayaan [Ministry of Education and Culture]. (2013). Kurikulum 2013, Paparan Wakil Menteri Pendidikan dan Kebudayaan R.I Bidang Pendidikan [The 2013 Curriculum, Presentation of the Vice Minister for Education and Culture]. Jakarta.

Khalid, F. M. (1992). The disconnected people. In F. M. Khalid & J. O'Brien (Eds), *Islam and Ecology* (pp. 99–111). London and New York: Cassell.

Kumar, K. (1988). Origins of India's "textbook culture". *Comparative Education Review, 32*(4), 452–464.

Leigh, B. (1991). Making the Indonesian state: The role of school texts. *Review of Indonesian and Malaysian Affairs, 25*(1), 17–43.

Lucas, A.M. (1972). Environment and environmental education: Conceptual issues and curriculum implications. PhD thesis, The Ohio State University, Columbus.

Mangunjaya, F. M. (2011). Developing environmental awareness and conservation through Islamic teaching. *Journal of Islamic Studies, 22*(1), 36–49.

Masri, A.-H. B. A. (1992). Islam and ecology. In F. M. Khalid & J. O'Brien (Eds), *Islam and Ecology* (pp. 1–24). London: Cassell.

Mendikbud (Kementrian Pendidikan dan Kebudayaan) [Ministry of Education and Culture]. (2012). *Dokumen kurikulum 2013*.

Mendikbud (Kementrian Pendidikan dan Kebudayaan) [Ministry of Education and Culture]. (2013). *Kurikulum 2013, Kompetensi Dasar, Sekolah Menengah Atas (SMA)/ Madrasah Aliyah (MA) [2013 Curriculum, Basic Competencies, Senior High School/Islamic Senior High School].*

Monbiot, G. (2015). Indonesia is burning. So why is the world looking away? *Guardian (online)*. Retrieved from www.theguardian.com/commentisfree/2015/oct/30/indonesia-fires-disaster-21st-century-world-media.

Munggoro, D. W., & Armansyah, A. (2008). *Menjadi Environmentalis – Itu Gampang! Panduan untuk Pemula (Becoming an Environmentalist – It's Easy! A Guide for Beginners)*. Jakarta: WALHI.

Negus, Y. (1992). Science within Islam: Learning how to care for our world. In F. M. Khalid & J. O'Brien (Eds), *Islam and Ecology* (pp. 37–50). London: Cassell.

Nomura, K., & Abe, O. (2005). The environmental education network in Indonesia. In K. Nomura & L. Hendarti (Eds), *Environmental Education and NGOs in Indonesia* (pp. 125–137). Jakarta: Yayasan Obor Indonesia.

Nomura, K., & Hendarti, L. (Eds). (2005). *Environmental Education and NGOs in Indonesia*. Jakarta: Yayasan Obor Indonesia.

Parker, L. (2002). The subjectification of citizenship: Student interpretations of school teachings in Bali. *Asian Studies Review, 26*, 3–37.

Parker, L. (2016). Religious environmental education? The new school curriculum in Indonesia. *Environmental Education Research.* doi:10.1080/13504622.2016.1150425.

Parus. (2005). Department of National Education. Partnership for Environmental Education in Primary and Secondary education. In K. Nomura & L. Hendarti (Eds.), *Environmental Education and NGOs in Indonesia* (pp. 65–76). Jakarta: Yayasan Obor Indonesia.

The Quran: A New Translation (2005). (M. A. S. Haleem, Trans.). New York: Oxford University Press.

Raihani. (2007). Education reforms in Indonesia in the twenty-first century. *International Education Journal, 8*(1), 172–183.

Rice, G. (2006). Pro-environmental behavior in Egypt: Is there a role for Islamic environmental ethics? *Journal of Business Ethics, 65*(4), 373–390.

Robottom, I., & Norhaidah, S. (2008). Western science and Islamic learners: When disciplines and culture intersect. *Journal of Research in International Education, 7*(2), 148–163.

Saniotis, A. (2012). Muslims and ecology: Fostering Islamic environmental ethics. *Contemporary Islam, 6*, 155–171.

Seta, A. K., & Mochtar, I. N. E. (Eds). (2014). *Pendidikan untuk pembangunan berkelanjutan di Indonesia: Implementasi dan kisah sukses (Education for Sustainable Development in Indonesia: Implementation and Success Stories).* Komisi Nasional Indonesia untuk UNESCO (KNIU) and Kementerian Pendidikan dan Kebudayaan.

Setia, A. (2007). The inner dimension of going green: Articulating an Islamic deep-ecology. *Islam and Science, 5*(2), 117–150.

Stevenson, R. B. (2007). Schooling and environmental education: Contradictions in purpose and practice. *Environmental Education Research, 13*(2), 139–153.

Stevenson, R. B., & Evans, N. S. (2011). The distinctive characteristics of environmental education research in Australia: An historical and comparative analysis. *Australian Journal of Environmental Education, 27*(1), 24–45. doi:10.1017/S0814062600 000057.

Stibbe, A. (2004). Environmental education across cultures: Beyond the discourse of shallow environmentalism. *Language and Intercultural Communication, 4*(4), 242–260. doi:10.1080/14708470408668875.

UNESCO. (n.d.). Learning to live together sustainably (SDG4.7): Trends and progress. Retrieved from https://en.unesco.org/gced/sdg47progress?language=en.

Wahyuni, (2013). FSGI presses govt to evaluate new school curriculum, *Jakarta Globe,* 5 August 2013. Retrieved from www.thejakartaglobe.com/news/fsgi-presses-govt-to-evaluate-new-school-curriculum/.

7   Is anyone responsible for the environment in Yogyakarta?

Introduction

This chapter is the first of the ethnographic chapters, analysing such EE as there is in schools in the central Javanese city of Yogyakarta. The city is generally known as the centre of traditional and contemporary Javanese culture, and is a hive of cultural activities: music, wayang and theatre performances, literature, batik-making and painting. It is the second major tourism destination in Indonesia, after Bali. Attractions include UNESCO World Heritage sites such as Borobudur and Prambanan Temples, and many come as cultural tourists, to see the Kraton (palace) and dance and wayang performances, and to learn something of the arts and crafts. It is known as the "city of students", and is a real education hub, with more than 100 institutions of higher education. Young people from all over Indonesia come here for higher education. A further, related, claim to fame is the large number of NGOs in the city. It is for these last two reasons in particular that our project went to study EE in Yogyakarta: we expected that with the combination of thousands of educated young people and perhaps hundreds of NGOs there would be an abundance of EE programmes for us to study, both within schools and without.

We were disappointed. There is very little happening formally in schools generally in Yogyakarta: there is the Adiwiyata Programme in some schools, and this chapter examines two Adiwiyata Schools, called here The High School and The Islamic School. There are also ephemeral efforts by environmental NGOs and individual environmentalists, working outside schools for the most part. This chapter and the next mainly use field notes of observations to describe what we call "hollow environmental education" in Yogyakarta. We found that many of the activities are not educational and only amount to a meaningless performance – hence "hollow" – in order to win Adiwiyata "points"; that students do not learn about environmental sustainability and the complex interactions between the environment and human society in Adiwiyata; that there is a preoccupation with status in the programme, to the extent that the environmental aims have been hijacked; and that there are many missed opportunities that a revamped Adiwiyata could productively exploit. We argue that nobody in government or in the education system is taking responsibility for EE.

We begin with analysis of the context: the administration and governance of the program, and the lack of training of teachers in EE. Then we introduce the two schools, examine the motivation for schools to join Adiwiyata, the organisation of the Adiwiyata Programme in our two case studies, and the system of fake documentation and accountability that has developed.

The heart of the section on EE in Yogyakarta is the dispiriting ethnographic descriptions of how EE is taught in three different classes. This is presented at the beginning of Chapter 8. Then we examine why it is like this – why teachers seem to lack commitment to teaching, how they use Adiwiyata for promotion opportunities, and why students are not more active. We follow up on students' stated desire for more hands-on activities, and move on to considerations of "student voice" and power, the importance of having students do EE in such a way that they feel enabled and competent, and not downhearted, and finally the necessity to move away from meaningless performance of EE towards meaningful participation by students. Some of the shortcomings – such as teacher insecurity and lack of knowledge, related to lack of training – are ubiquitous across the education system; others pertain more to the particular needs of EE.

The political and governmental context

The Special Region of Yogyakarta (DIY) is unique among provinces in Indonesia in its undemocratic leadership: an inherited Sultanate. This special status is due to its historic role during the war for independence (often called the Indonesian National Revolution 1945–1949) from the Netherlands, when it hosted the rebel nationalist government from January 1946 to December 1948. The Sultan of Yogyakarta at the time, Sultan Hamengko Buwono IX, was a supporter of the newly declared President Sukarno, and the award of the special status was granted in recognition of the special role played by Yogyakarta in securing independence. The current incumbent is Sri Sultan Hamengko Buwono X.

In Yogyakarta, the power of the Sultan goes far beyond his position as Sultan and Governor. Hamengku Buwono IX was a highly respected and much loved leader. He held power in both the Javanese sense,[1] as a spiritual and fatherly leader, and in the Western sense of economic and political power, in and beyond Yogyakarta – he was Vice President of Indonesia after Hatta. Hamengku Buwono IX was seen to care for the little people (*rakyat*), and under his guidance, Yogya became more democratic (Monfries, 2007). His son, Hamengku Buwono X, became Sultan of Yogyakarta after his father's death in 1988 and was the democratically elected Governor of Yogyakarta before the law was changed in 2012 to make the Governorship of Yogyakarta an inherited position (Colbron, 2016). The Sultan owns and controls vast amounts of land in the Yogyakarta region, in addition to his family's many business interests (Colbron, 2016, p. 78). He is very powerful, politically and economically. In recent years he has been seen to favour concern for his business interests and those of his family over concerns for the little people, with many land disputes arising as Yogyakarta has focused on tourism and development (Biennale Jogja XIII

Equator #3, 2015; Suryani, 2016; Wicaksono, 2016; Yanuardy, 2012). Under the leadership of the current Sultan, there has been no serious attempt in Yogya to embrace any kind of sustainable development or eco-tourism, despite the growing discontent of the Yogyakartan people (see Figure 7.1).[2]

Figure 7.1 Banner protesting the building of the Uttara apartments in Yogyakarta.
Photo credit: Kelsie Prabawa-Sear.

The Special Region is divided into one city administration (Kota Yogyakarta), and four districts (*kabupaten*): Sleman, Kulon Progo, Gunung Kidul and Bantul. Our two schools are in the City and Bantul. H. Hariyadi Suyuti has been the democratically elected Mayor (head of the Kota Yogyakarta) since 2011. While the previous Mayor (Herry Zudianto, Mayor 2001–2011) had implemented a minor greening and street-scaping programme, the current Mayor, with the Sultan, has pushed for unbridled development and tourism. The position of any Mayor of Yogyakarta would be constrained by the extra-ordinary power of the Sultan – and we will see, in Surabaya, the importance of pro-environment leadership, that is not happening in Yogya. Yogyakarta is still lacking the basic infrastructure needed to sustain the DIY of 3,457,491 people (BPS, 2010). There is no city-wide sewerage system, piped water or waste management. All of the new hotels, in addition to the thousands of households and boarding houses, draw water from wells, depleting ground water; effluent waste is discharged into septic tanks or waterways; and commercial waste is largely unmanaged – burnt or piped to the nearest waterway. Household waste is burnt, dumped, or paid for privately to be dumped at the nearest tipping site.

Decentralisation has been occurring in Indonesia since 1999, and education is one of the many portfolios that has been decentralised. However, there have been many problems associated with the process and there has been some claw-back, including in the Ministry of Education and Culture. One of the responses of local governments under decentralisation has been "a new tendency for the local governments to exploit the local resources even more intensively, including water, land and other physical assets to maximize their own income" (Firman, 2010, p. 400). Often, tasks have been devolved to provincial or district levels of government, but with no budget to fund the new responsibilities. For instance, Waste Law No. 18/2008 requires local governments to organise environmentally responsible landfill but to source funding locally (Meidiana & Gamse, 2011, 23). Another problem has been that there has been "no effective cooperation among the bordering municipalities and districts in Indonesia's metropolitan regions" (Firman, 2010, 401), and departmental silo-isation, where governments at local level sequester their resources and are disinclined to collaborate with other departments. Decentralisation has thus been a patchy affair, with a general tendency to "become inward-looking", and where the quality of leadership has often been identified as a key factor in a region's success or otherwise (Firman, 2010)).

Moving to EE, we see a classic case of ministerial silos. In DIY, the Adiwiyata Programme is administered by the provincial and district levels of the Environmental Agency (Badan Lingkungan Hidup, BLH), part of the Ministry of the Environment and Forestry. The Environmental Agency has officials at the provincial level, who seemed inactive with regard to Adiwiyata, and there was one official for Adiwiyata at each of the *kabupaten* offices of the Environmental Agency. The Environmental Agency has no official role in schools or in the education system. One government official suggested that in other provinces, the Ministry of Education and Culture was more supportive of the Adiwiyata

Programme than the Ministry in Yogya, and encouraged participation by state schools. She said that a lack of support from the Ministry of Education and Culture in Yogyakarta was the reason there were more Islamic schools than SMA participating in the programme. (As noted above, the Islamic Schools are administered by the Ministry of Religion, rather than the Ministry of Education and Culture). Any NGO that wants to access schools in any formal and wide-spread way would require the support of the Ministry of Education and Culture and, ideally, the Environmental Agency. Thus, the Agency which, from the outside, is responsible for EE, and for training teachers in EE (see below), is actually not acknowledged as such by the Ministry of Education and Culture.

In Yogya we saw no evidence that the Environmental Agency and the Education Office (Dinas Pendidikan) work together. The role of the Environmental Agency in EE is clear, in that there are officials assigned to Adiwiyata, and they supply the equipment for Adiwiyata Schools, but there is actually no assigned role for the Education Office with regard to EE or Adiwiyata. The nature of their "partnership" is unclear to the officials in the two offices. Despite this, and a lack of any evidence to suggest that they have ever worked co-operatively, when asked, Education Office staff reported cooperating with the Environmental Agency.

The only mention of Adiwiyata in the Ministry of Education's website is a notice from 2014 that Anies Baswedan (then Minister of Education and Culture) and Siti Nurbaya (Minister for the Environment) announced that 498 Adiwiyata schools had been awarded National status (Maulipaksi, 2014). Anies mentioned that raising environmental awareness was a problem for all layers of society, and commented that 498 schools might not seem many, when there are 208,000 schools in Indonesia (indeed), but even from such a small number, the message can be spread. Since there is no mention of what Adiwiyata is, it's a standalone and mysterious notice in the Ministry's website. It stands as a symbol of the Ministry's lack of attention to EE and shows how very marginal the Adiwiyata Programme is in the education system.[3]

Government roles and responsibilities

Aside from the deleterious positioning of the main EE programme for schools *outside* the education system, there is a gaping hole where in-service training of teachers should be. The Environmental Agency is responsible for training teachers in EE, but for various reasons this does not happen. This problem is addressed immediately below under "Training".

There was a further, local problem in Yogyakarta. Two government officials claimed that the Environment Agency at the DIY level was restricting the number of schools that could move up the Adiwiyata rankings and that this was why DIY had so few schools ranked at the higher levels.

There are still only a few [highly ranked schools] in Yogya, not many compared to how many schools we have. 300 schools and about 15 [highly

ranked]. Only 15, so that's not balanced. If we compare to Surabaya, East Java, they have lots [of schools] that are Adiwiyata *Mandiri* [ranked as Independent, the top rank]. But Yogyakarta [Province] has been restricted, so each district can only have one primary school, one middle school and one senior high school.

(8 April 2015, Bu Ani, Yogyakarta City Environmental Agency Office)

The reason for this restriction was said to be the "budget". We were assured by various people that this simply meant that the person in charge was using the money for something else. It seems there was a bottleneck at the point where schools were ready to attain the highest Mandiri level. The Environmental Agency has a responsibility to provide resources (both human and infrastructural, such as training workshops and shredding machines and compost bins), to the highest level schools. So by having few schools at the apex of the Adiwiyata pyramid, the Agency did not have to spend money. We interviewed the official who was responsible for Adiwiyata at the DIY level at the beginning of fieldwork.[4] Amongst other things, we asked him about the prospect of having NGOs work with schools to deliver EE (as in the Surabaya case described in Chapters 9 and 10). He said,

They [NGOs] will ask schools [for money] and actually it's we who don't have too much [money]. And then schools don't have ongoing funding so they will cut ties with the NGOs and then the process of learning and teaching about the environment will also be cut.… We also cannot be sure of the capabilities of NGOs.

(29 October 2014, DIY Environmental Agency Office, Yogyakarta)

He went on to explain that NGO workers lack management skills and often lack the background (or education) needed to teach EE and described it as a "difficult" situation. He knew that we were also conducting research in Surabaya and suggested more than once that schools and NGOs there have access to "a lot of money" from the government and private sources, unlike schools in Yogyakarta, which were "pure" in their approach. He asked us to note the source of their funding and told us not to only focus on the results. He advised we examine the processes because if the funds were to dry up, the results would too. We did duly take note of funding arrangements when researching EE in Surabaya and found that schools there were not required to pay anything to work with the NGO. It was evident that many of his comments were most likely to have been offered in justification for Yogyakarta's poor result in the Adiwiyata Programme. It seemed that in fact the poor results were partly due to the restrictions that he himself had put in place and had little to do with what would be the best for the schools participating in the Adiwiyata Programme.

Training and leadership

As mentioned above, the Environmental Agency had officials assigned to the Adiwiyata Programme, and it was the responsibility of the officials at the *kabupaten* level to train teachers in EE. However, the officials themselves had no environmental or education background or formal training. The two officials in charge of the Adiwiyata Program in the City of Yogyakarta and Kabupaten Bantul, where our two studied schools are located, reported that they did not have sufficient expertise to train teachers and instead tried to facilitate experts from the universities to come and speak at any training session. The Yogya City official reported that she preferred to work with individual schools rather than hold training workshops. Officials from the Environmental Agency require permission from the Ministry of Education and Culture before holding Adiwiyata training sessions for teachers in state schools. There was no training held for teachers during the 12 months of our fieldwork. Teacher training run by the Environmental Agency is not automatically recognised by the Ministry of Education and Culture (and therefore does not necessarily count for career advancement for teachers). According to officials at the Environmental Agency, this is a significant barrier to the recruitment of teachers to the Adiwiyata Programme.

Finally, it is worth noting the teachers' response to the lack of leadership, training and support from government. Pak Hendra,[5] an English teacher at The Islamic School, explained the need for leadership as follows:

> The school needs support from the government. If there is no government support, the school is not motivated. For example, they [the Environmental Agency] needs a specific plan for Adiwiyata. A five year plan – they must have 10 or more schools and offer them support. It needs to come from the Mayor, then delegated to the City-level government, Environmental Agency, etc. The policy must come from the Mayor. He has the authority and power to make policies. If he doesn't have the policies, staff just wait. It's the character of the Yogya people to follow the leader, so initiatives must come from the Mayor. The former Mayor had more concern for the environment and good planning – how to make the Yogya environment better. For example, gardens in the streets [street-scaping] and some parks. He showed that he had a programme to make the environment better, then the people give support. [That's the] character of Yogya – support and follow the leader. Do what the leader says. I saw a big difference between the former Mayor and past ones.
>
> (Pak Hendra, English teacher, interview 21 February 2015)

Pak Hendra believed that the 30 years under Suharto's rule had resulted in people having come to rely on the government. He finished by saying that people "are not used to thinking for themselves or taking action. [It] will take a long time to fix this. Adiwiyata depends on [the] Principal. If [the] Principal has

[a] good programme, others will follow. Strong leadership". We come back to this issue of leadership in the chapters on Surabaya.

Without the support of the Ministry for Education and Culture, the Environmental Agency is very limited in its access to schools and teachers and hence its capacity to be effective in EE. The Ministry for Education and Culture, it seems, is content to leave EE as the responsibility of the Environmental Agency – but they are hamstrung. With almost complete silence around environmental responsibility in the curriculum and textbooks, and complete lack of interest by the Ministry as well as the local Education Office, the possibilities for EE in senior high schools in Indonesia are extremely limited.

Fieldwork and selection of schools

It should be noted that we chose the two schools as case studies not in order to test hypotheses or to carry out controlled comparisons of EE programmes. In Yogya, it was very difficult to find schools that were doing any EE, and in the event, there was hardly any choice.

Our first port of call was the Provincial Office (DIY) of the Ministry of Education and Culture. The official, a former principal of one of Yogya's best public schools, was not sure of the best environmental schools, but suggested seven schools to visit. We visited and spoke to the principals of four of those schools (who said not to bother with the other three as they had no EE programmes). Only one of the schools seemed to have an environmental programme running. Each school principal offered reasons as to why their school was not a strong environmental school: one had had a person outside the school running an EE programme but that ended due to mismanagement and had not been re-established; another focused on outdoor education, which mainly consisted of a camp each year; and the other was not an Adiwiyata school. These were prestigious schools (two were famous private schools), but none was an Adiwiyata school. It is our impression that these schools are not Adiwiyata schools because they are such prestigious schools: they do not need the Adiwiyata imprimatur to attract students. We asked each principal if they knew of any other environmental schools in addition to those on our list. They could not think of any schools to suggest.

Our next point of call was the Provincial Office (DIY) of the Environment Agency where we met with officials responsible for the Adiwiyata Programme at the provincial level. They provided a list of schools that were currently participating in the Adiwiyata Programme. The list for 2014 contained only two primary schools, two junior high schools and two senior high schools. Each type of school had one school in Yogyakarta city and one in Bantul District. We contacted the two schools at senior high school level, a state Islamic day school (MAN, Madrasah Aliyah Negeri), called here The Islamic School, and a general state school (SMAN, Sekolah Menengah Atas Negeri), called here The High School. Both were willing to participate in our research. Thus, the schools were selected because they were currently participating in the Adiwiyata Programme;[6] they were willing to have us conduct research at the school; they were reasonably close to the city (within a 45 minute drive for the researcher); and they were different types of schools (Islamic and general).

Observations at the two Yogyakarta schools were carried out over an 11 month period. School visits were always prearranged with the contact teacher in an effort to capture any environmental actions or lessons going on at the school. Because the schools took a very ad hoc approach to their environmental activities, and they happened so rarely, it was pointless to hang around at school, waiting. It was often a week or more between activities and in some cases a month or more. The problem with this relatively efficient way of working was that the researcher had to rely on the teacher knowing that something was planned and remembering to invite the researcher. On more than one occasion the researcher learned after the fact that the school had undertaken an activity that would have been relevant to her research, but the teacher had forgotten to inform her. This happened less as time went on, but this problem shows the value of classic long-term fieldwork, hanging around, waiting for things to happen, and also the researcher's position of limited power which she just had to accept. At both schools the researcher gave her mobile phone number to environmental club students (with the teacher's permission) so that they could inform her of upcoming events in an effort to minimise missed opportunities, but with no results.

Some weeks the researcher would visit the schools once or twice and other weeks not at all, depending on their schedules. She only observed classes with environmental content, as defined by the teachers. She often visited on Friday and Saturday mornings when clean ups and environmental activities were scheduled. Because the schools ran their clean-ups at the same time, she had to divide her time between both. In addition to observations, she ran focus group discussions (FGDs) with students and teachers at both schools, recorded interviews with some student leaders at both schools and held informal conversations with numerous students. The informal conversations were not audio recorded but she took notes where possible.

Introducing the two schools

The Islamic School is situated on a main road in the highly developed and commercial city centre. The city centre is hot and crowded, and polluted. The school has a reputation as an old and respected educational institution. Even though *madrasah* are under government control, it is commonly the case that they have a distinctive history, which shapes their contemporary identity (Parker & Raihani, 2011, p. 717). Many have sprung up from within the community, initially as a Qur'an-teaching course or similar. Many teachers at MAN (and other levels of *madrasah*) feel like second-class citizens because they perceive that they are under-resourced and under-supported compared with SMAN and other schools under the Ministry for Education and Culture.

However, The Islamic School is not like that. Its institutional history goes back to the early days of the Republic, and for some years it was a state teachers' training college for girls to become Religion teachers. It is a more reputable institution than our second school, The High School, and, because of the trend to send children to Islamic schools, is a popular choice, even among the city's elite, e.g. a judge sends his child to this school. This is a school with a strong commitment to education, demonstrated when our researcher was asked to provide books for the students, and when they organised a Q&A session on EE for teachers with both authors. This school committed to the 2013 Curriculum and stayed with it, as they thought that teachers in the school were coping with its new demands.

There are around 580 students and 60 teachers in the school. A typical inner-city school, The Islamic School has no green areas. There are a few trees in pots. The school administration is housed in a colonial-style building from the Dutch era; classes are held in the newer buildings, which are three storeys high. The mosque is separated into upstairs (females) and downstairs (males), as there is not enough room to fit everyone on one level. The canteen cannot seat all of the students, so students eat their lunch and snacks in the classrooms. There is no outside area for students to sit, and the only open space is a paved area at the front of the school, which is in full sun. This means that the opportunities for outdoor, nature-based learning are very limited. In addition, most students come from the main city of Yogyakarta so are rarely exposed to the natural environment.

The area of the school has recently expanded, with government funding. There will be an underground parking area with either a sports field or a meeting/assembly area on top. A condition of the government funding was that the school would build a dormitory, and this will open shortly. The Islamic School decided that it would be for around 60 girls, as girls are "easier to manage, and cleaner than boys" (Teacher, 18 April 2015). The student gender balance in this school is already something like 8:3 in favour of girls, and presumably this will mean an even more "female" student population.

The High School is a medium-sized school of approximately 600 students. It falls into the Bantul District but is much closer to the city than it is to Bantul. The socioeconomic background of the students is mixed, placing it in the middle socioeconomically. Students come from all five districts in DIY, but more because of its position than its academic reputation. Being in Bantul District automatically means it is of somewhat lower status than a City school. It is situated amongst rice fields and houses, local shops and stalls. This is known as a green part of the DIY. The school has pleasant grounds, with trees, sports fields, and space to grow gardens, sort rubbish, make and use compost. While this looks the better school, educationally The Islamic School is much superior. Here there was no concern when teachers would leave school early or be absent from class. When the new 2013 curriculum was being introduced, they changed the curriculum half-way through the term, leaving students high and dry. When our researcher wanted to provide a token of thanks for participating in the research, they only suggested giving food and drinks to the students (not books). In other words, the commitment to education at The High School left a lot to be desired.

Joining Adiwiyata

Since decentralisation began in 1999, the education system has been decentralising, and in the process, moving towards school-based management (SBM), privatisation and marketisation. This process was largely driven by the World Bank, and means that local governments now pay much of school budgets, including salaries. The decentralisation of education has been contested, not least for the impact on the poor and needy, as schools and universities have often raised fees as a response. Various academic studies have been conducted on SBM (e.g. Bandur, 2008; Bjork, 2006; Kristiansen & Pratikno, 2006; Parker & Raihani, 2011), but the trend for schools to become commodities, each with their market niche, has been less studied, though see Darmaningtyas and Panimbang (2009), and Kartono (2009).[7] The neoliberal agenda means that each school has to "establish a unique difference in order to highlight their strengths and to give the students a reason to choose" a particular school (Kusumawati, 2010, p. 1). Amirrachman *et al.* argue that

> as a result of the educational marketisation that paralleled decentralisation, parental choice seems likely to become the preserve of the richer, more influential, or middle class parents. Poorer, less influential, or working class parents are likely to have more limited choices.
>
> (Amirrachman *et al.*, 2008, p. 40)

And, one could add, rural and remote parents have even more limited choices.[8]

Both of the studied schools were ordered to join Adiwiyata in order to build a reputation as a special type of school. Instead of being just a *madrasah* or an SMA, the schools now had a niche identity. The Adiwiyata Programme gave them the opportunity to gain prestige in a non-academic (and therefore less competitive) field. Neither school was a strong performer academically. Thus, by their own admission, the Yogyakarta schools were made to become Adiwiyata schools because they were not achieving highly in any other field. An unintended and unfortunate consequence is that EE is associated with the less-prestigious, academically-weaker schools.

From truants to environmental leaders

The environmental programme at The Islamic School came about by accident. The school was punishing late and truanting students by having them wear orange T-shirts whilst sweeping the school grounds and cleaning up litter. The unintended consequence was that this orange T-shirt group became "cool" (*gengsi*) and other students wanted to join the group. This group entered a school environmental competition and was awarded first prize on the basis of its work in keeping the school clean. Soon after this, The Islamic School was instructed by the Department of Religion to continue to build its environmental programme in order to build its reputation as a "green *madrasah*", and it joined the Adiwiyata Programme.

Forced participation

In 2011, The High School was told by the District Head (*Bupati*) that it would become an Adiwiyata school in order to make a name for itself. A teacher described the decision as being forced on the school, whether they wanted it or not ("*mau, nggak mau, terpaksa*"). At that time, the school was in a state of disrepair and the grounds were dry and dusty, with no trees. In the years since being forced to join the programme, the school had made itself "clean and green". At a school learning exchange day, the Sociology teacher explained this transition to visiting teachers and students. He first showed photographs on a big screen that depicted a dusty school with no greenery, grounds of dry, compacted, swept dirt and a canteen that was a shed of some metal sheets held up by wooden stakes in a dirty, dusty outdoor environment. He then showed pictures of the school as it currently is, with a new canteen, a new sick bay, and grass and many trees where before there had been dirt and dust. "We did this in three years," he proudly announced. "We were forced." He told us that now the school is quite famous.

The Bantul District falls within DIY and despite a lack of support for Adiwiyata at the provincial level from both the Ministry of Education and Culture and the Environmental Board, almost all of Bantul's schools (approximately 550 according to the representatives of the Environmental Agency in Bantul) are involved in the Adiwiyata Programme, with many at the National level (waiting to be allowed to move to the Mandiri level). In interviews, teachers from three Bantul schools noted that the District Head (*Bupati*) was highly supportive of the Adiwiyata Programme and expected all schools to participate in it.

Many schools around the world take up environmental learning programmes because of one or two committed individuals, who are environmentalists. This is not the pattern at all with Adiwiyata: it seems most schools join because they are told to, and most teachers who are made responsible for the Adiwiyata Programme in their school do it because they are told to.

Organisation of Adiwiyata in schools

At The Islamic School, Ibu Eni was appointed to be responsible for the Adiwiyata programme and is known as the leader of the Adiwiyata Team.[9] Ibu Eni was always present at the more prestigious environmental events such as a learning exchange between teachers from another school, visits from German researchers and an NGO seminar held at a hotel, but rarely participated in the hands-on activities such as the weighing of recyclables or clean-up mornings. Early on, we were under the (incorrect) impression that the Adiwiyata Team was made up of student representatives. We gradually realised that the team was composed entirely of teachers and the students were the cadets (*kader lingkungan*). There was a clear distinction between the roles: the team (adults) made the decisions and compiled documentation for Adiwiyata reporting and the cadets (students) did the physical work. Year 12 students were not permitted to participate in the programme as they needed to focus on their studies (for the National Examination). All student leaders (OSIS) were required to participate as "agents of change" (organisers) for the Rubbish Day event (described in Chapter 8). Despite having very few trees and therefore hardly any organic waste (besides that generated from the canteen), the school has a shredding machine and a composting area. It also has a greenhouse behind the carpark. These are requirements of the Adiwiyata programme, yet in our many visits to the school we never saw them being used. However, the school has a Rubbish Bank (Bank Sampah) and they weigh and count the recyclables. (Several schools had unused greenhouses and shredding machines; one had an unused and very expensive water-purifying machine provided by Adiwiyata.)

At The High School, the Adiwiyata Programme is open to year 10 and 11 students. We were told that it includes environmental cadets (*kader lingkungan*), nature lovers (Pencinta Alam) and an anti-illicit-drug group (Gerakan Anti Napsa dan Psikotropika). The School has almost the same facilities as The Islamic School, but the Bank Sampah is not well established and is inactive. The programme is coordinated by one of the school's biology teachers, Ibu Widiya. Ibu Widiya somewhat reluctantly led the programme, demonstrating very little enthusiasm for the programme and offering many reasons why the students were not actively involved in environmental projects. As part of the Adiwiyata programme, the students are meant to undertake weekly activities including composting organic waste, maintaining the plants in the greenhouse, sorting and weighing recyclables and tending to the fish pond. Due to a lack of interest in Adiwiyata as an extra-curricular activity, these activities were scheduled to be undertaken during the Saturday morning clean-up sessions (*Sabtu bersih*). During a focus group discussion with the Adiwiyata students, we asked if they had meetings and made any kinds of plans for environmental actions. The students looked left and right and seemed uncomfortable with the question before one student answered, "We have lots of meetings and make plans but *action* is only an expression". We never saw any students undertaking any environmental activities in our time at the school but we did witness some discussions about doing things.

We were puzzled as to how the School had attained National Adiwiyata status, and how they had "greened" their campus. Apparently they had managed to get a big loan, as the principal knew someone who could lend them money. They used those funds to build a new canteen, sick bay (Usaha Kesehatan Sekolah), prayer room, etc. and the greening was done at the same time. Students were required to bring plants from home. The school had a gardener, whose status was similar to that of an "office boy", and who was paid out of the school budget. It seems they did not have a special Adiwiyata budget. Usually schools would make a proposal to the principal to ask for money for something they needed. The achievement of National Adiwiyata status is still something of a mystery.

The school has a superior website, and includes a tab for Adiwiyata, and a YouTube clip of their "Engine Off" initiative (motorbike riders have to cut their engines on school grounds to reduce air pollution on campus). This does not actually happen in everyday life, but the clip is interesting for its representation of gender relations and surveillance. Most of the motorbike riders are males. As they ride in through the front gate they are met by a *piket* (monitoring picket) of girls, who do not actually do anything but are there as a reminder to the boys.[10] The Principal's welcome to the Adiwiyata page begins (in Indonesian): "Welcome to [the School], a National Adiwiyata School, and a healthy school, which enculturates a love for the environment and behaviour that is clean and healthy." He outlines his school's preparedness to be awarded "National" status in the Adiwiyata hierarchy. The webpage says that it has an "education curriculum based on the environment". There is a page on Outdoor Learning with photos of students outside, clustered around a banner promoting "Outdoor Learning"; at the Merapi Museum, looking at a model of a volcano;[11] and in a shadehouse. There is a page with a nice piece on conservation by WWF. Yet, in the various lists of extra-curricular activities and organisations at the school, and in the page on Awards (Prestasi), Adiwiyata is not mentioned. It's a rather uneven presentation of the school's commitment to the environment and to Adiwiyata.

Adiwiyata and ersatz accountability

Both of our Yogyakarta schools had achieved "National" level Adiwiyata status at the time of field research. This meant that they had progressed through various levels of the programme (school, district or city, and province) and were working towards attaining the highest level: the "Independent" level, Adiwiyata Mandiri. In order to have attained the "National" level, they had to have scored at least 90 per cent in meeting the standards outlined in the Adiwiyata Programme Manual, as assessed by a judging committee. In order to have met this standard, amongst many other things, both schools had to have demonstrated that 50–70 per cent of their teachers had prepared lesson plans related to environmental management and protection (Kementerian Lingkungan, 2013, pp. 23, 27).

Since both schools had achieved the national level, we were expecting to see a strong commitment to the environment and environmental education by the staff and students. Instead, we found an ad hoc approach that focused on reward and prestige over action, where one or two teachers controlled and ran the programme, and the role of the students was reduced to being the labour force to carry out activities or make up participation numbers that would score the school Adiwiyata points. Like the general education system in Indonesia, the Adiwiyata Programme has a very strong focus on scoring, competition and prizes, status and rank (*prestasi*), and assessment (Parker & Nilan, 2013, especially Chapter 5, pp. 92–95). Unlike the general education system, the Adiwiyata Programme focuses on assessing teachers' capacity to meet the standards, rather than student capabilities. We were puzzled as to how the schools could have achieved such a high rank when there was so little happening in the schools.

Documentation is unavoidable in Indonesia and is an aspect of everyday life. For example, if a guest comes to stay at one's house for more than one night, the host is required to report this person (with a copy of their ID) to the head of the neighbourhood (*RT, rukun tetangga*), head of the ward (*RW, rukun warga*) and head of the village (*dukuh*).[12] At each level, paperwork will be filled out and questions asked. Under the New Order, such reporting and documentation was often seen by scholars as a dense net of surveillance of the population and part of the larger project of social control. It seems that many of these systems have remained in place, perhaps through unthinking inertia, and perhaps also in order to justify the huge numbers of civil servants in Indonesia.

Documentation is often used in an attempt to demonstrate that one is not cheating or behaving in an immoral or corrupt manner. In order to move up the Adiwiyata rankings, schools are required to provide documentation proving every detail of their claims. Various teachers explained that while the Adiwiyata documentation was burdensome, they understood that it was required to be sure that schools did not cheat their way up the Adiwiyata rankings. During field work, the researcher was almost always photographed when visiting schools, and many of the schools mentioned that this would go in their Adiwiyata documentation. After fieldwork, when looking at our photographic documentation of Adiwiyata activities, we noticed that in nearly every one of our photos there is at least one other person, and sometimes several, taking a photo, also for "*dokumentasi*".

Teachers shared with us files and reports outlining their Adiwiyata Programme that often did not match our observations. Mas Rudi, from an environmental NGO in Surabaya, was on an Adiwiyata jury for East Java. He told us that he was a good judge because he could tell which schools were really doing what they reported and which just had "monuments to compost" (shredding machines and composting areas that were rarely used, as in The High School). Teachers did not leave it to students to take responsibility for documentation. While students did have to count (e.g. bags of compost made, the number of students participating in each event, etc.) and take photographs, they did not have any control over the actual documentation that went to Adiwiyata.

Various teachers complained about the heavy documentation requirements of the Adiwiyata Programme. At first we were puzzled as to why the students were not more involved in the documentation of their activities. Other teachers later suggested (somewhat apologetically) that the Adiwiyata teachers did not want the students involved in documentation because this would have meant that they could not report it as they wanted to (for instance, by embellishing the numbers). By handing over responsibility to students they would be foregoing opportunities to make themselves look good. After a few months of observing and interviewing teachers from various schools, we came to believe that this was indeed the case. For some teachers, this becomes an obsession with documentation, reward and rank (*prestasi*). After conducting an interview with a teacher lasting 40 minutes, the researcher received a text message from the teacher, asking for a certificate from the researcher's university in Australia so that she could use it in her application for promotion. When told that we could only provide a letter of thanks, she made her dissatisfaction clear. Even after we had sent the letter of thanks, she continued to send messages expressing her dissatisfaction. Another teacher asked us to provide her with a table listing all of our visits to the school and any photos of students doing things so that she could include it in her next Adiwiyata documentation submission. Slowly we realised that the point of the Adiwiyata Programme for the two schools was to advance up the rankings. Doing this provided opportunities for teacher career advancement and increased prestige for the school.

Long and others have noted the obsession with *prestasi* (prestige, rank, status) in Indonesia – among students, teachers and civil servants (Long, 2007, 2013; Parker & Nilan, 2013). Long traces it back to the popularity of the sociology of achievement in Indonesia – in particular the book, *The Achieving Society* by McClelland, 1961 – as a sort of blueprint for how Indonesia might produce achievement-oriented citizens and develop its human resources (Long, 2013, pp. 179ff.). This approach involves encouraging people to create "achievement-related fantasies" and instigating "activities aimed at producing achievement". The motivation to achieve is enhanced when people experience pleasure at having met "standards of excellence" that had been "impressed" upon them by authoritative or admired outsiders – or by the shame of failing to meet those standards. The ubiquity of *lomba* (competitions) in Indonesia from the 1970s (Long, 2013, p. 179; Parker, 2003, Chapter 6) suggests that we can trace the genealogy of the *prestasi* obsession to before the neoliberal audit culture. The *prestasi* obsession entails taking on the externally-imposed standards as a personal goal, the excitement of competition and of measurable improvement, as well as the pleasure (or pain) of winning or meeting a standard (or not). In many contexts, the achievement of *prestasi*, rather than the content of the work, becomes the goal. We suggest that this is what was happening with many of the Adiwiyata teachers.

Nevertheless, with its huge demands for paperwork and documentation, the Adiwiyata Programme is a classic example of the contemporary neoliberal "audit culture" (Strathern, 2003b). It seems the documentation requirement was designed to be a preventative of corruption. It was set up with a lack of trust as the implicit basis. Given the everyday ubiquity of corruption in Indonesia, it is perhaps not surprising. "Audits are needed when accountability can no longer be sustained by informal relations of trust alone but must be formalized, made visible and subject to independent validation" (Power, 1994, pp. 9–10). In creating a new programme out of nothing, if seems as if the creators deliberately built in a massive documentation mechanism as a substitute for trust. Power suggests that the popularity of the audit, way beyond its original field of financial accounting, means that "its spread actually creates the very distrust it is meant to address" (Power, 1994, p. 10). He says it would be "far-fetched to say that audit creates the very pathologies for which it is the prescribed treatment" (Power, 1994, p. 11), but in the case of Adiwiyata, where the requirements for documentation were set up *prior* to the event, staff were allowed time to "create their own reality" (Strathern, 2003a, p. 289), which was a parallel, "bottom-drawer" file of false documents.

Conclusion

This chapter has set the scene for Chapter 8, where we will enter schools in Yogyakarta and show how teachers taught about the environment and implemented the Adiwiyata Programme. This chapter has shown that the bureaucratic context in Yogyakarta is inimical to the fostering of good EE. The Education Ministry does not consider itself responsible for EE; the Environment Ministry is responsible but there is no avenue by which they can enter schools, train teachers and enable EE to happen. There is a lamentable lack of leadership by the Sultan and the Mayor because they are only interested in profitable development and tourism. We feel that the problems identified in the previous sentence could be overcome if these two leaders were only more motivated to foster social justice and environmental sustainability. There are also problems of lack of training and expertise, corruption and "passing the buck".

The Adiwiyata Programme, established and run by the Ministry of the Environment and Forests, has potential. Its activity-based approach to EE, examined in Prabawa-Sear (2018), Tanu and Parker (2018) and the next chapter, has much to recommend it. However, the way it has been institutionalised – i.e. schools are forced to participate, teachers are forced to run it – undermines its ethos. Its adoption by principals of schools that do not perform well academically, means that it has become a marketing tool. Further, the design of the Programme, with its hierarchy of success, and emphasis on numbers as a measure of performance or Key Performance Indicators (KPIs) – the number of students attending an event, the number of staff who had incorporated EE into their syllabi, the number of bags of compost made) – means that the motivation of teachers has nothing to do with the environment. The numbers were transformed from mere numbers into resources for prizes and rewards (Harper, 2003, pp. 46–47). Thus, the very mechanism of accountability, documentation, became the mechanism for cheating. The documentation also indexed the extent to which the Adiwiyata Programme was being hijacked for non-environmental ends: in Yogya, instead of teaching teachers and students how to take responsibility for the environment, it had become a mechanism for school marketing and *prestasi*, and a way for teachers to gain credit towards promotion.

Notes

1 See Anderson (1990 [1972]).

2 The Javanese phrase *Yogya ora didol* (Yogya is not for sale) can be found all around Yogyakarta – on banners, street art, T-shirts as well as a hashtag on social media posts. It is also common to see banners strung up at the entrance to neighbourhoods stating that "This neighbourhood (*kampung*) rejects hotels and apartments".

3 However, this is not a problem unique to Indonesia, or to the era of decentralisation. The positioning of EE outside of the Ministry of Education and Culture (and therefore the formal education sector) is an issue that is well documented from several countries, and there are many arguments for and against this positioning. See Rose (2011), Teamey (2007) and Wade (1996) for more on this.

4 Fortunately he has now moved on.

5 All names are pseudonyms, in compliance with Human Ethics regulations and commitments.

6 We selected Adiwiyata schools over non-Adiwiyata schools, even where one non-Adiwiyata school seemed to have a strong environmental programme. We felt that it was important to understand the strengths and limitations of the national programme and the role that government played in delivering it. We also knew that we would have the opportunity to explore EE programmes in schools that were not high-achieving Adiwiyata schools in Surabaya. This meant that the one school suggested by the Ministry of Education and Culture that was running an EE programme was not selected.

7 The marketisation of education in Indonesia has attracted some scholarly response in relation to higher education in Indonesia, but not much in relation to secondary education. For higher education, see for example Susanti (2011) and Welch (2007).

8 Rosser also argues that there has been some push-back to the neoliberal reform of the education system, by "predatory bureaucrats, populist politicians and other elite actors but also everyday actors such as parents, students, NGO activists and nationalist intellectuals" (Rosser, 2016).

9 She has asked to have it passed to somebody else but was told she had to keep doing it.

10 This reflects a broader gendering of school life in Indonesia, where girls are constructed as good and boys allowed and even encouraged to be bad: girls should be quiet and compliant, boys should be noisy and brave (Parker, 1997, pp. 504ff.). The idea that girls should be the moral guardians is also quite prevalent, for example, in the shift to wearing the Islamic headscarf (*jilbab* or *hijab*), many people note that girls should dress modestly in order to help boys control their sexual urges; if a girl is raped, she is at least partly responsible for leading the boy on; and so on. See Riyani (2016, pp. 160–168) on the core values of *malu* (shyness) and *pasrah* (compliance) for women. Brenner (1999, 2011) argues that women are viewed as vessels of morality. She argues that in New Order discourse, the state, the family and organised religion were closely bound to one another and viewed as the three most sacred institutions of Indonesian society (Brenner 1999). In this way, public discourse often attributes perceived failures of morality to a breakdown of the family unit and a failure of women to carry out their roles as wives and mothers (Brenner 1999). She highlights the Indonesian saying that "women are the pillars of the state. If women are good, the state will also be good, but if women are ruined, the state will be ruined as well" (*Amanah*, cited in Brenner, 1999, p. 31). Instead of focusing on the government's failure to address these issues, responsibility for the nation's wellbeing was placed on family institutions and, in particular, on women's ability to uphold public morality (Adamson, 2007; Bennett & Davies, 2015; Brenner, 1999).

11 In the photo of students at the museum, they are looking at a papier-mâché model of a volcano. The Merapi Museum is about the eruption of Gunung Merapi, an enormous active volcano just to the north of Yogyakarta. This is akin to the famous

Leunig cartoon of a person looking at a sunset on television when there is a spectacular sunset occurring in real life just outside the window. www.leunig.com.au/works/cartoons.

12 Elsewhere in Indonesia, the *dukuh* level does not exist.

References

Adamson, C. (2007). Gendered anxieties: Islam, women's rights, and moral hierarchy in Java. *Anthropological Quarterly*, 80(1), 5–37. doi:10.2307/4150942.

Amirrachman, A., Syafi'i, S., & Welch, A. (2008). Decentralising Indonesian education: The promise and the price. *World Studies in Education*, 9(1), 31–53.

Anderson, B. R. O. G. (1990 [1972]). The idea of power in Javanese culture. In *Language and Power: Exploring Political Cultures in Indonesia* (pp. 17–77). Ithaca, NY: Cornell University Press.

Bandur, A. (2008). *A Study of the Implementation of School-based Management in Flores Primary Schools in Indonesia*. PhD, The University of Newcastle, Newcastle, NSW.

Bennett, L. R., & Davies, S. G. (Eds). (2015). *Sex and Sexualities in Contemporary Indonesia. Sexual Politics, Health, Diversity and Representations*. Abingdon: Routledge.

Biennale Jogja XIII Equator #3. (2015). Anti tank "Sabda Warga". Retrieved from www.biennalejogja.org/2015/programmes/exhibition/artists/anti-tank-2/?lang=en.

Bjork, C. (2006). Transferring authority to local school communities in Indonesia: Ambitious plans, mixed results. In C. Bjork (Ed.), *Educational Decentralization: Asian Experiences and Conceptual Contributions* (pp. 129–147). Dordrecht Kluwer Academic.

BPS (Badan Pusat Statistik). (2010). Sensus Penduduk 2010 (2010 Census). Retrieved 24 August 2016 http://sp2010.bps.go.id/index.php/site/tabel?search-tabel=Penduduk+Menurut+Wilayah+dan+Agama+yang+Dianut&tid=321&search-wilayah=Indonesia&wid=0000000000&lang=id.

Brenner, S. (1999). On the public intimacy of the New Order. *Indonesia*, 67, 13–38. doi:10.2307/3351375.

Brenner, S. (2011). Private moralities in the public sphere: Democratization, Islam, and gender in Indonesia. *American Anthropologist*, 113(3), 478–490. doi:10.1111/j.1548-1433.2010.01355.x.

Colbron, C. (2016). The Sultan of development? *Inside Indonesia*, Jan–June 2016(123). Retrieved from www.insideindonesia.org/the-sultan-of-development?highlight=WyJzdWx0YW4iLCInc3VsdGFuIisInN1bHRhbidzIl0%3D.

Darmaningtyas, E. S., & Panimbang, I. F. (2009). *Tirani Kapital dalam Pendidikan (The Tyranny of Capital in Education)*. Yogyakarta: Damar Press/Pustaka Yashiba.

Firman, T. (2010). Multi local-government under Indonesia's decentralization reform: The case of Kartamantul (The Greater Yogyakarta). *Habitat International*, 34, 400–405.

Harper, R. (2003). The social organization of the IMF's mission work: An examination of international auditing. In M. Strathern (Ed.), *Audit Cultures: Anthropological Studies in Accountability, Ethics and the Academy* (pp. 21–53). Florence: Taylor & Francis.

Kartono, S. (2009). *Sekolah Bukan Pasar: Catatan Otokritik Seorang Guru (School is not a Market: Autocritical Notes of a Teacher)*. Jakarta: Penerbit Buku Kompas.

Kementerian Lingkungan Hidup Republik Indonesia. (2013). *Pedoman Pelaksanaan Program Adiwiyata*. Retrieved from www.menlh.go.id/Peraturan/Permen%2005%20tahun%202013_tentang%20Pedoman%20Pelaksanaan%20Program%20Adiwiyata%20oke.pdf.

Kristiansen, S., & Pratikno. (2006). Decentralising education in Indonesia. *International Journal of Educational Development, 26*, 513–531. doi:10.1016/j.ijedudev.2005.12.003.

Kusumawati, A. (2010). Privatisation and marketisation of Indonesian public universities: A systematic review of student choice criteria literature. In: University of Wollongong, Research Online.

Long, N. (2007). How to win a beauty contest in Tanjung Pinang. *Review of Indonesian and Malaysian Affairs, 41*(1), 91–117.

Long, N. (2013). *Being Malay in Indonesia: Histories, Hopes and Citizenship in the Riau Archipelago.* Singapore: Asian Studies Association of Australia in association with NUS Press and NIAS Press.

Maulipaksi, D. (2014). 498 Sekolah Terima Penghargaan Adiwiyata Nasional 2014. Retrieved from www.kemdikbud.go.id/main/blog/2014/12/498-sekolah-terima-penghargaan-adiwiyata-nasional-2014-3655-3655-3655.

Meidiana, C., & Gamse, T. (2011). The new Waste Law: Challenging opportunity for future landfill operation in Indonesia. *Waste Management & Research, 29*(1), 20–29.

McClelland, D. C. (1961). *The Achieving Society.* Princeton: van Nostrand.

Monfries, J. (2007). Hamengku Buwono IX of Jogjakarta: From Sultan to Vice President. *Life Writing, 4*(2), 165–180. doi:10.1080/14484520701559653.

Parker, L. (1997). Engendering school children in Bali. *Journal of the Royal Anthropological Institute, 3*, 497–516.

Parker, L. (2003). *From Subjects to Citizens: Balinese Villagers in the Indonesian Nation-state.* Copenhagen: Nordic Institute of Asian Studies Press.

Parker, L., & Nilan, P. (2013). *Adolescents in Contemporary Indonesia.* New York: Routledge.

Parker, L., & Raihani. (2011). Democratizing Indonesia through education? Community Participation in Islamic schooling. *Education Management, Administration and Leadership, 39*(6), 712–732.

Power, M. (1994). *The Audit Explosion.* London: Demos.

Prabawa-Sear, K. (2018). Winning beats learning: Environmental education in Indonesian senior high schools. *Indonesia and the Malay World, 46*(136), 283–302. doi:10.1080/13639811.2018.1496631.

Riyani, I. (2016). *The Silent Desire: Islam, Women's Sexuality and the Politics of Patriarchy in Indonesia.* PhD, The University of Western Australia.

Rose, P. (2011). Strategies for engagement: Government and national non-government education providers in South Asia. *Public Administration & Development, 31*(4), 294–305. doi:10.1002/pad.607.

Rosser, A. (2016). Resisting marketization: Everyday actors, courts and education reform in post-New Order Indonesia. In L. Rethel & J. Elias (Eds), *The Everyday Political Economy of Southeast Asia* (pp. 137–156). Cambridge, UK: Cambridge University Press.

Strathern, M. (2003a). Afterword: Accountability … and ethnography. In M. Strathern (Ed.), *Audit Cultures: Anthropological Studies in Accountability, Ethics and the Academy* (pp. 280–304). Florence: Taylor & Francis.

Strathern, M. (Ed.) (2003b). *Audit Cultures: Anthropological Studies in Accountability, Ethics and the Academy.* Florence: Taylor & Francis.

Suryani, B. (2016). Warga Parangtritis Siap Tolak Pembangunan Hotel (Citizens of Parangtritis Ready to Reject Hotel Development). *Harian Jogja (Jogja Daily)*, 29 February 2016. Retrieved from www.harianjogja.com/baca/2016/02/29/warga-tolak-hotel-warga-parangtritis-siap-tolak-pembangunan-hotel-696067.

Susanti, D. (2011). Privatisation and marketisation of higher education in Indonesia: The challenge for equal access and academic values. *Higher Education: The International Journal of Higher Education and Educational Planning, 61*(2), 209–218. doi:10.1007/s10734-010-9333-7.

Tanu, D., & Parker, L. (2018). Fun, "family", and friends: Developing pro-environmental behaviour among high school students in Indonesia. *Indonesia and the Malay World, 46*(136), 303–324. doi:10.1080/13639811.2018.1518015.

Teamey, K. (2007). *Whose Public Action? Analysing Inter-sectoral Collaboration for Service Delivery. Literature Review on Relationships between Government and Non-state Providers of Services.* Retrieved from Birmingham: www.birmingham.ac.uk/Documents/college-social-sciences/government-society/idd/research/non-state-providers/literature-review.pdf.

Wade, K. S. (1996). EE teacher inservice education: The need for new perspectives. *Journal of Environmental Education, 27*(2), [no page numbers].

Welch, A. R. (2007). Blurred vision?: Public and private higher education in Indonesia. *Higher Education, 54*(5), 665–687. doi:10.1007/s10734-006-9017-5.

Wicaksono, P. (2016). Tolak Pembangunan Hotel, Aktivis Yogya Mandi Kembang (Rejecting Hotel Development, Jogjan Activist Anoints himself with Flowers). *Tempo,* 5 February 2016.

Yanuardy, D. (2012). *Commoning, Dispossession Projects and Resistance: A Land Dispossession Project for Sand Iron Mining in Yogyakarta, Indonesia.* Paper presented at the International Conference on Global Land Grabbing II, Cornell University, Ithaca, NY. Retrieved from from www.academia.edu/12288247/Commoning_Dispossession_Project_and_Resistance_A_Land_Dispossession_Project_for_Sand_Iron_Mining_in_Yogyakarta_Indonesia?auto=download.

8 Hollow environmental education in Yogyakarta

Introduction

The 2013 Curriculum states that students should be able to show "an attitude of being a part of the solution to various problems ... in the social and *natural environments*" (Mendikbud, 2013, p. 7, emphasis added).[1] Despite this competency objective, the way that the Curriculum is currently used means that it fails to offer students the opportunity to involve themselves in the deepest problems of society. Giroux argues that the key to students acquiring the knowledge, skills, and ethical vocabulary necessary for the "richest possible participation in public life" is participation in the deepest problems (Giroux, 1999, pp. 146–147 citing Havel, 1998, p. 46). The Adiwiyata Programme could enable such participation, but teachers would have to be trained and made responsible for facilitating this. At present, as we saw in Chapter 7, Adiwiyata is not incorporated into the Curriculum; it has limited take-up; and is a long way from fostering student learning about how to solve environmental problems, as required by the Curriculum.

In this chapter, we use classic educational ethnography to show what is happening in schools in Yogyakarta – both in classes where teachers are using the Curriculum, and in and around schools, where students are doing Adiwiyata activities. The education system in Indonesia is still based on exams, rankings and results for each subject; and to that end, the usual pedagogy is reading and rote learning, so that "facts" can be transferred and retained. We examine three classes in particular: a Craft and Entrepreneurship lesson, a Biology lesson and a Geography lesson. To anticipate, in the Craft lesson, we see that when teachers stray from the traditional pedagogy, they run into problems – mainly because of their own lack of knowledge, confidence and pleasure in teaching. In contrast, the Biology class is an example of a more confident teacher following the curriculum to allow students to leave the classroom and explore the natural world. The third class is the best lesson we saw during fieldwork: it shows a smart, knowledgeable Geography teacher encouraging students to ask questions that go way beyond the textbook topics. The three examples show the crucial importance of the quality of teachers.

Later we report on a student-led environmental event, Rubbish Day, at one of the schools. We use that example to examine the possibilities for the amplification of "student voice" in EE in Indonesia. We discuss the discursive impact of the fact that many teachers are public servants first, and teachers second; and the power of the social value of *sungkan* (respectful politeness) among students. These combine to work against students exercising initiative, suggesting innovations or critiquing their lessons or teachers.

Three "environment" classes

It was very difficult for the researcher to get teachers to commit to a time when she could observe their class. Her aim was to observe any classes that were related to "the environment". This meant that she was open to observing any class in any subject on any day as long as it was somehow related to "the environment" (*lingkungan* or *lingkungan hidup*). It should not have been a difficult request as both schools reported integrating environmental issues across the curriculum as part of the Adiwiyata Programme requirements. On various occasions, teachers apologised to her in advance of the class for the "boring" teaching, "*monotone*" approach and "theory with no practice" that she would witness. Teaching "*monotone*" consisted of reading large amounts of information to students and having them repeat it back, sometimes with the inclusion of some written activities (usually simple comprehension questions to be answered). The teachers acknowledged that they felt uncomfortable having her observe them because they were not "yet" teaching the way they should (as defined by the Education Department in line with the new curriculum). So there were problems with both the content and the pedagogy.

The mushroom fiasco

The researcher observed five Prakarya (Craft and Entrepreneurship) classes at The High School. In four of these classes, students were told to go outside to the mushroom hut. This is an account of one these lessons.

> One by one, students brought their offcut of a log (from which the mushrooms would grow) to the teacher. (Each off-cut was shrouded in a plastic cover.) The teacher inspected each log and students were instructed to spray them. Occasionally the teacher would tell a student to break off a dry bit of the log or to open the plastic cover a little bit more. Not once were the students told *why* they should do something. On the odd occasion that a student asked a question, the teacher would avoid answering it.

STUDENT 1: How come they [mushrooms] grow differently Miss?
TEACHER: Yeah
STUDENT 2 (TO TEACHER): There is no sign of life.
TEACHER: [silence]
STUDENT 3: Why is it that it can go mouldy Miss? [Asking about a log that had gone slimy and mouldy]

TEACHER: Yes, it can. (*Ya ... bisa.*)
STUDENT 3: What's in the sprayer Miss?
TEACHER: Water.
RESEARCHER TO STUDENT: Why is this dry?
STUDENT: I don't know.

The students demonstrated very little understanding of what they were doing or why, and the teacher appeared to have no desire to explain it to them. Students who had successfully grown a mushroom were instructed to pick and weigh it and when we returned to class all students were told to note down any changes in their exercise books. Most students wrote nothing. We felt that what we were observing was quite far from EE, but were aware that our interpretation of EE could be vastly different from those of the teacher and the students. In order to clarify this, we asked the teacher and some students how these activities were related to EE.

RESEARCHER TO STUDENT: Does this activity have a connection to the environment?
STUDENT (MALE): Yeah, we clean the dry bits off [the log] on to the ground, then that gets cleaned up.

This student understood that the ground is the environment. We asked the teacher if there was a connection between this activity and the environment. She said that there was a connection because the logs were made from wood offcuts. During another class a few weeks later, we asked her how growing mushrooms was related to EE. She said, "We learn how to use nature in our area".

(14 January 2015)

Observations of these lessons, combined with conversations with the teacher and students, gave us the strong impression that there was not an issue of differences of interpretation about what EE was. This teacher (an IT specialist) really had *no idea* what EE was and was there teaching Prakarya because someone had to be. We were not convinced that anyone actually believed that this lesson had anything to do with EE. Much to our bemusement, *using natural resources* had become EE. This is also evident in school textbooks and the curriculum. Fortunately, in Surabaya we found a "mushroom hut" which was more successful (see Figure 8.1).

With the exception of the Biology and Geography classes described below, other classes that we observed at both schools were as follows: the teacher disseminated information to students in a classroom, often having students read the relevant passages in textbooks, and asked for an occasional response from (mostly male) students to indicate that they understood. The teacher would say something like, "The natural environment is very import-" and students would reply "-ant." The teacher would then nod with satisfaction at the students

Figure 8.1 Successful mushroom hut in Surabaya school.
Photo credit: Danau Tanu.

having demonstrated their "comprehension". This complete-the-word technique of teaching and "feedback" has been ubiquitous in schools in Indonesia for decades (Parker, 1992 pp. 62–66). It is a feature of rote learning pedagogy and, of course, does not demonstrate comprehension.

Despite teachers showing our researcher documents outlining the subjects where sustainability was integrated, no teacher was able to demonstrate how they integrated sustainability into their lessons. The lessons that we observed were lessons taught straight from textbooks and did not integrate sustainability into science or non-science subjects. Both the Head of Curriculum and the English teacher at The Islamic School explained how sustainability could or should be integrated (the theory), but no one was doing it in practice. What was outlined in the curriculum did not match what was happening in the classroom. This is not only a problem in Indonesia. Mokhele (2011) describes a similar situation in South Africa, where, in line with international recommendations and provisions,[2] environmental learning was to be integrated into all subject areas in primary school. Despite this policy and curriculum change, many schools all but ignored the environmental learning mandate in the curriculum (Mokhele, 2011, p. 78). This was exactly what we observed in schools in Yogyakarta.

Environmental education is difficult to place within a formal education system. It often falls through the gaps as it is not a discrete subject. Where it is applied with an integrated or cross-curricular approach, it often ends up being nowhere, as seen in the Yogyakarta and South African examples, and, to a lesser extent, in Australia too (Dyment, Hill, & Emery, 2015; Mokhele, 2011; Shumacher, Fuhrman, & Duncan, 2012).

The *not so exceptional* Biology class

There was one class at The High School that was different to the others. This class was in some ways "exceptional" in that it encouraged students to find answers in nature.

> The "exceptional" Grade 10 Biology class started as others had. The teacher asked the researcher to introduce herself and informed the class that she could speak Indonesian. During the brief introduction the researcher explained that she was there to observe and would try not to disrupt their class. During the explanation, some girls squealed with delight at her presence (and possibly language ability) and a female student called out "You're beautiful!". These kinds of antics usually came from the boys, but with two-thirds of the class female, it seemed this class had some brave young women in it. The researcher moved to the back of the class where students could not watch her and she could focus on the class activities. There were 29 students in the class, all wearing uniforms: 19 girls, 17 wore the hijab (*jilbab*). They sat in same-gender pairs at the old, hard wooden tables and chairs.

The teacher, Ibu Dian, began the lesson on eco-systems. She reminded the students that they are an Adiwiyata school (probably for our benefit) and asked them to discuss the important parts of an eco-system. There were a few mumbles but no discussions going on. The students appeared to be waiting for Ibu Dian to tell them what the important parts of an eco-system were. Ibu Dian then wrote up the parts of an eco-system under the terms *biotic* and *abiotic*. Ibu Dian gave some explanation as she wrote, and often left words unfinished for the students to complete: "inter-" – "-action", "commun-" – "-ity". Part-way through her explanation, she asked the students, "Where are humans in this?". The students were silent and gasped when she pointed to the word "animals", and she said, *"Paham, ya"* (You understand, don't you), not so much asking the students if they understood, but more inviting them to say that they understood. Sometimes they would nod and say *"Paham"* (Yeah, I understand) back to her. Compared with other teachers that we observed, Ibu Dian was a little more engaging with her tone of voice and occasional light-hearted comment. She also tended to engage the female students more than other teachers who, despite having a majority of female students in the class, clearly focused their attention on male students. Ibu Dian also paused and invited students to focus on her again (*kembali lagi*) when they had stopped listening and had started chatting amongst themselves. Those techniques make Ibu Dian an engaging teacher by local standards. Even so, it was a struggle to focus in the hot, muggy after-lunch class.

Much to our delight, Ibu Dian announced that we would go outside to continue the lesson. The students all listened as she explained the task and where the groups of students were to go in the school grounds to find answers for their worksheets. She sent groups to the greenhouse, the fish-pond, the back pond/reflection area, near the mushroom hut, east of the library and in front of the assembly area. Students were asked to list on their worksheets what type of creatures they could see, what role the creature plays in the eco-system and how many they saw. According to the worksheet (which accompanied the set lesson in the textbook), the aim of the exercise was to observe the interaction of the *biotik* and *abiotik* components of an eco-system. However, the worksheet did not require the students to list which were *biotik* and which were *abiotik*, so it was not a particularly well-thought-out worksheet. The researcher followed a group of four girls who enthusiastically filled in their sheet and told her that they found it "refreshing" to go outside for lessons and that it helps them to understand (*langsung mengerti*). The researcher then chatted with Ibu Dian, asking her the benefits of taking learning outside. She said, "It's fresher and cooler" for the morning classes, but hotter for afternoon classes. We were not surprised by her identifying the temperature as the main benefit of being outside but were somewhat disappointed that she had not identified any education-related benefits such as increased levels of interest, engagement with nature, inquiry-based learning or the like. Upon returning to the classroom, the students sat in groups and discussed their answers. Ibu Dian moved around to each group, checking their work and commenting before handing out marked test scores and dismissing the class.

(29 April 2015)

Upon reflection, this class stood out from others. It was different in that Ibu Dian facilitated active inquiry into nature and engaged all of the students in the class. The researcher had seen other classes move outside (such as the mushroom class above) but the students had not been encouraged to seek answers from, or related to, nature.[3] She had also seen some other engaging teachers, but unlike Ibu Dian, they tended to focus on the boys, not the whole class, and they taught *about* the environment. With the exception of a Geography teacher at The Islamic School, we never witnessed a teacher encouraging students to identify problems, find solutions or even consider that they might have a positive role to play in environmental management. The particular Curriculum competence that the class addressed was this one:

"3.7 Describe the diversity of genuses and species in an ecosystem through observation." (Mendikbud (Kementrian Pendidikan dan Kebudayaan) [Ministry of Education and Culture], 2013, p. 110). It appears in the government-approved textbook as Chapter 9 (Sulistyowati *et al.*, 2013). Although the topic of ecosystems lends itself beautifully to a discussion of conservation or of human intervention in nature, the lesson was only *about* nature, not *for* nature.

A *"quite good"* Geography class

This was a Grade 10 Geography class on Greenhouse Gases (GHG). In Chapter 6 on the 2013 Curriculum, we saw that Geography was the one subject that did convey some of the complexity and interactions between humans and the natural environment. This lesson was the best Geography lesson we saw, and is notable for the amount of student questioning that was not only allowed but encouraged by the teacher, Ibu Yuli.

She began by listing greenhouse gases on the board: CO, CO_2, NO_2, SO_2, HF, HCN, HN_4, then draw a diagram of the earth, the atmosphere and the sun's rays. She explains that "simply" the atmosphere keeps the temperature right, but these GHGs are getting thicker in the atmosphere. The sun's heat gets trapped. This is called the "greenhouse effect".[4] She tells the students to read the textbook and asks, Why are these gasses getting thicker? Where do the carbon emissions come from? The students answer, Transport and industry.[5] This answer of course distances the problem from their own lives, but fortunately Ibu Yuli adds that they also come from houses. She says that in a glass house, it's hotter than the outside temperature, and that's why we call this phenomenon the "greenhouse effect".

She says there are various phenomena related to climate change: El Niño, which can cause a long dry season, drought, which has an effect on harvests, and burning, causing smoke; La Niña, which can cause increased rainfall, floods and erosion, bringing more water-borne illnesses and disease. She says it's not just a physical problem, it is also social. She gives the example of a tree that fell in February near the hospital (the connection to GHG is not clear). On the board she writes, "Consequences – changing climate. Disasters, sea levels rising". The information on the board is not well arranged, but this is complex and difficult to capture diagrammatically.

A boy at the back of the class (who had screamed when the researcher walked in) asked about rivers under the ocean. Then he looked at the researcher to check that she was watching him ask an impressive question. Ibu Yuli asked him if he saw it on a TV show. Yes. Ibu Yuli says, It's not really connected to this topic but I'll explain it anyway. She draws a diagram on the board and explains it's about the Java Sea. She explains the difference between salt and fresh water; like oil and water, they don't mix, and fresh water sits lower than the other, making a type of water stream. But don't think that it's like the rivers you see when you do Scouts – it's different. She goes on to explain how the islands used to be joined and Papua was with Australia. The water near Sulawesi is deeper. The Java Sea is not too deep (30+ metres) compared with 1000+ metres. So we could get that plane out (that had recently crashed) from the Java Sea, but if it had been near Sulawesi, we couldn't. (She was referring to a flight that had crashed into the ocean going from Surabaya to Singapore.) Ibu Yuli is only speaking to the boys. The boys are answering and she is making eye contact with them. She is standing right at the desk of the front row (where girls sit) and is looking over them to speak to the boys.

She goes on to explain how climate change is making rice more expensive. She says that unhulled rice rots because it cannot dry out as we are having less sun and more rain. It needs the sun to dry out. We are waiting on technology so we can still dry the rice even if it's raining. We should be able to produce four crops a year, but we are only getting two crops. So we need to import rice from overseas. She also talks about if we have a long dry season it causes respiratory problems (from the smoke), diarrhoea and skin diseases.

Ibu Yuli moves on to carbon emissions, and says that what we have to do is reduce carbon. If only one person reduces carbon, "*sami mawon*" (she uses Javanese for), it'll just be the same. She asks them to write this in their books: What must you do to decrease carbon emissions? The students start writing, but they seem to mostly be writing notes on what the teacher has just said, which is in the student book and the textbook. She then asks them to share their answers. The class clown (who had screamed and asked about the river) offers and comes forward, in bare feet. He is being silly. She asks him to try again, to be more formal. He does. He says,

1   He'll ride to school because petrol is so expensive. She asks why, environmentally. Him: to reduce carbon.
2   Knalpot (motor exhaust) – he'll use standard petrol. Ibu Yuli adds that he should keep it in good condition so that it doesn't smoke.
3   He will not burn rubbish.

The teacher confirms, If you can ride your bicycle, do so. The second volunteer comes up (another boy). Also no shoes (they are going to pray after this class).

1   Reduce transport
2   Lights
3   Use less electronics
4   Turn off lights/use LED
5   Plant vegetation
6   Rubbish
7   Less cigarette smoking
8   Use alternative energy (but he can't explain what it means).

Others add compost. The teacher suggests solar (photovoltaic) cells – like at the traffic lights.

It feels like this student is reeling off a memorised list and has no real connection to it – perhaps copied from the book.

Talking about smoking cigarettes, Ibu Yuli says, Of course there is no smoking at The Islamic School, but as to what happens outside, I don't know. Especially for boys, try to quit! She also tells the students, if you have to go by motorbike, go the shortest way. Don't just go around and around (for entertainment). The girls are laughing and pointing at classmates who do this. (It's what boys do on Saturday nights in Yogya.)

A student asks the teacher the difference between LPG and natural gas. Ibu Yuli says there is LPG (liquid petroleum gas) and LNG (liquid natural gas). She explains that gas can also come from hotspots. She asks for another volunteer. A boy calls out "A girl!". A girl comes up and says,

1   CFS – this is the ozone layer. Ibu Yuli explains that it is in air-conditioning, perfume and fridges, and says not to buy perfumes in cans.
2   Reduce rubbish – gas from burning. Industry needs filters.

Ibu Yuli picks up a pencil. What does it take to produce this? Where does the wood come from? Trees. The middle bit? Carbon. So to make a simple thing like this is a complex process. It takes energy. It's not about having the money to buy more. We have to think where it comes from. Don't be selfish (*egois*). There will be another generation.

(18 April 2015)

The first point this class highlights is that this is a teacher who is confident of her own ability to handle questions, and encourages students to ask questions. Not many teachers in Indonesia are confident of their knowledge, and this passage shows that students can ask questions about anything, including questions that are not about the topic at hand – demanding a high knowledge base in their teacher. Ibu Yuli said, after class, that she encourages students to ask questions, *even though she might not know the answer*. Then she can encourage them to find the answer. This is a smart, knowledgeable and confident teacher, and she understands the new way of teaching that the government wants. Second, at several points she tries to make the subject matter connect to the students' lives – smoking, riding motor bikes round and round – or to recent news items – the downed plane, the man felled by a tree (though that was rather mysterious).

Given that this lesson is the best lesson we saw, the way it was gendered is quite concerning. The two main examples of everyday relevance were smoking and motorbike riding, which are boys' activities. The boys demanded and received more attention than the girls, even though they were the numerical minority. The teacher talked over the girls, directing her attention and her speech to the boys at the back. It is well established in the scholarly literature that boys in school are rewarded for asking "big" questions, for skim reading, for making grand generalisations, for talking "big" and being brave, which means they can get away with not having done the close reading and hard work, while girls typically are more industrious, read carefully and thoroughly, and tend to ask "smaller" questions on topic. The one girl who asked a question asked quite a small and careful one, showing that she was thinking about the topic and what she knew and didn't know (Giroux, 2011; Paule, 2015). The fourth point we would make is that the teacher's final admonition, to not be selfish and to consider the future generations, was not well connected to the lesson and was brief, but it sounds like a direct reference to Sustainable Development (SD). It echoes the Brundtland Report's definition of SD: "Sustainable development is development that meets the needs of the present without compromising the ability of future generations to meet their own needs." This was the only time that we heard such teaching in schools in Yogyakarta.

Student "voice" in EE

Over the last decade or two, there has been a real swing in academic study, in disciplines such as anthropology, sociology, as well as education, towards making children the focus of study, as an important social group in their own right (James, 2011 (2001), p. 2 online version). In this "new paradigm" of childhood studies (James *et al.*, 1998), children are no longer regarded as incomplete, inadequate or "becoming" adults, they have begun to be treated as human beings who can interpret the world for themselves and be "agents of change". Sometimes children became participants in research, researching themselves, or at least collecting data such as photographs of themselves and their everyday lives.

The shift was echoed in Development Studies, and in development and aid programmes, where "the child" has become not only the object focus of attention but also a "subject" with agency. The potential for children to not only reproduce society but also to change society probably began with the youth culture movement in the West in the 1960s. The 1989 UN Convention on the Rights of the Child is both a manifestation of, and a catalyst for, much of this work (United Nations, 1989). In these studies and programmes, the language of child empowerment is upbeat and motivational: children's "voices" are heard, they are "agents of change" and "active champions"; they are autonomous, flexible and adaptable.[6] The child has become an "active producer of knowledge, engaged in interaction with the adult world at all levels. Children ... are participatory and interactive subjects" (Hultqvist & Dahlberg, 2001, p. 8; Rudduck, 2002). In EE there has been a similar swing (e.g. Barratt Hacking, Cutter-Mackenzie, & Barratt, 2013; Gough, 1999; Prabawa-Sear & Baudains, 2011). The rhetoric of the empowered, action child is evident in the beginning of the following account of an EE event in Yogyakarta.

A student-led event: Rubbish Day

At The Islamic School, the only time that the students appeared to be afforded any agency was the Rubbish Day (*Hari Sampah*) event. This is an annual event as part of the Adiwiyata Programme at The Islamic School. It's also on the OSIS work programme, so student leaders know they have to do it and when, and are left to organise it.

> The Rubbish Day event was planned by an organising committee of 30 students, including the OSIS team, who referred to themselves as agents of change (*agen perubahan*). The committee planned the day's activities, obtained permission from the local police to walk around the block as part of a campaign, and submitted a proposal for funding which was met by the school. The funding covered the cost of audio equipment, stage hire and catering. The students had also planned to hold a talk-show with a local NGO but this was cancelled and replaced with students and teachers singing because the NGO's appearance fees were beyond the event budget.
>   On the day, male *agen* wore pink shirts and the female *agen* wore orange, while the rest of the students were in their PE uniforms. Students told me that this was because the supplier of the new pink shirts had only sent short-sleeved shirts, and since the school required girls to cover their arms (part of their *aurat*, in Islam), the girls had to wear the (old, long-sleeved) orange ones. The student organisers had name badges and walkie-talkies and were easily identifiable among the crowd. Various students and teachers proudly informed me that the students were "trusted" (*dipercaya*) to run this event. These symbols (special uniforms, badges and walkie-talkies) and the "agents of change" language suggested that the Adiwiyata student organisers had some authority, but as the morning proceeded, it was clear

that they had very little, and that this event was really about students performing, and in the process scoring some points for their Adiwiyata record.

It took over 40 minutes for students to divide into class groups in preparation for a "healthy walk" (*jalan sehat*). Many teachers were not in attendance (including Ibu Eni, the teacher in charge of Adiwiyata), and those who were there stood around chatting and asking to have photos with [the researcher]. The student organisers were repeatedly asking their peers to gather in class groups to no avail and on two occasions a student organiser requested assistance from teachers over the loud speaker to round up the students into class groups. Eventually the principal came to the stage and restored order. It appeared that no one was interested in taking orders from the student organisers, and teachers had no intention of helping them. Students set off in their class groups for the healthy walk. I asked some students why they were walking and they told me that they did not know. Some classes were led on the march by a student holding an A4 piece of paper with a message on it, but I could not read the messages amongst the sea of students. One might have expected, on Rubbish Day, that the walk was a picking-up-rubbish walk, but no, it was just a walk.

After the walk, there was a concert, and students, teachers and canteen vendors all joined in performing songs that had no relevance to the environment. During this time (approximately two hours), some classes participated in a rubbish relay where students ran to put rubbish in the correct bin (organic and inorganic) and a few students competed in a "reuse" competition, where students had to make something creative by reusing "rubbish" such as plastic bottles. The day ended with students eating together and announcements of door prize winners. (The catering was provided by the school canteen, but unfortunately all the food containers and utensils were disposable.) By the end, there were no teachers to be seen and students were hanging out in groups far from the activities. Before we left, one student said (in front of the main student organisers), "I want to go home. We've been here since 7 o'clock." The organisers were sweaty and no doubt very tired after a long morning in the blazing sun. I felt this way and assumed that most of the students did too. This event would score well for the Adiwiyata Programme, as all 580 students had "participated".

(Field notes, 21 February 2015)

Almost two weeks after the Rubbish Day event, the researcher organised a focus group discussion with the student organising committee to evaluate the event. Only two boys, of the 30 in the organising committee, came. One of the students, Imam, was reflective, and showed some disappointment at what had been achieved that day. He acknowledged that the *agents of change* had "made some progress" (compared with previous events) but had run an event that was "not very interesting" and "had [literally produced] a lot of rubbish". He said several times that students were "less than enthusiastic", and also that they were lazy and didn't care. Imam noted that they had tried to make it more interesting than previous Rubbish Days, getting students out on the "Healthy Walk", and providing prizes (*rewards*) for sorting rubbish, etc. But he said that it was "hypocritical to have a day to reduce rubbish and then to produce more." Imam clearly felt that despite their best efforts, the agents of change were "part of the problem, not part of the solution". As we noted in Chapter 6, one of the objectives of the 2013 curriculum for senior high schools is to have students who are responsible and pro-active and "part of the solution" to environmental problems (Kementerian, 2013, p. 7).

> Yeah, what can you do Miss? Actually it's quite difficult. We have to relate to those outside so sometimes our intentions are good but they're not taken well by the canteen people. Because if it's us [students] doing the reminding, sometimes they'll question, who are you [to say such things]?
>
> (Imam, student organiser, 15 March 2015)

These students had dedicated a lot of time and energy to running the event. They had shown good initiative and planning, they had considered "external relations", in getting police clearance for the walk, and sought funding. Clearly Imam had a holistic vision of what Rubbish Day should look like – and making more rubbish was not part of it. The absence of message in the "healthy walk" was really unfortunate – obviously the students should have been collecting rubbish, and to that end there should have been some instructions before the event by student leaders, different routes for different classes, and coordination of bags for disposal of the rubbish. That was a missed opportunity. The singsong was another: a golden opportunity to invite along an environmental NGO which could have taught them some "green" songs, or the student leaders could have found some on the internet. And so on.

Two weeks later, Imam was clearly disappointed. He felt let down, and that the students had been unsupported. Of course, teachers might have wanted not to interfere so that students could be more autonomous. At the start of Rubbish Day, there were many teachers visible, clustered undercover (out of the sun). Some asked for photos with the researcher before most disappeared. One English teacher stayed, and chatted at length with the researcher. He noted the missed opportunities (that the songs should have been about the environment, and that the point of the healthy walk was to pick up rubbish and this should have been made clear). A few male teachers came back to join in the singing on stage after the long walk. Generally, however, the teachers were not visible and even when asked over the loud speaker, they had not helped. The message that the staff gave through their lack of support was that this event did not matter. It seemed that the staff and principal were content to write off the morning as a morning for the students and an opportunity to get Adiwiyata points.

The dispiriting Rubbish Day event illustrates several problems with EE in schools in Yogyakarta. Here we will discuss three: the problem of power relations and student "voice"; the problem of negative messages; and the problem of ritual performance.

Student voice and power

The difficulty of getting the students assembled at the beginning of Rubbish Day symbolises the issue of power. All the talk of children's rights and the need to attend to children's voices is great in theory but this little vignette brings us down to earth in the reality that, in the Indonesia context at least, children have very little power and no authority. They were using their voices, literally, but no one was listening because they had no authority.[7] Clearly it was the school principal who had authority – not the student leaders. These are not young children: they are 16–18 years of age. In Indonesia, girls of this age can marry (16 is the minimum age for girls; 19 for boys), but the point is not really their actual age but their underdog positioning in the school hierarchy. The point is that in schools, they are the subordinates, the subalterns, and teachers are the superordinates, and that basic hierarchy and structural inequality cannot easily or quickly be upturned.

The students had been "trusted", and given permission to run the event, but the students were not afforded any kind of power. This was the only time we saw students lead an environment-related activity in schools in Yogyakarta. Beforehand, it was celebrated by staff and students as an action-oriented activity, a special event where students were "trusted" to make the decisions and run the activities. The word "trusted" implies the hierarchy, and the top-down movement from the school authorities, and also implies that the authorities believed there was some risk involved, though it's not clear what this risk might entail – some sort of disorder, perhaps. But one event does not erase long-held traditions of power. Traditions of power in education are well documented (Freire, 1996; Hayward, 2000; Karabel & Halsey, 1977), and, as we have already seen, Indonesia's education system is steeped in traditions of power, with students at the receiving end (Bjork, 2005, 2013; Leigh, 1991; Parker, 1992; Parker & Raihani, 2011). More broadly and philosophically, several scholars have drawn attention to the related political problem: that the institutionalisation of EE in schools, as socially conservative institutions, contradicts the socially transformative goals of EE (e.g. Gruenewald, 2004; Jickling, 1992; Stevenson, 2007).

As Barrett *et al.* note:

> Decentering power and embracing democratic processes in the classroom not only takes away the predictability of a tightly planned, teacher directed lesson, but it also risks opening up uncomfortable conversations and unsettling privileges that have become normalized within dominant environmental education narratives. Far from being "natural" or "normal", the dominant notions and processes of teaching and learning have been historically constructed to support certain political and social agendas. Over time however, these dominant beliefs and practices have become so "commonsense" that questioning of philosophy or the specifics is seldom introduced in the daily lives of teaching.… [T]eachers, and students understand and

enact these dominant pedagogical beliefs and self regulate their thoughts and behaviours to conform to them.

(Barrett *et al.*, 2005, p. 516)

They argue that engaging in action-oriented activities in schools challenges dominant conceptions about the organisation and transmission of knowledge, creating, for most teachers and students, contradictions with standard approaches to teaching and learning (Barrett *et al.*, 2005). This is exactly what Rubbish Day could have been: a contradiction wherein the authority to lead was temporarily vested in students; the venue was outside the classroom; and learning was by doing (ideally, practical pro-environmental behaviour). This sort of contradiction is one that is encouraged by both the Adiwiyata Programme and the 2013 curriculum in that they call for action-oriented approaches to learning. Barrett *et al.* (2005) acknowledge the difficulties experienced by both teachers and students in redefining and assuming unfamiliar roles that challenge the traditions of education. They argue that the roles of "powerful teacher and submissive student carry both the authority of law and the weight of tradition, despite educational theory and educational practice purporting a counter theme of independent learning and critical and creative thinking" (Barrett *et al.*, 2005, p. 514).

   The issue of "student voice" is one of agency and power. The simple difficulty experienced by student leaders in creating orderly class lines, even when equipped with loud speakers, shows the need for researchers to remember and acknowledge the real power of entrenched authorities and conservative social forces (Mizen & Ofosu-Kusi, 2013). Imam was cognisant of some of the difficulties, and to some extent of the ways the students were dependent (e.g. upon the canteen), and made vulnerable because of the very innovation that their Rubbish Day symbolised. Mizen and Ofosu-Kusi (2013) show that just going out and finding young people's voices is not enough; nor is it sufficient to "amplify" their voices (Beattie, 2012): we also have a duty not to downplay the structures and norms that keep them from being "agents of change". Further, giving voice to children is not automatically going to give them power. We would argue that in this context, students were trying to give voice and exercise agency – they had sought permission for the event, ordered in the uniforms and walkie-talkies, and so on – but to expect them to be empowered, as "agents of change", "shaping" transformative change, is too much of a burden to place on young shoulders, especially without the support of the staff.

The importance of hope

The English teacher thought that the Rubbish Day event was good for keeping the students interested, but seemed to think that that was about all it was good for. Imam's conclusion was that the event was "hypocritical" and "not very interesting"; it seemed that the Day had exhausted the student leaders, who had worked hard to organise the event, had somewhat depleted their motivation and enthusiasm, and had probably had no environmentally educational effect whatsoever.

There is some literature on the problem which says that it is very easy to slide into a depressing, "doomsday" discourse in EE. While it may be realistic, it is perhaps not helpful for the environment movement to use "scare tactics" or a relentless stream of environmental disasters and problems to prod children into caring for the environment. Nagel (2005), for instance, argues that contemporary young people are being raised on a diet of negative discourse surrounding the environment and sustainable development, which instead of scaring youth into action, has created a generation of children who are not only apathetic about environmental issues, but are "lost in a confused muddle of learned hopelessness" (Nagel, 2005, p. 71). While we do not consider that this picture fits the Indonesian case – indeed, it is the very absence of environmentalism in public discourse that motivated this research project – the disappointment experienced by Imam and his co-student workers gives us pause. Teaching about the environment needs to be done in such a way that students feel that their projects are do-able and ethically pure (not hypocritical), and that they can make a difference. EE needs to be enabling, not discouraging. In this case, support from the teachers, in particular, could have enabled the student leaders to do so much more with their Day.

Ojala (2011) highlights the importance of hope in pro-environmental behaviours and suggests that if students are feeling unsupported in their environmental actions, their levels of hope are likely to fall, contributing to apathy (Mokhele, 2011). While Imam in particular reported feeling unsupported in environmental actions, most young people in Indonesia appear hopeful and enthusiastic (Nilan et al., 2011; Parker & Nilan, 2013). In Indonesia, because of the very low level of environmental awareness, and the immensity of the task, we think it vital that those who are working in this area, especially young people, be able to work with like-minded others, in do-able, positive projects, that have defined, real-world outcomes.

It seems a shame to us that the principle environmental activity in which young people in Indonesia are engaged, is cleaning up rubbish. We are reminded of the talk given by Kirsty Albion, the National Co-Director of the Australian Youth Climate Coalition "Empowering Young People To Save Our Climate" (Monday, 3 November 2014, AAEE Conference, Hobart). She impressed upon her audience that there are two things that young environmentalists in Australia hate doing: stuffing envelopes and picking up rubbish. This work is not positive, it can be seen as mindless and demeaning, and is a terrible waste of youth potential. In contrast, she was able to tell the audience about how young people organised a 328 km. Walk for Solar from Port Augusta to Adelaide to convince the government to support the conversion of the decommissioned coal-fired power plant to solar, and the successful campaign to get the Commonwealth Bank to stop funding coal ports near the Great Barrier Reef. This brings us to the third issue raised by Rubbish Day: the lack of meaning in the event.

Meaningless performance

It would have been impossible to work out the name or the aim of the Rubbish Day event just by observing the activities. If the aim was to run an educational, action-based event, it failed. While it succeeded in getting students out of the classroom, and having the students do some exercise (which is, of course, beneficial in itself), this cannot count as EE. What we saw was, at best, environmental lip-service and ceremony that met the requirements of the Adiwiyata Programme. Certainly the number of participants was high, and for this reason the Adiwiyata system would deem it a successful activity. However, there was no environmentally-meaningful activity, lesson or learning.

Educationists in general, and perhaps especially in EE, have stressed the importance of the *meaningful participation* of students (e.g. P. Hart, 2000; R. Hart, 1992, 1997).[8] They bemoan tokenism and point out different ways students can "participate". While we support meaningful participation by students, we would also point out that in Indonesia there is a well-worn tradition of "meaningless performance", for instance, of Development initiatives (Parker, 2003, pp. 136–146) as well as in education. Bjork noted the focus on school rituals such as the flag-raising ceremony (Bjork, 2005, pp. xi–xii, 91–94). This can be seen as empty ritual and ritualistic compliance but such performance is often not so much meaningless as meaningful but in unintended directions. An example is a beauty contest in Riau, which was meant to be contested on merit but was actually judged on "race"/ethnicity (Long, 2007). Cleaning activities have been conducted by students in Indonesian schools for decades, and it seems quite fraudulent to claim these as EE. They offer no environmental educational benefit to students, but nowadays, in Adiwiyata schools, provide an opportunity for Adiwiyata points and teacher career advancement. Like flag ceremonies, they are an opportunity for teachers to demonstrate loyalty and commitment without having to do anything that is too challenging or time consuming.

One of the Adiwiyata schools in Yogyakarta asks students to separate waste, and they have different-coloured bins for organic and non-organic waste.[9] However, because of a lack of municipal waste services, the separated waste is collected by a private contractor, who combines the organic and non-organic waste and disposes of it. The students can see the contractor combining the non-organic and organic waste just outside the school gate. Not only is this situation failing to achieve any positive environmental outcome, it is hypocritical and demonstrates to the students that it is not what you do that counts, it's what you report that you do (to Adiwiyata judging panels). Students know that there are no separated waste services available in their region and without a word being said, students are being taught that positive environmental actions do not really matter.

The neo-liberal nature of the Adiwiyata Programme, with its prioritisation of numbers and documentation above all else, facilitates this loss of meaning and purpose, and allows it to become an exercise in how to gain prestige by doing as little as possible. Unfortunately, just having the facilities (shredders and composting areas, greenhouses, recycling bins) and having large numbers of students attend events such as *Sabtu Bersih* and Rubbish Day (even if they did not participate) are enough to meet the requirements of the programme. Like the example of the school that separated its waste only to have it combined again to be disposed of, these activities were demonstrating to students that behaviour and actions do not matter – "performance" does not become "practice" (West *et al.*, 2007). What matters is the documentation, which "proves" that the school is doing the right thing, even when it's not. Continual exposure to such superficial, if not fraudulent, behaviour must impact on students' understanding of the importance of EE and their environmental behaviours. It would not be surprising if students, like teachers, came to see Adiwiyata as little more than an opportunity from which to gain prestige, prizes and the like. Clearly the aims of EE have been lost in the competition to gain a higher Adiwiyata status, prestige for the school and points towards teacher promotion.

Teachers: public servants first?

Both the 2006 and 2013 Curricula encourage active learning, but what happens in school is the same old rote learning. Much of the continuing reliance on top-down rote learning and textbook teaching is the flow-on effect of a broader education system that focuses on knowledge retention and examination scores above all else. Although the Minister of Education and Culture has announced that, from now on, the national examinations will not determine school graduation – individual schools will – the exams are still there and students and teachers alike still consider them the main instrument for assessing and ranking student academic performance (and therefore the quality of teachers and schools). The effect of these external exams flows back down the grades, constraining teachers to stick to passing on examinable information that students have to memorise, and discouraging them from being experimental, from letting students ask questions, to follow their curiosity or express themselves. This means that all those desiderata in international EE curricula, such as UNESCO's – critical thinking, being part of the solution, problem solving, systemic thinking – do not see the light of day in Indonesia.

We have noted along the way that one of the reasons for the persistence of outmoded pedagogy is teachers' lack of confidence in their own ability and knowledge base: they fear being found to be ignorant. Generally speaking, younger teachers seem more confident, more willing to be open to students' questions and less inclined to stick to the textbook come what may.[10] We think this is a result of teachers gradually being better educated and therefore feeling more confident about their knowledge. The history of mass schooling in Indonesia is a recent one: it really only began with primary schools in the 1970s. So many older teachers often only have an equivalent of senior high school level qualifications, such as a Diploma. Indeed a World Bank report states that "Before 2005, around 25 percent of the teachers had failed to go beyond high school" (Chang *et al.*, 2014, p. 18). The Teacher and Lecturer Law No. 14/2005 aimed to fix this, stipulating that teachers must have the equivalent of a four years' Bachelor's degree. This has meant considerable upgrading, as it was calculated that 60 per cent of teachers did not meet this standard. Another important component of this Law was the doubling of teachers' salaries. However, this ambitious and well-funded reform has not (yet) resulted in any improvements in Indonesia's performance in international tests such as the PISA and TIMMS.[11] Nevertheless, the qualifications and training of teachers are immensely important, and have a direct impact on student outcomes. There are two aspects to teacher quality: both subject knowledge and pedagogical skill, and it seems that both are still lacking in Indonesian schools.

Bjork (2005) writes that he was unprepared for what he saw in six Indonesian Junior High Schools (SMP) during fieldwork conducted in the late 1990s: he was not surprised by the poor quality of facilities or the poverty of some students, but he was surprised by the "framing of professional responsibility" and teachers' lack of attention to teaching and learning. Similarly, the World Bank noted the poor motivation of teachers, which it linked to their low workload, low pay and poor qualifications (Chang *et al.*, 2014).

No doubt there is also the "inertia" factor: many teachers just want to minimise the amount of work that they have to do. When the 2006 curriculum was introduced, many teachers felt lost: they had no idea how to teach if they could not teach from the textbook, and they thought that reconceptualising everything they did would be a lot of work – so they continued to teach from the textbook. The 2013 curriculum provides just the outcomes or competencies that are required of students at each level; it therefore requires teachers to devise their own syllabi and lesson plans. We have frequently asked teachers for these and have been stalled or fobbed off time and again. The usual thing is to resort to the textbook.

Related to the lack of motivation and commitment is the important fact that most teachers in Indonesia are civil servants. This has been discussed in Chapter 5, where we describe the "public service mindset" of teachers, which is characterised by values such as loyalty and obedience, and a professional preoccupation with rising up the hierarchy over time. A World Bank report pointed out that

> this focus on income and status was linked to the realization that a large percentage of the teachers in 2005 had been hired during the large expansion of the system in the 1970s–80s with a relatively low level of initial education; little preservice education; few opportunities for later systematic in-service upgrading; and exposure to a school culture that rewarded passivity and loyalty rather than proactive, innovative approaches to the improvement of student outcomes.
>
> (Chang *et al.*, 2014, pp. 24–25)

Even now, although the civil service is not a glamorous position that attracts young people in towns and cities (Nilan *et al.*, 2011, p. 718), the reality is that youth un- and under-employment is a big and long-term problem in Indonesia (Elfindri *et al.*, 2015; Manning & Purnagunawan, 2011, pp. 322–326; Nasir, 2015). Rural and remote youth still see the public service as the most desirable job (Nilan *et al.*, 2011).

> Permanent employment is generally in short supply in Indonesia. Unemployment and underemployment rates may be as high as 40 per cent of the workforce (Dhanani 2004; *Jakarta Post*, 17 December 2005). So to secure an earning position with stability and reasonable income opportunities, people are willing to make huge investments.
>
> (Kristiansen, Pratikno, & Ramli, 2006, p. 208)

Consequently, people pay large sums to get into the public service. Kristiansen *et al.* (2006) found that in West Nusa Tenggara, among civil servants who had comparatively recently joined up, the average payment to obtain a civil service placement was Rp.27.4 million, which approximated to two and one-half years' full salary for a starting official (dependent on education level) (Kristiansen *et al.*, 2006, p. 221). Of course this is a corrupt practice, as entry to the civil service, and to teaching, is supposed to be on the basis of merit.

While not all teachers are civil servants, there is a prevalent public servant mindset of obedience and follow-the-leader, often couched in terms of loyalty. This mentality is similar to what Pak Hendra was talking about when he mentioned the need for leadership in EE and Adiwiyata (Chapter 7). He said that people "are not used to thinking for themselves or taking action. [It] will take a long time to fix this…. If [the] Principal has [a] good programme, others will follow. Strong leadership". Bjork, in the following, was writing of the education system in the New Order, before democratisation and decentralisation, but in many ways things have not changed:

> State authority in Indonesia is, and has always been, so pervasive that few individuals question their lack of power in the schools. As civil servants they have learned to follow the directives of upper level officials, not dispute them…. The Indonesian government has ensured … that educators treat the civil servant identity as "superordinate".… One effect of that emphasis is that teachers have not established an identity for themselves separate from that applied to all civil servants, or a distinct set of professional standards.
>
> (Bjork, 2005, pp. 105–106)

In many ways, the civil service, and, by extension, the teaching fraternity, still feel quite feudal. The traditions of the *priyayi* (the aristocratic elite who made up the pre-independence bureaucracy) (e.g. Antlov & Cederroth, 1994; Sutherland, 1979), or of the *abdi dalem*, the "servants of the Sultan" (Moertono, 1968), with their humble attitudes, ethos of service, docility, attitude of *pasrah* (submission), and hyper-sensitivity to social hierarchy, still prevail, despite efforts to reform both the bureaucracy and teachers. These authority structures, in which knowledge/power only flow downwards, mean that teachers, and ultimately students, are only meant to receive (*menerima*), not to ask or to push (see also Liem *et al.*, 2009; Maulana *et al.*, 2011; Parker, 1997). Maulana reports that in teacher–student relationships in schools in Indonesia, there is "a very high power distance index, indicating a high level of inequality of power and wealth within the society", "a very low index regarding individualism," and

> pervasive cultural values ... such as paternalism and respect for older individuals, implicitly regulating interactions between the young and the old.... Order and neatness are maintained by the elders and the younger generation is expected to follow the rules. This conservative situation allows the gap to grow and forms a directing–following interactional pattern between the two generations, which is reflected in the school system as hierarchical and monotonous.
>
> (Maulana *et al.*, 2011, pp. 37–38)

Students feeling *sungkan*

Considering this feudal mentality among the teaching fraternity and, to some extent, the wider society, it is not surprising that students also feel the weight of expectations of docility, submission and acceptance. Returning to EE and Adiwiyata, we can see how the teaching hierarchy and the general education culture combine to inhibit opportunity for students to engage in positive action. In fact, students reported a reluctance to carry out environmental activities because of cultural barriers referred to as *segan* and *sungkan*. These words (Indonesian and Javanese respectively) are both used to describe one's reluctance to do something on account of another person's higher status.[12]

> *Sungkan* refers to a feeling of respectful politeness before a superior or an unfamiliar equal, an attitude of constraint, a repression of one's own impulses and desires, so as not to disturb the emotional equanimity of one who may be spiritually higher.
>
> (Geertz, 1959, p. 233)

To feel *sungkan* is to be civilised and mature: "to be able to perform the social minuet with grace" (Geertz, 1961, p. 114). Students used these terms to explain why they would not approach teachers with ideas for environmental programmes or to ask for help when their school decided to revert to the old 2006 curriculum. This idea of *segan* is not only relevant for students and teachers, but is still evident across all levels of Javanese society. A child should not tell an elder what to do and a lower-ranking public servant should not openly question a decision from above. Respect for elders and superiors is an important and cherished part of Javanese culture and a significant barrier to Indonesian (particularly Javanese) schools changing from a top-down education system to a student-led, critical enquiry approach. No matter how often teachers are told (by the Department of Education) that they should use the "new" approach to education, while society still holds strong values around respect for position and *segan* and *sungkan*, there will be a limited uptake in schools.

Academic versus *"praktek"* EE

Teachers made a clear distinction between "academic" activities, which were assessable, quantifiable and examinable, and "non-academic" activities, which were not assessable, quantifiable, nor likely to be in exams. Non-academic activities were therefore not valuable. Environmental actions associated with the Adiwiyata Programme (composting, recycling, etc.) were always considered "non-academic" by teachers and therefore to lack real value. Such activities took time away from academic pursuits, which is why Grade 12 students were never allowed to participate in Adiwiyata. This meant that EE was an unvalued "add-on" for teachers who were already struggling with an extremely crowded curriculum and a change in curriculum. While we obviously feel that EE activities are indeed very valuable, the way that the activities were conducted offered little educational value for students, as the three classes described above show. It is not surprising that usually a school principal had to appoint someone to take responsibility for Adiwiyata in a school.

Praktek

As part of focus group discussions (FGD), we asked students what they thought made effective EE. The most common response was *praktek* (practical or hands-on activities). Various students reported wanting less theory and more practice.

> In my opinion, the most effective way is to go straight down (*langsung turun*) – practice (*praktek*) and not just theory. This whole time we have seen too much theory. All theory, but when's it time for practice?
> (Akmal, FGD among members of the Environment Club,
> The Islamic School, 10 December 2014)

Akmal unselfconsciously indicates the hierarchy of knowledge: theory is "above" and practice is "below". This is absolutely congruent with the way most university students in Anthropology classes and those doing service work in Indonesia[13] conceptualise fieldwork: they will "descend" (*turun*) to the village to do fieldwork (and usually hope that it doesn't last too long). In the Indonesian education system, book learning is elevated, high status, certain and reliable; learning from observation and experiment is low status, questionable, uncertain, and requires validation.

> Knowledge is claimed when the pupil can replicate the teacher's knowledge … and the student is then *ipso facto* capable and clever. The knowledge

itself is not questioned or analyzed: it is accepted without proof other than that it is being taught. Underlying this directional and hierarchical system of knowledge transmission is the assumption that "those who know", and therefore control the flow of knowledge, have the necessary qualifications to cope with this knowledge, which is a kind of authoritative power. This attitude towards knowledge and learning is not unique to Bali, and is manifest in many aspects of educational practice throughout Indonesia.

(Parker, 2003, pp. 236–237)

While students made clear their desire for more hands-on learning (*praktek*), we were forced to question what this meant in different contexts. There were few opportunities to observe classes that were *praktek*, but of the few that we saw, it appeared that leaving the classroom was the defining feature of *praktek*. The Biology class described above was one. While this was more hands-on than reading a list of insects and their features from a textbook, it was a biology lesson and not environmental education. Like most of the *praktek* classes that we observed, this class could have become an environmental lesson, but the opportunity was missed.

In a chapter on teacher training and effectiveness in Indonesia, Bjork describes how 57 per cent of teachers in his survey reported using *student-driven* and *active* approaches to teaching. Bjork had seen very little evidence of this in classroom observations. The teachers reported using workbooks in class, requiring students to complete more revision exercises and assigning homework more regularly as student-centred teaching techniques (Bjork, 2013, p. 54). None of these examples is an example of student-driven or active approaches to teaching. Similarly, our concepts of hands-on or *praktek* in regard to EE were quite different to those of students and teachers. Our concept of student-driven, hands-on learning in senior-high-school level EE would begin with students identifying an issue and working on a project to address that issue. Ideally, they would use competencies such as critical thinking, imagining future scenarios, conducting research, making decisions in a collaborative way, understanding causes and consequences, and problem solving, and demonstrate values such as environmental sensitivity, care, responsibility, and social and environmental justice. Clearly, Indonesian understandings of active learning and *praktek* are extremely limited.

Theory

While we appreciate that inner-city students need more exposure to the natural world and to opportunities to experience "nature", and so we support the students' calls for more hands-on EE, we also question the standard and applicability of the theory that they are taught. The same student, Akmal, went on to qualify his argument, and admitted that, actually, they need both theory and *praktek*. He explained that, "If we know the theory, then we know to add some *tetes tebu* [drops from sugarcane] [to compost] so that the micro-organisms will live again [*hidup lagi*] and that can be used to "provoke" microorganisms that are in the compost in a more useful way." *Tetes tebu* is used in a fermentation process in some instances, and we assume the student is referring to this. It is quite concerning that he thought that it could bring dead microorganisms to life but at the same time it was good that he understood something about composting (the existence of microorganisms). We do not know if he understood the environmental benefits of composting, but his knowledge was more *science* than *environmental*. Good EE would have students composting because it is a productive and sustainable way to deal with organic waste.

This example was one of many where students who were active in a school's Environmental Club or Adiwiyata Programme failed to understand quite straightforward concepts such as the environmental benefits of composting, how waterways are connected in the neighbourhood (when talking about drainage and *biopori*), why their mushroom log was dry, why they were even growing mushrooms, or how their environmental actions are related to climate change. Quite often they had learnt some kind of "theory" or scientific explanation, but failed to understand its relevance to them, their actions or to broader environmental issues. This situation suggests that not only do students need more practical opportunities, but also they need more in-depth and systemic learning that links behaviours to issues via theory, and demonstrates the interconnectedness of human behaviour and environmental issues. This can only be achieved where teachers have the knowledge and skills to facilitate such learning. Understanding the inter-connectedness of the environment and the impact of humankind's action on the environment is a fundamental first step to understanding environmental issues. It is clear from our fieldwork that even the most enthusiastic senior high school students lacked this understanding.

Conclusion

This chapter has focused on "real life" in schools. It has demonstrated how traditions of rote learning pedagogy, the "public service mentality", the strongly felt social hierarchy and teacher inadequacies combine to work against effective environmental education in schools in Yogyakarta. Many of these issues are issues for education generally in Indonesia, and are not specific to environmental education. However, this chapter and the preceding one have also been critical of the Adiwiyata Programme, showing how the obsession with documentation, performance and rank, means that the more worthwhile aspects of the programme – such as its focus on pro-environmental practice – are undermined. The chapter has shown that the integration of environmental education into the Curriculum – one of the criteria for high rank in Adiwiyata – has not happened. Teachers have no idea what EE is, let alone how to integrate it into the Curriculum. Many think that it is anything *about* the natural environment: growing a mushroom on a timber offcut is therefore EE. Admittedly, the value of the mushroom exercise was hard to fathom. Nevertheless, the teacher's contribution to this lesson was appalling: not only was she ignorant, she was failing to engage with the students and the content of the lesson. Fortunately, there are some teachers – often the younger, better-trained ones, such as the Biology and Geography teachers above – who are enthusiastic and engaged. The fact that most teachers are public servants first, and teachers second, has been shown to be a major obstacle to good teaching practice in Indonesia generally, and for EE in particular.

A real concern is that many of the efforts deemed to be EE have no real environmental value. Often the performance of EE is a sham: rubbish is sorted by students, only to be re-combined by the garbage contractor, in full view of the students. The meaningless performance of pseudo-environmental behaviours eats away at the integrity of a commitment to the environment, and potentially has a dispiriting effect on young people. This is another way in which EE in Yogyakarta is "hollow".

Young people in Indonesia are enthusiastic, often idealistic, and optimistic, and represent a great resource for the future. The story of Rubbish Day is a sad story: the capable young leaders were trying hard to make a relevant and inspirational event. It was a real shame they lacked the support and guidance of the teachers. We can see that with good leadership and support, there is a lot of potential for young people to become committed and knowledgeable environmentalists.

Notes

1 This is one of the overarching social competence objectives for Grade X–XII in the 2013 Curriculum. See Chapter 6.
2 Mokhele (2011, p. 78) reports that the change to integrated EE policy came about as a result of international recommendations and provisions, "particularly those originating from the Johannesburg World Summit for Sustainable Development (WSSD) held in 2002".
3 There is a significant literature on the benefits of being in nature for students, and on how this can help to build environmental concern (Nisbet, Zelenski, & Murphy, 2009; Petra et al., 2013). This has also been documented for Chemistry students in senior high schools in Indonesia (Kusmawan et al., 2009). Students in these city schools in Yogyakarta and Surabaya get little exposure to "nature".
4 In Indonesian, the term "greenhouse" is literally "glasshouse" (*rumah kaca*) rather than greenhouse. This sometimes causes problems in understanding, e.g. people think that we are using too much glass in our buildings and this causes the air to heat up and so we have the greenhouse effect.
5 The relevant competence in the 2013 Curriculum is: "3.5 analyse the dynamics of the atmosphere and its influence on everyday life" (Mendikbud (Kementrian Pendidikan dan Kebudayaan) [Ministry of Education and Culture], 2013, p. 132). The topics of Greenhouse Gases, climate change and global warming are not itemised in the 2013 Curriculum, but they are in the textbook for Geography Grade X. It has only a few sentences:

> The activities of humankind and progress in technology and industry are speeding up climate change. Waste and smoke from transport and industry burning fossil fuels is the main cause of the increase in greenhouse gases in the atmosphere. Apart from that, the cattle industry, waste from cattle, and heaped plants (? *tumbuhan yang menumpuk*) play a part in the increase in GHG.
>
> (Tika et al., 2014, p. 182)

Then it mentions cattle excrement producing methane, and that the agriculture system is also to blame because of the use of inorganic fertiliser "as well as because of the changes in land use rules and forestry" (*sic*).

6 Recently, this movement has particularly focused on girls. If one Googles "empowering girls", one can see the range of international NGOs and NFPs that are involved in "girl" projects. One example is "The Girl Effect": www.girleffect.org/.

7 Despite all the talk, this problem is apparent in the Global North too (O'Boyle, 2013).

8 The journal *Children, Youth and Environments* is replete with articles on this topic (e.g. West *et al.*, 2007).

9 These are quite common in schools, but it is important to always check and see, (a) if they are being used, and (b) if the rubbish is being sorted correctly.

10 Although not especially young, Ibu Dian is a fairly new teacher, having begun in Agriculture.

11 Indeed, Shaeffer, an expert on education in Indonesia and co-author of the World Bank Report on the Teacher Law of 2005, reported to UNESCO that "Ultimately, the granting of certification and the doubling of income did not significantly have an impact on the 'professionalisation' of teachers, on teacher competencies, or on learner outcomes" (Shaeffer, 2015, slide 10).

12 Heider identifies a cluster of emotion words around "respect", "shyness", "shame", etc., including *segan*, which leads to withdrawal and avoidance of engagement (Heider, 1991, pp. 85, 302–309).

13 In Indonesian universities, students have to do a practicum semester, usually for two months, called Kuliah Kerja Nyata (KKN) or Student Service Learning, during which they "descend" to a village, ostensibly to help the community with Development (Windred, 2017).

References

Antlov, H., & Cederroth, S. (Eds). (1994). *Leadership on Java: Gentle Hints, Authoritarian Rule*. Richmond, Surrey: Curzon Press.

Barratt Hacking, E., Cutter-Mackenzie, A., & Barratt, R. (2013). Children as active researchers: The potential of environmental education research involving children. In R. B. Stevenson, M. Brody, J. Dillon, & A. E. Wals (Eds), *International Handbook of Research on Environmental Education* (pp. 438–458). New York: Routledge.

Barrett, M. J., Hart, P., Nolan, K., & Sammel, A. (2005). Challenges in implementing action oriented sustainability education. In W. L. Filho (Ed.), *Handbook of Sustainability Research* (Vol. 20, pp. 505–534). Frankfurt am Main: Peter Lang.

Beattie, H. (2012). Amplifying student voice: The missing link in school transformation. *Management in Education*, 26(3), 158–160. doi:10.1177/0892020612445700.

Bjork, C. (2005). *Indonesian Education: Teachers, Schools, and Central Bureaucracy*. New York: Routledge.

Bjork, C. (2013). Teacher training, school norms and teacher effectiveness in Indonesia. In D. Suryadarma & G. W. Jones (Eds), *Education in Indonesia* (pp. 53–67). Singapore: ISEAS Publishing.

Chang, M. C., Shaeffer, S., Al-Samarrai, S., Ragatz, A. B., de Ree, J., & Stevenson, R. (2014). *Teacher Reform in Indonesia: The Role of Politics and Evidence in Policy Making. Directions in Development*. Retrieved from Washington, DC: World Bank. doi: 10.1596/978-0-8213-9829-6.

Dyment, J., Hill, A., & Emery, S. (2015). Sustainability as a cross-curricular priority in the Australian Curriculum: A Tasmanian investigation. *Environmental Education Research*, 21(8), 1105–1126.

Elfindri, Soebiakto, B., Harizal, & Rezki, J. F. (2015). Youth idleness in Indonesia. *Asian Social Science, 11*(13), 251–259.

Freire, P. (1996). *Pedagogy of the Oppressed* (new rev. edn). London: London: Penguin.

Geertz, H. (1959). The vocabulary of emotion: A study of Javanese socialization processes. *Psychiatry, 22*(3), 225–237.

Geertz, H. (1961). *The Javanese Family: A Study of Kinship and Socialization*. New York: The Free Press of Glencoe.

Giroux, H. A. (1999). Schools for sale: Public education, corporate culture, and the citizen-consumer. *The Educational Forum, 63*(2), 140–149. doi:10.1080/001317299 08984404.

Giroux, H. A. (2011). *On Critical Pedagogy*. New York Continuum.

Gough, A. (1999). Kids don't like wearing the same jeans as their Mums and Dads: So whose "life" should be in significant life experiences research? *Environmental Education Research, 5*(4), 383–394.

Gruenewald, D. A. (2004). A Foucauldian analysis of environmental education: Toward the socioecological challenge of the earth charter. *Curriculum Inquiry, 34*(1), 71–107.

Hart, P. (2000). Searching for meaning in children's participation in environmental education. In B. Jensen, K. Schnack, & V. Simovska (Eds), *Critical Environmental and Health Education Research Issues and Challenges* (pp. 7–28). Denmark: Research Centre for Environmental and Health Education, The Danish University of Education.

Hart, R. (1992). *Children's Participation: From Tokenism to Citizenship*. Florence: UNICEF, Innocenti Essays #4.

Hart, R. (1997). *Children's Participation: The Theory and Practice of Involving Young Citizens in Community Development and Environmental Care*. London: Earthscan.

Hayward, C. R. (2000). *Power and Pedagogy*. Cambridge: Cambridge: Cambridge University Press.

Heider, K. G. (1991). *Landscapes of Emotion: Mapping Three Cultures of Emotion in Indonesia*. Cambridge: Cambridge University Press.

Hultqvist, K., & Dahlberg, G. (2001). Governing the child in the new millennium. In K. Hultqvist & G. Dahlberg (Eds), *Governing the Child in the New Millennium* (pp. 1–14). Milton Park, Abingdon: Routledge.

James, A. (2011 (2001)). Ethnography in the study of children and childhood. In P. Atkinson, A. Coffey, S. Delamont, J. Lofland, & L. Lofland (Eds), *Handbook of Ethnography* (pp. 1–22). London: Sage.

James, A., Jenks, C., & Prout, A. (1998). *Theorizing Childhood*. Cambridge: Polity Press.

Jickling, B. (1992). Why I don't want my children to be educated for sustainable development. *Journal of Environmental Education, 23*(4), 5–8.

Karabel, J., & Halsey, A. H. (1977). *Power and Ideology in Education*. New York: Oxford University Press.

Kementerian Pendidikan dan Kebudayaan (Ministry of Education and Culture). (2013). *Kurikulum 2013: Kompetensi Dasar Sekolah Menengah Atas (SMA)/Madrasah Aliyah (MA)*. Jakarta.

Kristiansen S, Pratikno, & Ramli, M. (2006). Buying an income: The market for civil service positions in Indonesia. *Contemporary Southeast Asia, 28*(2), 207–233. doi:10.1355/cs28-2b.

Kusmawan, U., O'Toole, J. M., Reynolds, R., & Bourke, S. (2009). Beliefs, attitudes, intentions and locality: The impact of different teaching approaches on the ecological affinity of Indonesian secondary school students. *International Research in Geographical and Environmental Education, 18*(3), 157–169.

Leigh, B. (1991). Making the Indonesian state: The role of school texts. *Review of Indonesian and Malaysian Affairs, 25*(1), 17–43.

Liem, G. A. D., Martin, A. J., Nair, E., Bernardo, A. B. I., & Prasetya, P. H. (2009). Cultural factors relevant to secondary school students in Australia, Singapore, the Philippines and Indonesia: Relative differences and congruencies. *Australian Journal of Guidance & Counselling, 19*(2), 161–178.

Long, N. (2007). How to win a beauty contest in Tanjung Pinang. *Review of Indonesian and Malaysian Affairs, 41*(1), 91–117.

Manning, C., & Purnagunawan, R. M. (2011). Survey of recent developments. *Bulletin of Indonesian Economic Studies, 47*(3), 303–332.

Maulana, R., Opdenakker, M.-C., den Brok, P., & Bosker, R. (2011). Teacher–student interpersonal relationships in Indonesia: Profiles and importance to student motivation. *Asia Pacific Journal of Education, 31*(1), 33–49. doi:10.1080/02188791.2011.544061.

Mendikbud (Kementrian Pendidikan dan Kebudayaan) [Ministry of Education and Culture]. (2013). *Kurikulum 2013, Kompetensi Dasar, Sekolah Menengah Atas (SMA)/ Madrasah Aliyah (MA) [2013 Curriculum, Basic Competencies, Senior High School/Islamic Senior High School].*

Mizen, P., & Ofosu-Kusi, Y. (2013). Agency as vulnerability: Accounting for children's movement to the streets of Accra. *The Sociological Review, 61*, 363–382. doi:10.1111/1467-954X.12021.

Moertono, S. (1968). *State and Statecraft in Old Java.* Ithaca, NY: Cornell University.

Mokhele, M. L. (2011). Integrated environmental teaching in South Africa: An impossible dream? *Perspectives in Education, 29*(4), 78–86.

Nagel, M. (2005). Constructing apathy: How environmentalism and environmental education may be fostering "learned hopelessness" in children. *Australian Journal of Environmental Education, 21*, 71–80.

Nasir, S. (2015). Jobless youth raise risk of Indonesia's "demographic bonus" turning into disaster *The Conversation*, 19 November 2015.

Nilan, P., Parker, L., Bennett, L., & Robinson, K. (2011). Indonesian youth looking towards the future. *Journal of Youth Studies, 14*(6), 709–728.

Nisbet, E. K., Zelenski, J. M., & Murphy, S. A. (2009). The nature relatedness scale: Linking individuals' connection with nature to environmental concern and behavior. *Environment and Behavior, 41*(5), 715–740. doi:10.1177/0013916508318748.

O'Boyle, A. (2013). Valuing the talk of young people: Are we there yet? *London Review of Education, 11*(2), 127–139. doi:10.1080/14748460.2013.79980.

Ojala, M. (2011). Hope and climate change: The importance of hope for environmental engagement among young people. *Environmental Education Research*, 1–18.

Parker, L. (1992). The creation of Indonesian citizens in Balinese primary schools. *Review of Indonesian and Malaysian Affairs, 26*, 42–70.

Parker, L. (1997). Engendering school children in Bali. *Journal of the Royal Anthropological Institute, 3*, 497–516.

Parker, L. (2003). *From Subjects to Citizens: Balinese Villagers in the Indonesian Nationstate.* Copenhagen: Nordic Institute of Asian Studies Press.

Parker, L., & Nilan, P. (2013). *Adolescents in Contemporary Indonesia.* New York: Routledge.

Parker, L., & Raihani, R. (2011). Democratizing Indonesia through education? Community participation in Islamic schooling. *Educational Management Administration & Leadership, 39*(6), 712–732. doi:10.1177/1741143211416389.

Paule, M. (2015). Dinosaur discourses: Taking stock of gendered learning myths. *Gender and Education, 27*(7), 744–758. doi:10.1080/09540253.2015.1093101.

Petra, Y. B., Mathias, S., Rainer, F., Dušan, K., & Markus, P. (2013). Different shades of green: Comparative study on nature relatedness and ecologic consciousness among South Korean, Swiss, and Czech students. *Central European Business Review*, 2(2), 7–18.

Prabawa-Sear, K. A., & Baudains, C. (2011). Asking the participants: Students' views on their environmental attitudes, behaviours, motivators and barriers. *Australian Journal of Environmental Education*, 27(2), 219–228.

Rudduck, J. (2002). The transformative potential of consulting young people about teaching, learning and schooling. *Scottish Educational Review*, 34(2).

Shaeffer, S. (2015). *The Indonesian Teacher Law of 2005: Intentions, Implementation, and Impact.* Paper presented at the 17th UNESCO-APEID International Conference, Bangkok, Thailand. www.unescobkk.org/fileadmin/user_upload/apeid/Conference/17th Conference/docs/PS_IV_1-_SHELDON_SHAEFFER_UNESCO.pdf.

Shumacher, S. L., Fuhrman, N. E., & Duncan, D. W. (2012). The influence of school culture on environmental education integration: A case study of an urban private school system. *Journal of Agricultural Education*, 53(4), 141–155. doi:10.5032/jae. 2012.04141.

Stevenson, R. B. (2007). Schooling and environmental education: Contradictions in purpose and practice. *Environmental Education Research*, 13(2), 139–153.

Sulistyowati, E., Omegawati, W. H., & Hidayat, M. L. (2013). *Biologi untuk SMA/MA Kelas X Peminatan Matematika dan Ilmu-Ilmu Alam (Biology for SMA/MA Grade X, Maths and Natural Sciences Stream).* Klaten: PT Intan Pariwara.

Sutherland, H. (1979). *The Making of a Bureaucratic Elite: The Colonial Transformation of the Javanese Priyayi.* Singapore: Asian Studies Association of Australia & Heinemann Educational Books (Asia) Ltd.

Tika, P., Amin, S., Arofah, A., & Hermanto. (2014). *Jelajah Dunia. Geografi. SMA/MA Kelas X (Roaming the World. Geography. SMA/MA Grade X).* Jakarta: PT Bumi Aksara.

United Nations. (1989). *The United Nations Convention on the Rights of the Child.* Retrieved from www.unicef.org.uk/Documents/Publication-pdfs/UNCRC_PRESS2009 10web.pdf.

West, A., Mei, C. X., Ye, Z., Na, Z. C., & Qiang, C. (2007). From performance to practice: Changing the meaning of child participation in China. *Children, Youth and Environments*, 17(1), 14–32.

Windred, S. (2017). Fighting apathy, seeking engagement *Inside Indonesia*, 29 July 2017.

9 A coordinated approach to environmental education in Surabaya

Introduction

Surabaya is a city that is pushing ahead with a rather aggressive approach to EE. This approach is quite proudly referred to as *paksarela* (forced volunteering) by NGO workers and government officials. We hypothesised that the holistic and compulsory nature of the approach to EE in Surabaya might be an example of environmentality in progress. To recapitulate from Chapter 2, "environmentality" borrows from Foucault's concept of "governmentality" (Dean, 1999; Foucault, 1991) and posits that a dense net of government regulations, sometimes complemented with NGO activity, around protection of the natural environment can foster the development of new, environmentally aware subjects, who are "responsibilised" and come to autonomously care for the environment (Agrawal, 2005a, 2005b). Agrawal defines environmentality as "the knowledge, politics, institutions and subjectivities that come to be linked together with the emergence of the environment as a domain that requires regulation and protection" (Agrawal, 2005a, p. 226). He states that environmental subjectivities come into existence when local people "come to care for, act and think of their actions in relation to something they define as the environment" (Agrawal, 2005a, p. 164).

After introducing the fieldwork, this chapter describes the approach to EE in Surabaya, beginning with the vital role played by the Mayor of Surabaya, Ibu Risma. Then we examine the cooperation among government agencies, an environmental NGO, called here TENGO, and schools in enforcing a city-wide approach to EE. Data are taken from interviews with heads of government agencies (Department of Education, Parks and Sanitation and the Environmental Agency) and TENGO staff; focus group discussions with school principals, teachers and students; and participant observation at various environmental competitions, education workshops and environmental activities over an 11 month period.

This chapter examines the processes used by the Surabayan Government and TENGO as part of Surabaya's large-scale EE program and considers these processes in the Javanese context with a particular focus on leadership, respect and power. Consideration of the energetic leadership of Bu Risma, and the intricate and coordinated net of services being mobilised by multiple government agencies and the ubiquitous TENGO, could lead one to conclude that Surabaya's municipal "clean and green" programme is an example of environmentality. However, the following chapter on EE in practice in and around schools in Surabaya suggests that "responsibilisation" for the environment has not occurred. The two chapters together form our second case study and together they question the applicability of the concept of environmentality in schools in Surabaya.

Fieldwork in Surabaya

The researcher's access to all government officials and schools in Surabaya was arranged by TENGO. Mas Rudi founded TENGO in 1999 as a result of his personal interest in the environment and his frustration at the ongoing issues of litter and waste in Surabaya. Dr Prabawa-Sear and Mas Rudi had a three-year working relationship prior to, during and after fieldwork. The Project also employed a second researcher, Dr Danau Tanu, to conduct intensive fieldwork in Surabaya for three months in 2015. The two fieldworkers often attended events and schools together. They found that this worked well, allowing more comprehensive observation of complex and busy events.

TENGO arranged access to all of the government offices cited below, schools and out-of-school activities in Surabaya. Of course this "hosting" or "sponsorship" was a double-edged sword.[1] The activities, events and schools that the researchers could attend were to some extent controlled by TENGO, though they were able to request particular schools, offices and so forth. At least one representative from TENGO was in attendance at each interview. Dr Prabawa-Sear knew she was being used by TENGO, as a photogenic white woman, to add glamour and appeal to their events. She was often photographed, and appeared in publicity for their events. On occasion, she was required to participate in events in ways that made her uncomfortable, e.g. as a judge, or as "expert" when she felt she was not an expert (e.g. a journalism workshop). On the other hand, she felt rather insulted at times at the presumption of her ignorance and naivety: she could clearly see when things were not going well, despite TENGO's attempts to portray themselves, and their cooperation with schools, as flawless. She felt she was quite capable of identifying and assessing the impact of TENGO on the interviewees and the interview data. In general, TENGO were really helpful. TENGO provided her with valuable insight into the relationships among TENGO, government departments and schools, and provided an ease of access that it is unlikely she would have had on her own. She was often introduced as a friend of TENGO, which was a positive thing: it made the interviewees feel more relaxed and made for more candid and valuable conversations (Figure 9.1). The social relationships with TENGO staff continued long after fieldwork. However, as one incident described in the following chapter shows, the sponsorship of TENGO was not 100 per cent positive.

Figure 9.1 Doing fieldwork in schools in Surabaya.
Photo credit: Danau Tanu.

The mayor, the government and EE

Surabaya began a greening programme in 2006, largely in response to major flooding.[2] Attempts were made to mitigate flooding by providing more green space to absorb rainwater. The programme received a large boost from the election of Tri Rismaharini as Mayor from 2010. She is affectionately known as "Bu Risma".[3]

The mayor

Bu Risma is now internationally known as an award-winning, can-do mayor with a deep commitment to making Surabaya "clean and green".[4] This policy, usually known by the English phrase, "Surabaya Green and Clean", is seen by some as a marketing tool, as Surabaya was perceived to lack a distinctive image. When asked why Surabaya adopted this clean, green image, government officials most often answered, "Because of Bu Risma", but if pressed would say, "Because we have nothing else". Jakarta is the business, media and political capital;[5] Yogyakarta is the "centre of Javanese culture"; Medan has Sumatra's forests and natural resources; cities in Kalimantan have mining, forests and plantations; Bali has its beaches and culture; but Surabaya only has a shipping port and mosquitoes. Surabaya has to be a green city in order to draw people in and to stop its own people from leaving.[6] As they see it, Surabaya has no choice but to be a green city.

Bu Risma has the training for, and a serious commitment to, improving the environment of Surabaya. She trained as an architect and became a public servant, eventually heading the Surabaya Municipal Parks Office (Fionna, 2017, p. 10). She is well known and respected by the Surabayan people for her hands-on approach (e.g. cleaning toilets) and her displays of emotion. At the mention of her name, people proudly recount stories of her anger (screaming at people who trampled plants at a park), her generosity (carrying bags of rice in her car for hungry people), her hands-on approach (going to schools unannounced and telling principals to implement EE programmes), her accessibility (being available to heads of government at all hours, by walkie-talkie),[7] her "clean" approach to governing the city (it is said she is the first non-corrupt mayor) and her insistence that heads of government are not only seen in their offices, but also work *in the field*. She has an almost legendary reputation, and was re-elected in 2015 in a landslide, garnering 86 per cent of the vote (Fionna, 2017, p. 23). However, she has also been described as a polarising and divisive figure. She is not seen as pro-business, so business people generally do not support her.

The concept of "forced volunteering", which is the hallmark of environmental practices in Surabaya, comes from Bu Risma. It means that orders for officials to participate in "green" activities flow down the government chain, beginning with Bu Risma, to heads of departments and agencies, who are made to participate in environmental activities such as the Car Free Day and to represent their departments at environmental events, to school principals, who must have environmental programmes at their schools. Some teachers are nominated (by principals) to lead environmental programmes and manage the student environmental cadets who carry out the work.[8] In Surabaya, forced participation and leadership emerge as two themes that are integral to the city's approach to greening.

In some ways, this style of leadership is quite fresh in Indonesia, and is particularly unconventional in the Java context: the quick and public expression of emotion, the valorising of "the field" and of action rather than passivity upturn ideal Javanese concepts of the leader.[9] However, there is also a tradition that contrasts the *alus* (refined, controlled) ideal behaviour in the inland centres of *priyayi* culture such as Yogyakarta and Solo[10] with the much coarser (*kasar*), what-you-see-is-what-you-get behaviour in the *pasisir* (coast, edges): the port and trading cultures of the coastal towns (e.g. Lombard, 1986; Vickers, 1987, 1997). Bu Risma's forthright style, single-minded commitment and unwillingness to compromise have sometimes triggered opposition, for example, from surrounding districts.[11] There is also a gender aspect to her leadership style. While the ideal Javanese male is a model of self-control, women are expected to be more emotional, more active and more impulsive (Djajadiningrat-Nieuwenhuis, 1992). Keeler describes how men are generally higher-status than women, and are expected to be weightier, more even-tempered and serious than women, but in forfeiting some of that status and ser-iousness, women "gain some freedom of action" (Keeler, 1990, p. 151). Bu Risma's can-do style thus echoes a real "ideal type" within Javanese society, albeit not at the very apex of Javanese society.[12]

Almost everyone in Surabaya – from government to NGOs, schools and the general public – agrees that the Surabayan government is doing a good job environmentally, and that this is a result of the leadership and example of Bu Risma. The head of the Environmental Agency echoed the remarks of various others when he explained as follows:

> Actually, the key is the example of our Mayor, who always goes into the field and the like. The community knows that the government doesn't only talk – they also join in (action). Every Friday we work together [*kerja bakti*] in the field. If there is a flood, all officials go to the field. So that gives the impression that the government isn't just all talk, but also takes action in the field. Secondly, it is always said that it is easy to build something but to take care of it is difficult, and that the government can't take care of things. The Surabayan government has proven itself – for example, the parks that have been maintained by the government. That's good, and it was the gov-ernment who made the parks in Surabaya open [free].
>
> (Interview, Head of Environmental Agency, 26 February 2015)

A coordinated government ...

From interviews with the Head and Deputy Head of the Office for Parks and Sanitation, the Education Office and the Environmental Agency, it became clear that Bu Risma was the reason that these offices were working together on EE and improved environmental infrastructure. Surabaya boasts 23 compost sta-tions for composting community waste and 60 parks and green spaces, making up over 23 per cent of the city's area.[13] Most other cities in Indonesia (like Yogyakarta) are yet to achieve the 2 per cent green space stipulated by law and have no composting stations, no government-run waste pick-up service and very few, if any, free parks. The Surabayan government sees many advantages in having green spaces for its people, including physical health, mental health, rec-reational and environmental advantages. Parks have CCTV and offer free Wi-Fi and varied themes (e.g. skateboarding, libraries, meditation and outdoor adven-ture activities), in an effort to encourage its people to use the parks in a variety of ways. Surabaya also holds weekly car-free community events, has a data-tracked waste system (with trucks tracked, weigh bridges and real-time report-ing), river restoration projects, extensive street-scaping, limits on signage (reducing visual pollution), bans on burning rubbish and numerous community EE programmes.

These programmes require the coordination and cooperation of multiple departments and agencies. As noted in the preceding chapters on Yogya, such cooperation is not a common practice in other cities under decentralisation. Each government official that we interviewed noted the importance of working with other departments and agencies. The Environmental Agency provided the administrative coordination for environmental projects while the Public Works Office, Agriculture Office and Office of Housing, Planning and Urban Development provided the technical skills.[14] Schools that were looking to obtain resources for their environmental projects could get shredding machines from the Environmental Agency, rubbish trailers from Parks and Sanitation and plants from the Agriculture Office. According to the Environmental Agency, they facilitated this for schools.

Whilst all the government officials noted the importance of their team effort, it was also evident that there was something of a hierarchy among the different agencies. The Education Office was clearly at the top and the Environmental Agency at the bottom. The other agencies sat somewhere in the middle. The Head of the Environmental Agency openly acknowledged that his Agency relied heavily on the support of the Education Office in persuading some schools to participate in Adiwiyata activities. He explained,

> To be honest, the schools are most afraid of the Education Office because they are below it. So ... we always work with them [the Education Office]. So if there is someone from the Education Office [involved] and there is someone [from the schools] who doesn't show up, we report them to the Education Office and the Office calls them and asks why they didn't come. If we [the Environmental Agency] are by ourselves, it's hard, so we always couple with them.
>
> (Interview with Environmental Agency, 26 February 2015)

Working well with an NGO

The Head and Deputy Head of the Department of Parks and Sanitation both raised the point that the government is limited in its ability to address Surabaya's environmental issues, and they acknowledged the role of education, media, business, citizens and NGOs. Chapter 7 showed the reluctance of government agencies in Yogyakarta to work with NGOs to support EE in schools. Surabaya, however, relies on the work of NGOs, TENGO, in particular, to introduce and support EE in schools. Teamey (2007) suggests that the more flexible organisational structure and dedicated agendas of non-state providers allow them to be perceived as more innovative, accountable and effective in terms of cost and delivery, while having greater knowledge of community needs than state providers. This observation holds true for the situation in Surabaya. From the interviews with the heads of government agencies, it was clear that these leaders were happy to hand responsibility for Surabaya's EE programme to TENGO, and it was partly due to the factors highlighted by Teamey. In addition to this, there was a perception that TENGO knew more of what needed to be done and how to do it.

Actually, we feel that there is no way that we could develop the [EE pro-grammes of] schools on our own. We don't have much power. Our techni-cal capacity is also limited. We don't have the capacity to up-date as quickly as them [TENGO]. We are also lacking personnel … We just have to support them and I think there are no negatives to this.…"

(Interview, Head of Environmental Agency, 26 February 2015)

TENGO was undoubtedly providing a service that the Surabayan government was not in a position to provide. TENGO were working with many schools, pro-viding training and competitions with support from private sponsors. They were also able to work with schools in a way that the government could not. The young staff at TENGO easily built relationships with students, and teachers and principals felt at ease with the TENGO staff as they were not employees of the Ministry of Education and Culture.

The Education Office

The opportunity to interview the Head of the Education Office, Pak Imran, came with no notice, and "on the fly". Dr Prabawa-Sear had been speaking at a seminar for junior high school teachers on integrating EE into the schools' health program. As they left the seminar, she was informed by her TENGO colleague that she could meet the Head of the Education Office, but needed to be quick. TENGO did not have the same close relationship with the Education Office that they had with the Environment Agency and Parks and Sanitation, and this was evident in the demeanour of her two TENGO colleagues. One sat silently, only nodding his head and offering a very humble handshake at the beginning and end of the interview, and the other, the head of TENGO, Mas Rudi, was much more polite and formal in his approach in this meeting than he had been in others. The researcher was hoping for some insightful conversations about Sura-baya's unique approach to EE and the policy behind it. Instead what she got was Pak Imran constantly attributing the EE programme to TENGO (much to Mas Rudi's obvious delight). He dismissed her questions wherever possible.

The [EE] programme came from TENGO. They were first, first with sociali-sation and grouping and the like. To be precise, the grand design came from TENGO. Then because we have the same vison and mission, we only had to join the pre-existing program. So we only needed to facilitate the connection.

(Interview, Pak Imran, Head of Education Office, 4 February 2015)

"Socialisation" (*sosialisasi*) is a key term in government discourse. It usually means the top-down transmission of a government policy followed by assumed internalisation of the government messages by the recipients of the education. For instance, government billboards exhort the "*sosialisasi*" of traffic rules to make traffic more orderly. Some of Pak Imran's praise for TENGO and reluctance to acknowledge his Office's role in the EE programme could be attributed to Mas Rudi's presence in the room, but it became clear that this reluctance was also due to the facts that Pak Imran was not particularly knowledgeable about the programme, and that his Office played only a small, supporting role in the delivery of EE to schools in Surabaya. This second point was highlighted later in discussions with the Head of the Environmental Agency when it was suggested that the Education Office's main contribution was that it offered the *"fear factor"* that scared schools into participating in the programme run by TENGO.

When the researcher asked Pak Imran if the Education Office had a policy relating to EE, he could not give a straight answer. He said that, "It seems that there is already a section of the grand design to build Surabaya, other than education, health, environment. It is already in the City level policy, not just in the Education Office." She enquired where she might find such a document (the grand design). He asked Mas Rudi if it might be on their (TENGO's) website, or maybe that of his own Office. Mas Rudi replied that it was surely there, to which Pak Imran agreed that it was "surely there". A search of both websites revealed, predictably, that it was definitely "not there".

Pak Imran was a busy man and fielded multiple phone calls during the 25-minute interview. He appeared to be uninterested, and his dismissive manner only changed when the researcher raised the prospect of teacher exchange and cross-cultural learning (in reference to the reward that TENGO offers teachers). He was not interested in discussing the value or otherwise of Surabayan teachers going to Australia as there was "no funding for that", but was very interested in the idea of Australian teachers and students coming to Surabaya because he had two guest houses where they could stay. He gave her his business cards for the guest houses and suggested that maybe she would like to stay there too. Later that evening, Mas Rudi was uncharacteristically friendly and chatty and informed her that Pak Imran had called him not long after the interview and asked him to come back immediately to talk to him and other officials in the Education Office about the opportunity for teachers to come to Surabaya. It seemed that this interview had given Mas Rudi the access to the head of the Office that he had been wanting for some time.

Office of Parks and Sanitation (DKP)

The interview at the Office of Parks and Sanitation was a contrast to the interview at the Education Office, reminding us that "the state" is neither monolithic nor homogeneous. The atmosphere was relaxed and somewhat jovial. It was attended by the Head, Pak Candra; his deputy, Pak Wanto; the man responsible for waste data, Pak Ipal (joined at times); the researcher and her colleague; and three TENGO colleagues (Mas Rudi, Rehan and Sumo). The TENGO colleagues made themselves at home, moving around the impressive office, slouching on the couches, helping themselves to food on the coffee table and using Pak Wanto's computer while he spoke with the researcher. At times Pak Candra left the room (to pray and attend to business).

With so many people in the room, all of whom were interested in the topics of conversation, the interview became a discussion among colleagues. The conversation started with the researcher asking Pak Candra questions. Over the course of an hour and a half, the conversation moved to various topics, including Surabaya's waste management approach and infrastructure (trucks, tracking systems, compost huts, landfill), parks and public spaces, working with other departments and agencies, the characteristics and expectations of Surabayan people regarding their city and neighbourhoods (high rates of public participation and awareness, high expectations of their government), and community education approaches.

The education programme about which Pak Candra spoke the most was the traditional markets programme.[15] He described it as a programme that aimed to "change *mindsets*" and make the markets "clean". Every day university students and school students (primary to senior high school) were expected to go to the local traditional market and change the "*mindset*" of the sellers so that they would sort their waste into organic and inorganic.

> Now, we are trying it at Grebeg markets with the children. They [the traders] will be embarrassed if small children are already moving [environmentally speaking]. They [the traders] will be embarrassed and then they must change.
> (Interview, Pak Candra, Head of Parks and Sanitation, 7 January 2015)

The markets programme was still in its initial stages during our fieldwork, so it was impossible to judge whether it was successful in introducing waste separation. What was particularly interesting was the focus on separating, rather than reducing waste first, and the tactic of embarrassing or shaming people into action, and using children as the tool to do that. The use of children as messengers and *shamers* is explored later in this chapter, including the opinions of some young people on their role in this and other programs.

The Environmental Agency

Dr Prabawa-Sear was introduced to Pak Mujirun, the Head of the Environmental Agency, by a TENGO colleague at the launch of the Sustainable Consumption and Production campaign. He was warm and friendly and invited her to come to his office. Pak Mujirun's answers to questions were considered and seemed quite honest. Unlike many other government officials with whom she spoke, Pak Mujirun was not afraid to admit that there were some challenges in facilitating EE in schools and in particular with the Adiwiyata Programme. These included its focus on documentation and the limited capacity of teachers and principals to integrate environmental learning across different subjects.

The interview with Pak Mujirun started with him outlining the importance of focusing on the low socioeconomic areas of Surabaya. He said that the housing estates of middle to upper class people were "of course good" and it was the "high density slum areas" that were the focus of the city's clean and green programme.[16] The theory, he explained, was that if the neighbourhoods of people with less money were good and clean, then those with money would feel embarrassed and clean up their act. The anthropological literature (e.g. Geertz, 1961; Guinness, 1986, 2009) emphasises that the avoidance of shame and embarrassment (*isin*) and the desire to maintain harmonious social relationships (*rukun*) are strong influences on behaviour in Javanese society so it was understandable that Pak Mujirun might consider avoiding embarrassment a good motivator.

While there is an obvious need for improved sanitation and infrastructure in slum areas, it was surprising to hear his dismissal of the need to work with the middle to upper class, who were already "good". It seemed strange to expect that the upper class would be inspired by or embarrassed by the improved "slums" (to which they would be very unlikely to ever go) to apply changes to their own housing estates. Finally, he had implicitly contradicted himself: if the more salubrious areas were already "of course, good", there was no need for an upward flow of environmentally responsible behaviour. Of course, in practice, there is an urgent need for more comfortable people to be more pro-environment in their habits, but the types of pro-environment behaviour required from them are different: they can deal, at a superficial level, with waste, by paying contractors to take it away, but there is no mention anywhere of policy to restrict their consumption of status goods as well as resources such as energy. Most strikingly, in his discourse, as elsewhere in offices, schools, curricula and textbooks, there was no identification of middle- and upper-class people as keen consumers – they produce significantly more waste than poorer people, and use more resources. There is a consistent silence around this topic in Indonesia, even within Indonesian environmentalism.

It is important to note that the government officials themselves are Javanese, and although they have a higher social position than sellers in the market, they are not of higher standing than Surabaya's wealthy and well-connected business people. If forced to confront those in positions of power and status, they would struggle with issues related to respect, speaking out of place and maintaining the appearance of social harmony through respecting the position of others (Geertz, 1989).

But is there a plan?

The Surabayan government's efforts to improve its city were unquestionably impressive, and even more so when examined in the context of Indonesia and the Global South. However, it became clear that there was no plan for officials to follow. At every agency, we asked for The Plan, or the "grand design", and while everyone thought there must be one, no one was able to produce it. As far as we could see, there were no documented policies related to Surabaya's greening and environmental efforts. This meant that it was difficult to understand how Surabaya had come to take the approach it did with some of its programmes (described in the next chapter). We came to understand that the approach taken was not so much *designed*, as it would be in developed countries – with strategic directions, policies and plans, impact studies, consultation, deadlines, monitoring and so on – as *built* on an ad hoc basis. It was built over time on the widely accepted Javanese tradition of leadership (albeit a female leader) and acceptance of directives from leaders. The Weberian idea of a bureaucracy as a rational planning body was nowhere in evidence.

This does raise the question, Are these efforts sufficiently entrenched to outlive the Mayor's second (and by law, therefore final) term as Mayor? The absence of an overall strategy or policy document is worrying, and highlights that the effort to "green Surabaya" might not be sustainable.

Further, upon closer examination, it became evident that most of Surabaya's environmental programmes lacked underpinning science or knowledge. The environmental explanations given for projects provided by government officials, teachers and NGO staff were baffling at times, and highlighted a simplistic and sometimes confused understanding of environmental concepts and environmental science. For instance, one of the senior officials at Parks and Sanitation explained that the Office was asking villagers to compost their organic waste so that

> There will be a reduction in diarrhoea flies, then from [the sale of] that compost they can buy plants. The compost that is used for fertiliser has lots of oxygen, so the village gets more oxygen and then respiratory diseases decrease, don't they?

The Adiwiyata Programme

The Adiwiyata Programme (as discussed in preceding chapters) is the national environmental education programme and is managed by the Environmental Agency at the provincial, city and district levels.

When asked about the challenges of integrating EE across the curriculum, Pak Mujirun admitted that, although it was compulsory for schools to integrate it, and therefore for all teachers to include it in their subjects, there were still some teachers who could not yet do it. This, he explained, was due to a lack of "*sosialisasi*", which was the responsibility of principals. He said that often one teacher would be given responsibility for the environmental programme and others would not feel the need to participate. This was evident during competitions: when teachers who were not involved in the EE programme were asked to participate, they felt confused or unsure. He explained that the government had to teach teachers that EE was everybody's responsibility – all teachers, students and parents.[17] This sentiment from Pak Mujirun did not match the message in schools, e.g. Grade 12 students were not allowed to participate in environmental activities.

The Adiwiyata Programme is not held in such esteem in Surabaya as it is in other areas of Indonesia. In Surabaya, teachers, TENGO staff and government officials all reported that Adiwiyata focused too much on documentation and reporting, whereas the Surabayan approach was to focus on action. As described in Chapter 7 for Yogyakarta, the plethora of paperwork required for documentation in Adiwiyata was considered a burden, and no one considered that it was worthwhile. The tendency of such management processes is to create a false system that looks like it enhances accountability but actually creates alternative systems (bottom-drawer files) designed to fool those in authority into thinking that the bureaucrats are performing well. Further, scholars of neoliberal audit cultures, such as Strathern (2003) and Shore (2008), point out the power of such accountability regimes to create particular bureaucratic subjectivities.

All of the schools that Prabawa-Sear visited in Surabaya (15 in total) were involved in the Adiwiyata programme, but instead of Adiwiyata being the only EE effort (as was the case with the schools in Yogyakarta), Adiwiyata was only a part of their EE programme. TENGO facilitated many non-Adiwiyata EE activities (outlined below). Despite having impressive environmental programmes, some of the schools that we visited had not achieved a high Adiwiyata rank – the opposite to the Yogya situation, where we could not work out how inactive schools could have attained such a high Adiwiyata rank. Most schools attributed their low Adiwiyata ranking to the excessive documentation and reporting required. This suggests that rather than the audit culture being successful in Surabaya, the teachers (and TENGO) resisted it. In Surabaya, Adiwiyata did not succeed in creating environmentality, mainly because its own processes were too onerous and there were alternatives available.

Pak Mujirun explained that the documentation required for Adiwiyata was, in some cases, more difficult than the environmental activities. He argued that the environmental outcomes were more important than the documentation, and if schools failed to do well because of a lack of documentation, then that was "ironic".[18] Rather than lobby for change at a national level, Surabaya was integrating Adiwiyata into their EE programmes but positioning Adiwiyata in the back seat and prioritising TENGO's Eco Schools programme in the front seat. Some of the Eco School activities are described in Chapter 10.

TENGO and EE

TENGO is based in a house in a new, upper-class housing estate on the east coast of Surabaya. The house is supplied by a supporter, Pak Edo. Pak Edo came to be a supporter of TENGO after his daughter was involved in TENGO activities at her primary school. Pak Edo is a very wealthy businessman and government employee. The researcher stayed in the guest room at Pak Edo's family home on each of her trips to Surabaya, at the insistence of TENGO.

At the time of fieldwork, there were six full-time staff (four male, two female) and two part-time staff, and, for a few months, three university students from Madura who worked as volunteers at TENGO. At the very end of fieldwork, one of the senior high school students joined the organisation.

Mas Rudi (38 years old) was the founder and autonomous boss of the NGO. Mas Rudi spent very little time in the office and staff usually did not know his whereabouts. When the researcher enquired, she was often told that he was picking up or dropping off his daughter at school or at the mosque praying or cleaning. Some days he came to the office in the late afternoon, other days not at all. Staff never left the office before 7.00 pm in case Mas Rudi returned to the office after evening prayers (*Maghrib*). Mas Rudi was responsible for all decision-making and planning, and all financial transactions, and he supervised all of the staff. He explained to the researcher that he was not accountable to anyone and that the Education Office often relied on his opinions regarding the performance of principals in schools.

Funding and support

The organisation received funding from various private sponsors in the form of cash and in-kind support (such as the house used as the office). The various government agencies all reported supporting TENGO in the following ways: providing venues and catering for TENGO events; providing prizes for schools, teachers and students (cash and environmental infrastructure such as water filters, shredding machines, rainwater tanks and solar panels); and making staff available to attend events. The government officials never reported it, but they also gave TENGO access to schools and a level of legitimacy that other environmental education NGOs do not have. As we witnessed in Yogyakarta, NGOs struggle to gain access to schools without the support of the Education Office. The presence of government staff at TENGO events gave the events and TENGO itself a level of importance and legitimacy. Bu Risma, in addition to other government representatives, often attended prize-giving ceremonies, making TENGO events a big ticket event for teachers eager to make a good impression and boost their (bureaucratic) careers.

At the time of fieldwork, TENGO was in the second year of a funding agreement with a private electricity supply company, which sells electricity to the government. This company (referred to as PES hereafter) supported TENGO as part of its corporate social responsibility (CSR) programme. The full details of the sponsorship agreement were not made available to the researcher, but PES reported supplying the Eco-mobile (a specially fitted van, used for free school incursions and community events – see Figure 9.2). The Eco-mobile was fitted with solar panels to supply power to the large monitor, a computer, and audio system; a fold-out library with books in Indonesian and English on various environmental themes; and a range of educational resources including compost bins and a huge (5 metre × 5 metre) environmentally-themed snakes and ladders game. The Eco-mobile was used at community events as a library and merchandise stand. The van was decorated with the sponsor's stickers and environmentally-themed cartoons. It was large, colourful and highly visible. The driver often struggled to squeeze it down narrow streets with low-hanging power lines in order to enter school courtyards. In addition to the Eco-mobile,

Figure 9.2 TENGO teaching students out of the Eco-mobile.
Photo credit: Danau Tanu.

PES sponsored the Cross-Cultural Exchange (CCE) trips to Perth, Western Australia (described in Chapter 10). The manager of the CSR program at PES, Mbak Oni, explained that PES chose to sponsor TENGO not only for their EE work, but also in recognition of their working relationship with the Surabayan government.

Leadership and power/knowledge

NGO scholars often suggest that non-state providers (not EE specific) tend to be less hierarchical, more democratic and flexible than governments (Eldridge, 1995; Hadiwinata, 2003; Teamey, 2007, p. 5). This was not the case with TENGO. From a distance, TENGO might appear to be flexible, in that the staff would work all kinds of hours and would fulfil whichever roles were needed at any time. It was an autonomous organisation, but upon closer observation, it was clear that this was not flexibility as much as it was inflexibility. Staff had no say over their working hours and spent 12 hours at work most days (though not necessarily working the whole time), and they struggled to have days off. They felt that they had to do whatever job they were assigned. There was no clear organisational structure other than the dictatorship of Mas Rudi. There was not a single instance that suggested TENGO was any type of democracy. There was, however, a clear social hierarchy, with Mas Rudi at the top, the older men in the middle, the younger men below them, and the two young female staff members at the bottom of the hierarchy. So strong was this hierarchy that the fear of being reprimanded by the boss appeared to be a bigger motivator and influencer of behaviour for the two young female staff than was concern for the environment, or anything else (including evening curfews at home). The following illustrates this point.

In preparation for the Cross-Cultural Exchange competition, Mbak Rani (19 years old) was sent to buy tea for the judges – a typical domestic task assigned to female workers. The researcher offered to accompany her as she wanted to buy herself a coffee. Rani took a container for the pre-made tea. The *warung* (streetside stall) did not have any tea ready, so Mbak Rani bought tea leaves and sugar. The seller put these items into a plastic bag. Mbak Rani did not refuse the plastic bag. As they entered the building (10 metres from the stall), she looked at the researcher with panic and said, "Oh no Kelsie, I have a plastic bag!" She quickly shoved the plastic bag into a bin in the entrance way, placed the tea and sugar in the container and continued walking. Mbak Rani was not concerned that she had taken the plastic for environmental reasons (it would have been more environmentally responsible to put it in her pocket to re-use later). Her concern was Mas Rudi.

It was not only the young female staff who were submissive to Mas Rudi. On various occasions, Mas Rehan (24 years old) insisted that the researcher and TENGO colleagues sit and wait for Mas Rudi to arrive or to send instructions via text message, even when they had not eaten for many hours or could be waiting indefinitely. The researcher was never able to convince him that they could go away briefly to get tea or coffee while they waited. The older staff were more likely to disappear, and one bragged that he was too old to be scared of Mas Rudi – which indexed the extent to which Mas Rudi's power was recognised within the organisation. The power relations in this organisation had a substantial impact on the researcher's access to information. She recorded three interviews with staff but felt that they were conscious of not speaking out of turn and fearful of telling her anything that they should not. She found it much easier for everyone if she observed and waited to ask questions when the time was right. She learned more about the running of the organisation from "being there" than from formal interviews (Borneman & Hammoudi, 2009). On occasions when she was socialising with TENGO staff after hours, they would ask her what she thought about how TENGO was run. She always tried to answer diplomatically but it was clear that they all struggled with Mas Rudi's management style and the long working hours, often for weeks at a time without a day off. The staff said that when they had voiced their dissatisfaction with their boss about the working hours in the past, he had told them that he was building their skills and stamina and that if they could work for TENGO, they could work anywhere, even if they did not have a university degree. The three younger staff members seemed to accept this explanation.

Despite the questionable management practices and lack of transparency around finances, TENGO is leading the way in EE in Surabaya, and Indonesia.[19] No other NGO has the same focus on EE and the reach in schools that TENGO has. This naturally leads to the question of the source of their ideas, knowledge and methodologies. According to Mas Rudi and the staff, the majority of the facts come from the internet. Mas Rudi and his staff reported that they used a combination of Mas Rudi's ideas and techniques that they (Rudi, Budi and Rehan) had learned from their experiences with Kids for Change, a Perth-based (Western Australia) ENGO. Ultimately, all information and facts had to be presented to Mas Rudi for his approval or to have been produced by Mas Rudi himself. All PowerPoint presentations, ideas for school activities, teacher training materials and workshops, competitions and judging criteria were created or produced by Mas Rudi. One of the most interesting training sessions was entitled "Mistaken Paradigms" (*paradigma keliru*). This PowerPoint presentation highlighted in detail all of the things that he felt teachers were doing wrong in their environmental efforts.

Foucault argues that power is not only about repression but also about the power to produce knowledge. Following Foucault, Rutherford (2007) argues that the power to produce knowledge about the environment is key in formulating the terms of its management. This was clearly true in Surabaya, where Mas Rudi was defining not only the environmental issues but also the solutions for both schools and government departments. In doing so, he had significant power and influence over how environmental issues were being managed.

TENGO and Kids for Change

TENGO and Kids for Change (KfC) have what KfC refer to as a "training partnership". The opportunity for this partnership came about in 1999 when Mas Rudi travelled to Perth for a Scouts Leadership conference. The CEO of Kids for Change, called Kit hereafter, was a presenter and facilitator at this conference, and spoke to Mas Rudi's group about the work of Clean Up Australia and Kids for Change. Rudi later contacted Kit and asked if he could come to a KfC conference to learn and be mentored by Kit. Kit and her KfC colleagues saw this as a good opportunity to develop a partnership and they worked to make this happen. Mas Rudi came to the next KfC conference, and for the next few years, with funding from various Australian groups, he brought groups of Surabayan young people to Perth for the annual KfC conference. Each year, the Surabaya group stayed on after the three-day conference, at Kit's house, and went to Perth schools that were working with KfC, met KfC sponsors, and undertook environmental projects such as river revegetation. According to Kit,[20] it was at about this point that Mas Rudi decided that he would like to start an organisation in Surabaya. Thus, TENGO and KfC were emerging not-for-profit organisations that decided that both organisations would benefit from such a partnership. The partnership has continued over 15 years, with groups going to Perth most years, and Kit self-funding four trips for herself to Surabaya and having one sponsored trip to Jakarta. The relationship between Kit and Mas Rudi, as seen during his visits to Perth, was something like a mother–son relationship, with respect underlying it. Kit does not speak Indonesian; Mas Rudi speaks English. When his team come to Perth, communication is a real issue.

The communication between the two organisations has been very limited, other than in preparation for visits. Kit explained that KfC

> don't really get an update from TENGO about what they do with the information that they glean when they are here, but it's really apparent when I visit Indonesia [that] the projects that roll out of the programmes [are those] that they have seen here.[21]

The researcher agreed with Kit that this was true: she had seen many projects and infrastructure in Surabaya that resulted indirectly from Perth visits. At one school, Mas Rudi had introduced a water re-use system that captured the water that had been used by the mosque congregation to cleanse themselves before prayer (*air wuduh*). This water, which Mas Rudi described as clean water with prayers, was pumped to the school's vegetable gardens. He said that this idea came to him when he was in Perth and saw how water was often re-used.

This project and many others may have originated in Perth at KfC, but there is no doubt that Mas Rudi's ability to take an idea and turn it into a reality is something to be admired, especially in an environment like a school in Indonesia, where traditions weigh heavily and change is often not easily accepted. Not only did Mas Rudi convince others of the value of these ideas, he sourced the necessary infrastructure, expertise and funding to make them happen. The projects that resulted from Perth ideas included rainwater tanks, solar panels, vertical gardens, plastic-free canteens and energy use monitoring and reduction programmes. These Perth ideas were evident in various schools: schools obtained the hardware as prizes for winning one of TENGO's various competitions. The Perth ideas were used in conjunction with environmental projects that are commonplace in Adiwiyata schools such as *biopori*, fish ponds, and pharmacy gardens. Kit described how she saw first-hand the results of Mas Rudi's visit to the Grove Library in Perth and its grey water project. She described it as

> but one example of how Rudi has taken a concept and applied it to the cultural setting in Surabaya. When I visited two years ago he took me on a tour and pointed out the initiatives that had been inspired by that visit.
> (See Chapter 10 for more on the programme.)

The partnership with KfC has been a rewarding one for TENGO and Surabaya, but the benefits to KfC are not so obvious. There have not been any Perth groups travel to Surabaya over the 15-year partnership, and, although Perth students and teachers interact with TENGO guests when they are in Perth, it appears that this is almost entirely a one-way cultural flow. By 2017, the partnership had come to an end, mainly because of the withdrawal of KfC.

Understanding and teaching environmental issues

While the TENGO staff saw environmental actions in place in Perth, they did not always understand the complexity and interrelatedness of environmental, social and economic issues. This indicated a simplistic understanding of environmental issues, which is perhaps more accurately described as "silo-ed knowledge": that is, it was not systemic and was usually limited to a single cause and effect. Evidence of this silo-ed knowledge presented itself numerous times throughout field research with the organisation, mostly when the staff (Budi, Rehan and occasionally Rudi, though not so often) were presenting at teacher or student seminars.

This lack of understanding of the complexities of environmental issues was not surprising, as Mas Budi and Rudi were the only ones with tertiary-level education and no one had studied environmental issues in any formal setting. It is concerning that the Education Office and the Environmental Agency had such confidence in TENGO that they felt they could "leave it to them", as the experts. Mas Rehan and Mbak Anisa, who facilitated approximately 90 per cent of presentations and education seminars with students, had high school education only. Mas Rudi presented to teachers at training workshops, with one exception. This exception was a last-minute decision that really highlighted that Mas Rehan lacked the knowledge and understanding of environmental issues needed to present to teachers.

A Community School in Yogya had approached TENGO to facilitate a full-day professional learning workshop for the teachers.[22] At that point, the school's EE programme was in its infancy (less than a year old) so it was thought that the timing was right. Teachers from The Islamic School (described in the previous chapter) were also invited to learn from TENGO and share their experiences in implementing EE in Yogya. Mas Rudi was supposed to come to Yogyakarta to deliver the training. He had explained via text message that Mas Rehan, the usual facilitator, "would be confused" if the audience were teachers, suggesting that he knew that Rehan's knowledge and experience were not suited to this audience of teachers at a British curriculum, independent community school. Despite these reservations, he sent Rehan and a university volunteer, Yoyok, to conduct the training. This was because Mas Rudi had to act as a judge for a clean canteen and toilet competition for Surabayan schools. The pre-planning and organisation was lacking, but unfortunately typical. Mas Rehan and Mas Yoyok arrived in Yogyakarta by bus at 2.45 am, were picked up by the researcher's husband, and had a few hours' sleep at their house before they made their way to the school.

The group of teachers was made up of two early childhood specialists, a music specialist, two maths specialists, and three science specialists (seven females and one male in total) from the Community School, plus the researcher; in addition there were two Islamic School teachers who joined for a few hours. The workshop was conducted in Indonesian. Mas Rehan began the workshop with a brief introduction and moved on to a PowerPoint presentation which outlined some of the main environmental issues that the world is facing. Although he is a confident speaker and quite charismatic, it was clear that he did not have an in-depth understanding of the issues. It was not long before one of the senior science teachers, Ibu Nini, raised her hand and challenged Mas Rehan's information. She rather assertively informed him that the "greenhouse effect" was a naturally occurring effect and should not be presented as a negative thing. The researcher was confused for a second, thinking that she was questioning the existence of the greenhouse effect, but quickly realised her point. Mas Rehan looked a little confused, so the researcher agreed with Ibu Nini and explained to the group that, yes, the greenhouse effect was indeed naturally occurring and very important in keeping the Earth at a liveable temperature, and that we should be careful to clarify that the issue is in fact the *enhanced* greenhouse effect. The greenhouse effect is not explicitly included in the current Indonesian school curriculum but the researcher had observed a Grade 10 Geography class learning about this at The Islamic School. She had also previously been contracted by the Western Australian Government to develop an education package for Grade 5 students on this topic. Clearly any environmental educator should understand the difference and comprehend the issue, particularly if he or she is teaching others about it.

This was not the only time during the workshop that Rehan's knowledge was found lacking by the other teachers. He kept the information as general as possible (at times relying on sweeping generalisations such as "the earth is feeling blue" (*bumi lagi galau*)), and utilised emotion more than fact. One of the resources he used was a video that showed one still picture after another of environmental disasters including melting icebergs, burning forests, arid land, floods, oil spills, displaced people, salt lakes and starving cattle. There was no explanation, just an emotive musical score with no lyrics and the name of the continent where each disaster had (allegedly) occurred. Mas Rehan later struggled to explain coral bleaching after showing a picture of colourless coral. He made mention of chlorophyll dying and that meant a loss of food for fish which impacts on the amount of fish available to humans. While this was not factually incorrect, it was a very small part of the issue, a part-understanding. There was no explanation of how or why coral bleaching occurs and the broader implications of it. He also explained to the group that Australians rely on rainwater and drink straight from their rainwater tanks without treating the water. The group were amazed and looked to the researcher for confirmation. She explained that some farming and rural families rely on rainwater but most city households are connected to the main water supply and some people use rainwater tanks for their gardens. Mas Rehan and TENGO's use of emotive images and music, and their part-understandings, appear to have been enough to satisfy most teachers in Surabaya, but the teachers at this school, and Ibu Nini in particular, were not impressed.

The workshop with the Community School teachers highlighted that TENGO was used to providing information to students and teachers at a superficial level – presenting the slogan and not the whole issue. From observations, it seemed that Mas Rudi had a greater depth of understanding of environmental issues than his staff, but their approach to EE was the same, no matter who was facilitating. The teaching resources that Mas Rehan used were those (Power-Point presentations and the information-less video) that Mas Rudi used with teachers in Surabaya.

Later, Ibu Nini and the researcher had quite a long discussion about the importance of providing a clear, factual picture when teaching about environmental issues, and of allowing students to form their own views. Ibu Nini felt that TENGO were not doing this, and were instead making generalisations (such as chemicals are bad, so avoid chemicals) and were relying on emotional motivators rather than fact-based education. This view of Ibu Nini's was one that we had not encountered among the 200 or so other teachers we met during field work. Ibu Nini agreed that this was not a common way of thinking amongst educators in Indonesia and felt that she had learned this from her time teaching the New South Wales curriculum in Jakarta at an international school. She said she had learned "the Australian way of teaching the science behind environmental issues, which makes it more balanced".

Training partners with contrasting approaches

Although TENGO and KfC had been training partners for 15 years, their approaches to EE could not have been more different. KfC facilitates a ten-step process with young people to envision a better future, to explore their local area and identify issues important to them (not only environmental issues – any issues), then they create projects to address the issues the young people feel most passionate about. This ten-step process includes identifying mentors and working with organisations and professionals to achieve the aims of their projects. This process encompasses the five components of education for sustainability[23] as outlined by the Australian Research Institute in Education for Sustainability (ARIES) (Australian Research Institute in Education for Sustainability, 2009, p. 3) and also fits with the educational approaches of the UN's Decade of Education for Sustainable Development (Australian Research Institute in Education for Sustainability, 2009, p. 2).[24]

In some ways, TENGO's approach resembled a more traditional approach to EE, as was popular in the 1980s in Australia (Gough, 2013; Wals & Dillon, 2013). Adults (government officials and Mas Rudi) identify the issue and plan how they would like the issue to be addressed by the schools. This approach usually means that students (and sometimes teachers) are given defined tasks in order to address an issue. The traditional market education programme is a good example of this. The Office for Parks and Sanitation identified the amount of organic waste being generated at traditional markets as an issue and the Office and TENGO designed a programme where students would begin by sorting, transporting and composting the organic market waste, and then somehow be responsible for influencing the behaviour of the sellers. As mentioned previously, this programme does not aim to reduce the amount of waste produced, but to encourage sellers to separate their organic waste from other types of waste so that students can collect it and compost it at their schools.

However, TENGO's approach with the Eco Schools programme was almost exclusively focused on *praktek* (practice). It is not at all clear that students understood why they were doing this work. Given the lack of interest in environmental issues in the Ministry of Education and Culture, and teachers' lack of knowledge and motivation, TENGO's approach made sense. In order to be eligible for certain awards, students had to have completed a minimum number of "green hours". The "green hours" part of the programme focused solely on students and teachers undertaking hands-on activities such as composting, working in the greenhouse, collecting organic waste from the market, making or repairing *biopori* and tending the school gardens.

Students always clamoured for more *praktek* and less *teori* (theory). TENGO was ill-equipped to provide the *teori* – the science knowledge – that should have underpinned the *praktek*. Teachers were similarly ill-equipped. Our researcher in Surabaya, Dr Tanu, asked one of the teachers, who was second in charge of the school's environmental club, which classes were ideal for observing EE. He repeatedly suggested maths and other unsuitable classes. Fortunately, one of the students, the leader of the environmental club and a high achiever, politely pointed out that maths classes do not have EE incorporated into their lessons and suggested choosing more suitable classes for the researcher to observe, such as sociology or natural science. Given this lack of teacher and NGO capacity, TENGO's focus on *praktek* was understandable.

Inverting the power flow: children taking on their elders?

We mentioned above that Pak Mujirun's plan for cleaning and greening Surabaya started with the "high density slum areas". Similarly, the market waste composting project started with young people, who were to "shame" adult, lower-class market sellers into separating waste. In both cases, model green behaviour was supposed to flow "upwards", effecting change in the wealthier suburbs and among market-sellers respectively. To us, these plans were fraught with sociological flaws: in Javanese society, children obey parents, and lower class people are subordinate to higher class people. Authority flows downwards, not upwards. Society follows the leader, but the leaders cannot be poor, uneducated market-sellers and children. This issue is discussed further in the next chapter.

While Javanese people are often concerned to avoid embarrassment (for themselves as well as for others), and rarely show anger towards children, children are also given very little voice and are certainly not expected to be involved in adult affairs, as discussed in the previous chapter (Geertz, 1959; Jay, 1969). A Javanese child is not raised to voice his or her opinion (or to have an opinion), or to correct his or her parents (particularly not their father). Geertz describes a childhood of "psychological preparations" in order to develop the "ability to sharply inhibit one's behaviour, to choose inaction rather than action" (Geertz, 1961, p. 150). The way a Javanese mother cares for her child (constant carrying and gently and physically modifying the child's behaviour), Geertz argues, is "all in the direction of encouraging a deeply passive attitude" (Geertz, 1961, p. 150).[25]

It may be argued that such values as respect, obedience and knowing one's place may have been hegemonic in the 1950s and 1960s but that society has changed much in the intervening decades in the direction of a lessening of respect, the dissolution of social hierarchy, the growing importance of individual autonomy and the growing likelihood of teenager rebellion. However, recent research into high-school youth unexpectedly revealed the social conservatism of young people in Indonesia (Parker & Nilan, 2013). We would argue from our research that, although there are definitely signs of loosening, and, particularly in the cities, there are young people who do rebel, the pattern of social hierarchy and the values that underpin it continue in school, where children are not encouraged to question or challenge, and the expectation is of docile, passive behaviour. Further, these days, children are dependent upon their parents for longer and longer, with long years of education and uncertain job prospects resulting in later marriage. At the same time, young people are no longer expected to help with paid work or income earning within the family, and yet have increased need for money (White, 2012).

The Head of the Environmental Agency in Surabaya enthused: "If little children can be examples – little children can reprimand parents – then it is sure to have a big impact and have direct effects.... Kids have a big influence".[26] This approach is clearly in contrast to the social norms of Javanese society. The normative pattern could be seen in a social media post by Animal Friends Jogja (AFJ). There was a picture of a kitten that a volunteer had saved from being drowned by three young children in a river in Yogyakarta. The post stated that when the AFJ volunteer asked the children why they were trying to drown the kitten, they replied that they did not want to and felt that it was wrong, but they had been told to do so by their father. They were relieved when the AFJ volunteer stopped them and took the kitten away (Animal Friends Jogja, 2016).

Students did not feel that they had much influence over others. In almost every focus group discussion (FGD), students described how they were mocked and bullied by classmates for their environmental efforts;[27] they were told that they smelt like rubbish; and were called names like beggars, pickers and the like.[28] Children are being expected to change societal behaviour, when they themselves occupy a subordinate position in Javanese families and in society as a whole.

In one discussion, two female Grade 12 students from a very prestigious private Catholic school shared their views on bullying and the environmental programme. One described being embarrassed by a teacher when encouraging her friend to turn off lights:

It's like being belittled by a teacher.... For example, I've had this happen: when you tell your friend to turn off the lights to conserve energy and the teacher says, "Oh, what are you doing? The school pays [for the electricity] you know". Like that! And in the end it's we who cop it.

(FGD, Catholic School, 13 October 2014)

For the virtuous student, the teacher's lack of environmental responsibility – if the school pays for electricity, there is no problem in leaving the classroom lights on – was compounded by the teacher's criticism of the student, so that in the end she "copped it". Similar stories were shared time and time again by students, suggesting that the methodology of children shaming others into action was indeed a flawed one. If these students were struggling to influence their peers and teachers, how could it be assumed that they could change the behaviour of parents and elders?

Dr Prabawa-Sear raised this issue with one of Surabaya's most celebrated Eco-students. This young man, Fajar, was one of the winners of the trip to Perth (described in the next chapter), and later went on to work with TENGO. His honest response confirmed our concerns about the strategy of expecting students to change the behaviours of those of higher social standing.

I'll be honest. In my family, I have already told my parents, my younger sister, my older sister, "Please, when you go shopping, take one of the shopping bags that I have set aside." But nothing changes – especially my parents. Maybe my older or younger sister might use them if I remind them as they are leaving, but not my parents. My mother says, "I'll just use this one, that one will smell and it's not enough for all the shopping." *Indeed, educating the family is not as easy as we say.... Especially in Surabaya. In our culture, children have to obey elders, so it's really, really difficult.*

(Interview in Perth, Fajar, 5 June 2015, our emphasis)

The apparent conflict between TENGO's approach and Javanese norms around social hierarchy was perplexing at first. It was difficult to understand why TENGO would encourage this approach. We think there are two explanations. The first is the Javanese value of conflict avoidance, and the value of *rukun*, or social harmony. It was suggested by the Head of the Environmental Agency when he explained that by using children as messengers, they could avoid confrontation that might occur between adults.

We know that in Surabayan society, and maybe Indonesian society, if, for example, we [adults] do something bad, like litter, and are reminded by another adult, well, then things could get messy. But if we are reminded by a child, no one will get angry with a child and will surely be embarrassed....

(Interview, Head of Environmental Agency, 26 February 2015)

Second, it seems that this approach was based on the model that KfC used: that young people must actually do meaningful, pro-environmental work. However, perhaps in an effort to make it less complex, the TENGO approach was missing some of the vital aspects of the KfC approach: that young people define the issues and decide on the projects themselves; that projects are age-appropriate and aim to empower young people; that projects are realistic in their goals; and they do not aim to change the behaviour of adults by simply telling them what to do. Not only was this approach of TENGO's actually far from that of KfC, but also it did not resemble the characteristics for Education for Sustainable Development (ESD) as promoted by UNESCO (2016).

Unpaid child labour?

Although the market waste composting programme will likely achieve some positive environmental outcomes, its value as an EE programme is doubtful. It does not include four of the five characteristics of ESD as defined by UNESCO: envisioning, critical thinking and reflection, systemic thinking, and participation in decision-making (UNESCO, 2016). Arguably, the one characteristic that it might be said to include is *building partnerships* – that is, for example, between schools and markets, or between TENGO and schools. This situation raises the question, if this programme does not meet UNESCO's definitional characteristics of ESD, and Indonesia offers no standards or definitions of its own for EE,[29] then on what basis can we label the traditional markets organic waste programme an environmental *education* programme? Further, and more concerning, is it not a government waste and composting programme that uses unpaid child labour? Using unpaid child labour for any length of time would be exploitative and unethical, and certainly against the spirit of any EE programme.

This programme's success relies heavily on students working physically hard, for free, to compost many hundreds of kilograms of organic waste, and on power relations: first, that TENGO has the power to compel students to actually perform this arduous work, and second, the presumption that adult market-sellers will be motivated to sort, transport and compost their own organic waste when they are shamed into it by seeing students doing the dirty work. It is highly arguable that students are performing this hard labour voluntarily. "Forced volunteering" is the name of the programme (see Figures 9.3 and 9.4).

As discussed in the previous chapter, we support EE programmes that meaningfully involve young people, from "go to woe", but they have to be done in a culturally sensitive and fair way. This contributes to the effectiveness of programmes, enhances motivation and helps to prevent apathy and hopelessness. We note again the optimism of young people in Indonesia, and think that this is something that could be harnessed to good effect with environmental events. TENGO-run events, such as the Yel-Yel competition, described in the next chapter, demonstrate to students that they are not alone in their environmental actions and that Surabayans are working on improving the environment together. Events such as these are very important in maintaining enthusiasm and hope for students and teachers alike.

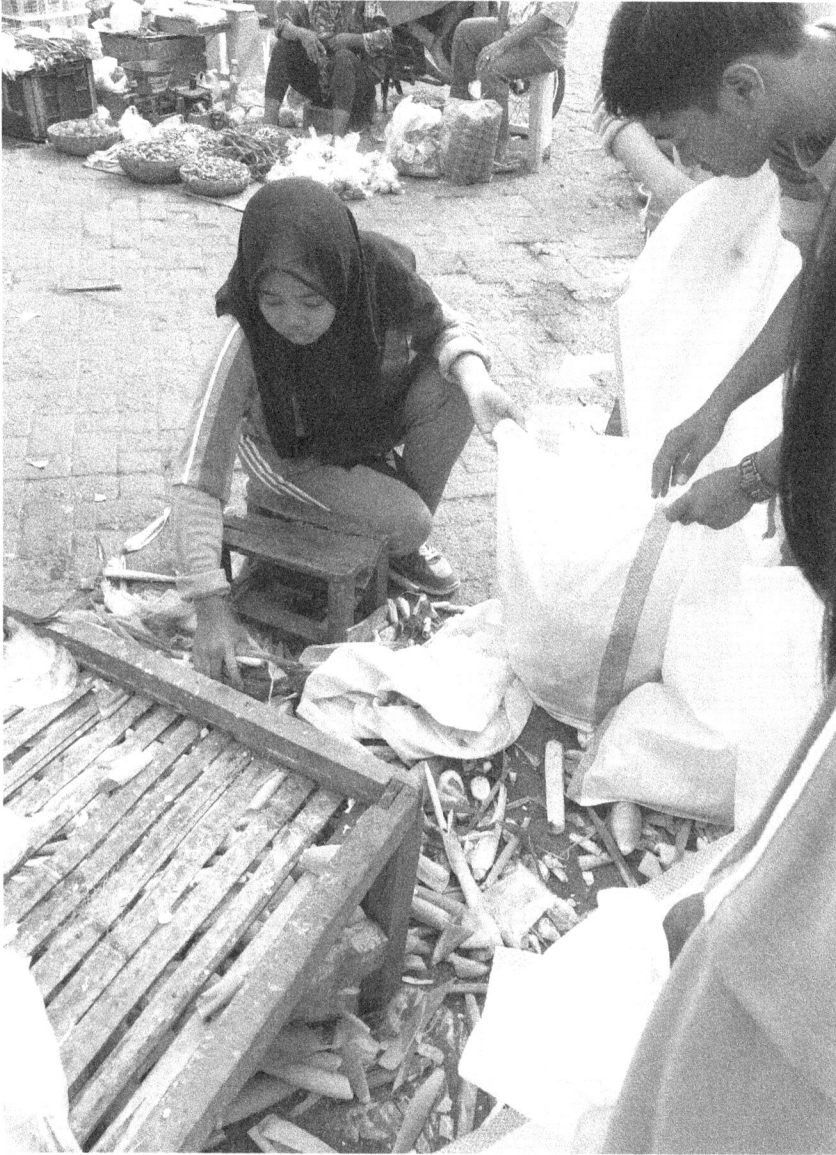

Figure 9.3 Students collecting organic waste in the market.
Photo credit: Danau Tanu.

Figure 9.4 Students carrying the organic waste to be made into compost.
Photo credit: Danau Tanu.

A model to emulate?

The approach taken by TENGO and the Surabayan government is held up as an example to others. Pak Wanto from the Office for Parks and Sanitation informed us that as well as having government representatives from all over Indonesia come to Surabaya to learn about their approach to environmental management, they have also had guests from Bangladesh, Laos, Malaysia and Timor Lesté. There is no doubt that Surabaya is a much "cleaner and greener" city now than in 2006, when it began its environmental programme. The main river smells like a living river; mangroves have been planted along foreshores; city parks are green, clean and well patronised; waste is being collected and sorted; city streets are comparatively clean and often lined with trees.

Despite the positive discourse from the government departments and TENGO, and the real environmental outcomes, it is hard to appreciate the approach to EE. Children are being made to do hard physical work, for free, and in the case of the market waste programme, they are being made responsible for EE via modelling ideal behaviour for the edification of uneducated market-sellers. In order that government officials can avoid telling other adults what to do – i.e. to avoid potential social conflict – children are being made to undertake the responsibility of adults, and, in the case of the traditional markets, the responsibility of the Surabayan government.

Conclusion: environmentality in Surabaya?

The Surabaya example is unique in Indonesia – and possibly the world – in its coordinated, whole-of-government approach to greening the city, in partnership with an ENGO. Is it environmentality, as conceptualised by Agrawal?

Birkenholtz notes that first, environmentality requires that the environment comes to be a thing in itself, as something that the state sees as needing protection and management (Birkenholtz, 2008, p. 83). We can see that Surabaya fits this first precept. Bu Risma has been largely responsible for identifying Surabaya's environment – previously taken for granted as dirty and chaotic – as in need of care and management, and for taking it up as a government priority. Rutherford (2007) noted that environmentality entails the production of new knowledge about the environment, and Birkenholtz's second precept was that new and different types of knowledge about the environment are shared. We have noted that two key individuals in Surabaya, Bu Risma and Mas Rudi, constructed new knowledge, in the form of isolating problematic aspects of the environment – waste, denuded coastlands, dirty rivers – and devised ways to "solve" the problems through local government services and projects – particular waste management practices, littoral regeneration, market clean-ups. However, we would not subscribe to the proposition that this new knowledge has been "shared". As we have seen, even among the government agencies and schools responsible for implementing these schemes, there is gross ignorance. The eventual targets of "socialisation" – first the students and lower classes, then their families and the higher classes, and finally the general population – are not made privy to new knowledge and instead are expected to do the unpaid physical work.

Notes

1 More information on this can be found in Prabawa-Sear (2019).
2 While the greening Surabaya programme is a new programme, there is a long tradition of government civic improvement programmes in cities around Indonesia, going back to colonial times. Colombijn has shown that these almost invariably favoured middle-class interests, for example, in hygiene and neatness, rather than improving the environment per se or the interests of the poor, because "kampong people lacked power" (Colombijn, 2013, p. 199).
3 "Bu" is short for "Ibu", which is literally "mother", but is a common way to address adult women.
4 *Fortune* magazine placed her number 24 in their list of the world's greatest leaders:

> Elected as Surabaya's mayor in 2010, Rismaharini has transformed her city of 2.7 million people into a new kind of Indonesian metropolis, one that celebrates green space and environmental sustainability. The city, long known for pollution and congestion, now boasts 11 richly landscaped parks and other green spaces. In some cases even cemeteries have been expanded and redesigned to absorb more water and reduce flooding, an ever-present risk in Indonesia.
> ("The World's Greatest Leaders – #24 Tri Rismaharini," 2015)

5 See Kusno (2011) for a description of Jakarta's approach to being a clean and green city.
6 Nevertheless, she has not been pro-tourism. For instance, luxury cruise liners laden with wealthy tourists wanted to dock in Surabaya but there were no city tours developed to cater to this demand.
7 However, this accessibility did not extend to our researchers. Despite many phone calls, requests for introductions to intermediaries, etc. we were never able to interview Bu Risma.
8 Most other teachers and students are not expected to participate unless there is a competition requiring big numbers to win.
9 There are some similarities to President Jokowi. Although he was a businessman before becoming President, and Bu Risma was a public servant, they both have the ability to incisively cut through to the main issue and to solve it, a reputation for can-do action and an image of being "clean" (non-corrupt).
10 See Chapter 8 on *priyayi* culture. Nevertheless, Surabaya is definitely seen as a Javanese city (Fionna, 2017, p. 5), unlike, say, Jakarta, which is often characterised as Indonesian (Fionna, 2017, p. 5), not least because immigrants to Surabaya are usually from East Java, while immigrants to Jakarta come from all over Indonesia.
11 These conflicts have impacted negatively in some areas of environmental management, e.g. waste management, where Surabaya needs the cooperation of surrounding areas.
12 The *alus: kasar* dichotomy is sometimes seen as a Javanese principle, and gendered, with maleness associated with *alus* behaviour and femaleness with *kasar*, but like many dualities, it is slippery, so scholars sometimes talk of Javanese *alus*: non-Javanese (sometimes Malay) *kasar*.
13 Interview with Head and Deputy Head of Parks and Sanitation, Surabaya, 7 January 2015.
14 Interview with Environmental Agency, Surabaya, 26 February 2015.

15 Traditional markets (literally, "wet" markets, as opposed to supermarkets and mini-markets) are generally held in an undercover market area each morning. Items sold include fruit and vegetables, meat and poultry, clothing and locally made snacks and drinks. Larger markets (like Grebeg) also sell non-electronic household items such as brooms, cooking utensils, kerosene cookers, etc.

16 Besides her environmental work, Bu Risma is famous for having closed down "Dolly", Surabaya's largest red-light district, as well as other brothel areas (Fionna, 2017, pp. 15–19). It is tempting to see the moral cleansing of the red light districts as a parallel with the "clean and green" policy – not unlike the moral crusades of the colonial era, when cleanliness and hygiene were invoked to justify the policing of prostitution and homosexuality (Bloembergen, 2011). However, Bu Risma was not particularly active in re-training or otherwise providing for the many sex workers rendered homeless and unemployed by the closure. It seems she has not been interested in job creation programmes or other projects that might improve the lot of the poor and vulnerable.

17 Lest the reader assume that this is a failing peculiar to Indonesia, or to education systems in the Global South, there is evidence that in Australia, the setting of EE as a cross-curricular priority for the national curriculum since 2010 has met various challenges, including that "Educators do not necessarily know how to teach in cross-curricular ways" (Dyment, Hill, & Emery, 2015, p. 1118). Several Global North countries have adopted the cross-curricular approach, e.g. the UK and Sweden, but one of the dangers is that while sustainability is supposed to be "everywhere" in the curriculum, it ends up being "nowhere" because it is nobody's particular responsibility to teach it. See also Hayes (2010).

18 Interview with Pak Mujirun, Head of the Environmental Agency, Surabaya, 26 February 2016.

19 Mas Rudi was awarded a Kalpataru Certificate by the President for his and TENGO's work for the environment. This is the premier environment award in Indonesia.

20 Interview with Kit, CEO of Kids for Change, Perth, 13 March 2015.

21 Interview with Kit, CEO of Kids for Change, Perth, 13 March 2015.

22 The School was attended by Dr Prabawa-Sear's daughter.

23 These are: envisioning a better future, critical thinking and reflection, participation, partnerships for change and systemic thinking.

24 These are: interdisciplinary and holistic learning, values-based learning, critically reflective thinking, multi-method approaches, participatory decision-making and locally relevant information.

25 In Java and Bali, women breastfeed their babies for long periods and carry them constantly. Babies are taught to reach out for and give things with their right hands only, by the mother gently pushing the left hand into inaction and pulling out the right hand. There is rarely any verbal admonishment. There is a strong expectation of docile behaviour, and this seems to work.

26 Interview with Pak Mujirun, Head of Environmental Agency, Surabaya, 26 February 2015.

27 The verb (to) *bully* or be bullied (*dibully*) is borrowed from the English, but is usually used to describe a less vicious or hurtful treatment than the English word. The Indonesian use of the word translates more accurately to "tease/be teased" in English. It is also used to describe being pressured to do something.

28 The issue of "being dirty" and being associated with rubbish is discussed more fully in Tanu and Parker (2018).

29 In an extensive literature review, we have not been able to find any documentation from Indonesia that defines or outlines what EE or ESD is. There is one line in the Adiwiyata Programme Guidelines that states that the Adiwiyata Programme is "built on principles of education, participation and sustainability" (Kementerian

Lingkungan Hidup Republik Indonesia, 2013, p. 5). What these principles are is not stated. Nomura (2009) explores Indonesia's transition from EE to ESD, but differentiates ESD and EE by the topics covered (ESD includes social issues) rather than the approaches used.

References

Agrawal, A. (2005a). *Environmentality: Technologies of Government and the Making of Subjects*. Durham: Duke University Press.

Agrawal, A. (2005b). Environmentality: Community, intimate government, and the making of environmental subjects in Kumaon, India. *Current Anthropology*, 46(2), 161–190.

Animal Friends Jogja (Producer). (2016). Instagram Animalfriendsjogja: Little Moses rescued on (*sic*) time before being drowned to (*sic*) a river! [Instagram post, 12 April 2016.] Retrieved from https://instagram.com/p/BEFqi2RD1g1/.

Australian Research Institute in Education for Sustainability. (2009). *Education for Sustainability: The Role of Education in Engaging and Equipping People for Change*. Retrieved from http://aries.mq.edu.au/publications/aries/efs_brochure/pdf/efs_brochure.pdf.

Birkenholtz, T. L. (2008). "Environmentality" in Rajasthan's groundwater sector: Divergent environmental knowledges and subjectivities. In M. K. Goodman, M. Boykoff, & K. Evered (Eds), *Contentious Geographies: Environmental Knowledge, Meaning, Scale* (pp. 81–96). Aldershot, UK: Routledge.

Bloembergen, M. (2011). Being clean is being strong: Policing cleanliness and gay vices in the Netherlands Indies in the 1930s. In K. Van Dijk & J. G. Taylor (Eds), *Cleanliness and Culture: Indonesian Histories* (pp. 117–146). Leiden: KITLV Press.

Borneman, J., & Hammoudi, A. (2009). *Being There: The Fieldwork Encounter and the Making of Truth*. Berkeley: University of California Press.

Colombijn, F. (2013). *Under Construction: The Politics of Urban Space and Housing during the Decolonization of Indonesia, 1930-1960*. Leiden: Brill.

Dean, M. (1999). *Governmentality: Power and Rule in Modern Society*. London: Sage Publications.

Djajadiningrat-Nieuwenhuis, M. (1992). Ibuism and Priyayization: Path to power? In E. Locher-Scholten & A. Niehof (Eds), *Indonesian Women in Focus: Past and Present Notions* (pp. 43–51). Leiden: KITLV Press.

Dyment, J., Hill, A., & Emery, S. (2015). Sustainability as a cross-curricular priority in the Australian Curriculum: A Tasmanian investigation. *Environmental Education Research*, 21(8), 1105–1126.

Eldridge, P. J. (1995). *Non-Government Organisations and Democratic Participation in Indonesia*. Kuala Lumpur: Oxford University Press.

Fionna, U. (2017). *Investigating the Popularity of Surabaya's Mayor Tri Rismaharini*. Retrieved from Trends in Southeast Asia, 2/17. Singapore: ISEAS Publishing.

Foucault, M. (1991). Governmentality (Rosi Braidotti and revised by Colin Gordon). In G. Burchell, C. Gordon, & P. Miller (Eds), *The Foucault Effect: Studies in Governmentality* (pp. 87–104). Chicago: University of Chicago Press.

Geertz, H. (1959). The vocabulary of emotion: A study of Javanese socialization processes. *Psychiatry*, 22(3), 225–237.

Geertz, H. (1961). *The Javanese Family: A Study of Kinship and Socialization*. New York: The Free Press of Glencoe.

Geertz, H. (1989). *The Javanese Family: A Study of Kinship and Socialization*. Prospect Heights, IL: Waveland Press.

Gough, A. (2013). The emergence of environmental education research. In R. B. Stevenson, A. E. Wals, J. Dillon, & M. Brody (Eds), *International Handbook of Research on Environmental Education* (pp. 13–22). New York: Routledge.

Guinness, P. (1986). *Harmony and Hierarchy in a Javanese Kampung*. Singapore: Oxford University Press.

Guinness, P. (2009). *Kampung, Islam and State in Urban Java*. Singapore: Asian Studies Association of Australia in association with NUS Press.

Hadiwinata, B. (2003). *The Politics of NGOs in Indonesia: Developing Democracy and Managing a Movement*. London: RoutledgeCurzon.

Hayes, D. (2010). The seductive charms of a cross-curricular approach. *Education 3–13*, 38(4), 381–387. doi:10.1080/03004270903519238.

Jay, R. R. (Ed.) (1969). *Javanese Villagers: Social Relations in Rural Modjokuto*. Cambridge, MA: MIT Press.

Kementerian Lingkungan Hidup Republik Indonesia. (2013). *Pedoman Pelaksanaan Program Adiwiyata*. Retrieved from www.menlh.go.id/Peraturan/Permen%2005%20tahun%20 2013_tentang%20Pedoman%20Pelaksanaan%20Program%20Adiwiyata%20oke.pdf.

Kusno, A. (2011). The green governmentality in an Indonesian metropolis. *Singapore Journal of Tropical Geography*, 32(3), 314–331. doi:10.1111/j.1467-9493.2011.00440.x.

Lombard, D. (1986). Réflexions sur le concept de "Pasisir" et sur son utilité pour l'étude des littératures. In C. D. Grijns & S. O. Robson (Eds), *Cultural Contact and Textual Interpretation* (pp. 19–24). Dordrecht & Cinnaminson: Foris.

Meek, D. (2015). Learning as territoriality: The political ecology of education in the Brazilian landless workers' movement. *The Journal of Peasant Studies*, 42(6), 1179–1200. doi:10.1080/03066150.2014.978299.

Nomura, K. (2009). A perspective on education for sustainable development: Historical development of environmental education in Indonesia. *International Journal of Educational Development*, 29(6), 621–627. doi:http://dx.doi.org/10.1016/j.ijedudev.2008.12.002.

Parker, L., & Nilan, P. (2013). *Adolescents in Contemporary Indonesia*. New York: Routledge.

Prabawa-Sear, K. (2019). The impact of culture on environmental education in Java, Indonesia. PhD thesis, School of Social Sciences, The University of Western Australia.

Rutherford, S. (2007). Green governmentality: Insights and opportunities in the study of nature's rule. *Progress in Human Geography*, 31(3), 291–307.

Shore, C. (2008). Audit culture and illiberal governance: Universities and the politics of accountability. *Anthropological Theory*, 8(3), 278–298. doi:10.1177/14634996080 93815.

Strathern, M. (Ed.). (2003). *Audit Cultures: Anthropological Studies in Accountability, Ethics and the Academy*. Florence: Taylor & Francis.

Tanu, D., & Parker, L. (2018). Fun, "family", and friends: Developing pro-environmental behaviour among high school students in Indonesia. *Indonesia and the Malay World*, 46(136). doi:10.1080/13639811.2018.1518015.

Teamey, K. (2007). *Whose Public Action? Analysing Inter-sectoral Collaboration for Service Delivery. Literature Review on Relationships between Government and Non-state Providers of Services*. Retrieved from www.birmingham.ac.uk/Documents/college-social-sciences/ government-society/idd/research/non-state-providers/literature-review.pdf.

The World's Greatest Leaders – #24 Tri Rismaharini. (2015). *Fortune*.

UNESCO. (2016). Education for Sustainable Development Unit (ESD): Characteristics of ESD. Retrieved from www.unescobkk.org/education/esd-unit/characteristics-of-esd/.

Vickers, A. (1987). Hinduism and Islam in Indonesia: Bali and the Pasisir World. *Indonesia, 44*(October), 31–58.

Vickers, A. (1997). "Malay identity": Modernity, invented tradition, and forms of knowledge. *Review of Indonesian and Malaysian Affairs, 31*(1), 173–211.

Wals, A. E. J., & Dillon, J. (2013). Conventional and emerging learning theories. In R. B. Stevenson, M. Brody, J. Dillon, & A. E. Wals (Eds), *International Handbook for Environmental Education Research* (pp. 253–261). New York: Routledge.

White, B. (2012). Changing childhoods: Javanese village children in three generations. *Journal of Agrarian Change, 12*(1), 81–97.

10 Environmentality in Surabaya?

The previous chapter described the strenuous efforts being made by government, in partnership with TENGO, to make a positive difference to Surabaya, by making EE compulsory and making the city "clean and green". Much of the credit must go to Bu Risma, the energetic mayor of Surabaya, for galvanising government officials into action and overcoming the obstacles of bureaucratic laxity, incapacity and silos. This holistic, all-of-government + ENGO approach is new in Indonesia, and suggests the possibility of environmentality in Surabaya. Has the institutionalisation of environmental concern in Surabaya proceeded to the extent that individual actors are "responsibilised" to act and care for the environment? In the previous chapter, we held over answering the question, in order to see if and how young people in Surabaya practise environmentalism. This chapter answers that question, first, by describing some EE events and then exploring the environmentalism of some individual students (and teachers). We begin with some environmental competitions.

Competitions

Competitions (*lomba*) are a very popular way to draw a crowd in Indonesia (Creese, 2014; Long, 2007, 2013) and have been used for many decades as a means of community education (Sears, 1989) and state-driven institutional agendas (Creese, 2014; Parker, 2003). Gade, talking about competitive Islamic performances in Indonesia, mentions the educative function of such performances: "the phenomenon of performance doubling as pedagogy" (Gade, 2004, p. 2).

TENGO, with the support of the Surabayan government and private sponsors, had a continuous cycle of competitions for schools, principals, teachers and students.[1] These included: Surabaya Eco Schools, Ecopreneur (eco-entrepreneur) (of the week, month and year), Eco teacher (of the week, month and year), Eco student (of the week, month and year), Eco principal (of the week, month and year), Yel-Yel Surabaya Eco Schools (dress-up and singing competition for high schools), Eco Schools jingle competition, Green hours, poster drawing, Princess and Prince of the Environment (for primary school students) and the Cross-Cultural Exchange Competition (for students, teachers and principals). The researcher attended two large events: the Yel-Yel Surabaya Eco Schools and the Cross-Cultural Exchange Selections, described below.

The Head of the Environmental Agency explained that his Agency supported the competitions because it was an opportunity to educate the audience: the parents. He said that, "When there is an event for students, parents definitely come. The way to educate parents is to educate children". While this approach might seem to make sense, its success depends on three factors: parents coming, the event being educational and the hoped-for ingestion of environmental messages by parents. The two major competitions that we witnessed, over five days, lacked these crucial factors, as outlined below.

Yel-Yel Eco Schools

Yel-Yel Eco Schools is a competition run for junior and senior high schools. Each school creates a song or chant and performs it with accompanying dance whilst wearing costumes made out of *"reuse"* (re-used) materials. The competitions were held in public spaces – Taman Flora Bratang (a public park) and Kenjeran Beach. Hundreds of students attended and performed each day (26 schools on Day One and 60 schools on Day Two), and there was a considerable crowd of passers-by (the general public) that gathered for the senior high school competition at Taman Flora Bratang. There were very few parents of students present; there were a few small groups of friends of the performers who cheered enthusiastically for their school.

While the competition was mostly highly entertaining (very colourful and loud, with many bright costumes – see Figure 10.1), the educational content was questionable. There were some large banners placed around the stage, but audience members were not able to get close to the banners while the competition was running. Some schools included an environmental message in their performance (the most common were "do not litter" and "conserve water") but many schools stuck to chanting lines such as "Surabaya Eco School!", "Surabaya clean and green!". The competition held at Kenjeran Beach (for junior high schools) had many more participants (we estimate 1000). Some schools struggled to fit all of their students onto the stage. Schools were congratulated for having large numbers of students participate. However, there were barely ten non-participants who constituted the audience. This was probably because one must pay to enter the beach and Indonesians sensibly tend to avoid the beach during the heat of the day. Each school group was accompanied by teachers, but like the Taman Flora Bratang competition, there were very few (five or so) parents in attendance.

The researcher had been co-opted to be one of the three judges. She was asked by Mas Rudi to give each school a score out of 100 based on their costumes and music. He said that all costumes and instruments must be made out of re-used material – if they were not, a score of 0 was to be given. The costumes for one performance consisted largely of yellow and blue plastic bags,

Figure 10.1 Yel-Yel dance competition.
Photo credit: Kelsie Prabawa-Sear.

which looked new, and two sets were suspect. The researcher asked Mas Rudi what he thought. On both occasions he said that it was fine. She was told to judge the music, but was unsure how to judge the music and was given no further direction on it. Mas Rudi was the second judge and gave scores on "compilation";[2] and a representative from PES, the main TENGO sponsor, gave scores on lyrics and movement. We did not discuss or share our scores and Mas Rudi collected them at the end of the day. Throughout the day it was announced that pictures were already being uploaded to Instagram, Facebook and Twitter and students were encouraged to upload their own photos and use hashtags.

When writing up her field notes after two long, hot days of judging the competition, the researcher summarised the successes and missed opportunities of the competition as follows:

Successes
- Community building: There was a great feeling of camaraderie and community. The large number of schools and students indicates to participants that they are part of something bigger, a Surabaya-wide approach to going green.[3]

- Indoctrination:[4] It felt very like indoctrination with the continual chanting of the slogans ("Surabaya Eco Schools", "Surabaya clean and green", "conserve water"). Over and over and over again. The kids sang with such gusto that it seemed that they were really enthusiastic about the message.

Missed opportunities
- Massive amounts of waste were created. Although the costumes were to be made from re-used materials, I have my doubts about the plastic bags. The number of bags used was crazy. There was also a lot of litter from the costumes and snacks. Rudi and Rehan told performers to clean up after themselves but this did little to stem the flow of litter.
- Lack of feedback from TENGO to participants. Feedback could have been on reducing waste in the performances for a start. Other ideas would be: kids design the production, write songs, make costumes, encourage difference (they all just love to be the same!!), share stories through performance – not just chant slogans.
- The messages were shallow. Performances tended to focus on repeating messages with little meaning: "conserve water", "conserve electricity". Not much focus on how to. About two schools included examples of how. Really felt like lots of chanting of messages (indoctrination) and little consideration of how or what. I would have loved to have seen an actual story of their actions.
- Overall, I got the impression that the *lomba* (competition) and the prestige that goes with participating and winning is more important than any environmental message. Perhaps the message was considered in the judging of the lyrics, but I'm not convinced that happened. It would be great to see a category of judgement for the message or theme. I would have LOVED to have seen a school that had a "minimise waste" theme and avoided all the waste!!

(Field notes, edited, 12 October 2014)

As mentioned above, contests have long been used in Indonesia as a way of drawing crowds together, and as a means of community education (particularly of government development programmes). For Indonesians, participation in a crowd and an excited atmosphere (*keramaian*) is a positive thing in itself. Although a contest might potentially create division, and promote self-interest and ill will, it means that many people are at least sharing an activity (and not doing something else), and looking inward, at the contest and the competitors – potentially creating a "web of competitive relations" and "a unit that matters": the Eco School movement (Long, 2013, p. 184). Through participation, the competition enables "particular kinds of achieving selves": students who belong to their school, to the Yel-Yel Competition, to the Eco School movement, and to the larger Surabaya "clean and green" movement, thus sustaining the Yel-Yel competition "beyond the moment" (Long, 2013, p. 184). The pity of it was, though, that the fun of doing things together crowded out the environmental content.

The winners of the competitions were not announced at the event, but were announced online later. While this meant there was no climax to the event, it was a clever tactic to draw the students and teachers to TENGO's social media feeds. There was no obvious dissatisfaction with this outcome, by students or teachers, who appeared to be used to this approach for announcing winners.

Cross-Cultural Exchange selections[5]

The "Cross-Cultural Exchange" (hereafter CCE) was TENGO's name for a group tour of students, teachers, a school principal, TENGO representatives and the TENGO founder/patron to Perth, under the auspices of KfC. The CCE selections event was a much smaller and less glitzy competition than the Yel-Yel Eco Schools competition. The event was held in a government building and only attended by competitors and judges. Only one parent came to watch their child compete for a chance to represent Surabaya in the "exchange" to Perth. Had any other parents (or families of teachers and principals) attended, it would have been an excellent opportunity for them to learn about the work for the environment that is happening in the schools.

In order to be eligible to compete, teachers and students had to have completed 50 "green hours" (in the past four months). Green hours are hours of work for the environment. In the case of students, the hours have to have been logged and signed off by a teacher. These hours can only be accrued by undertaking physical work (composting, planting trees, making *biopori*) and socialisation work (working with others to encourage them to be more pro-environment, working with neighbouring schools in one's own neighbourhood on a project). Mas Rudi was very clear that ceremonial activities did not count.

In addition to a three-minute presentation entitled "My Actions" (*Kiprahku*), students were required to submit an essay of the same title. Teachers and principals were only required to do the oral presentation. The competition was open to students of all ages. Two students, two teachers and one principal were selected as winners. The prize was a trip to Perth to attend a KfC conference, meet other environmental students and teachers, visit environmental projects and learn more about Western Australia.

The afternoon before the selections, the researcher had asked Mas Rehan if this was going to be a real selection process or if Mas Rudi would just pick whomever he wanted.[6] Mas Rehan smiled in acknowledgment and said, "We'll write notes on each one". She gave him a disapproving look to which he replied, "Maybe we could select three or so" (implying that Mas Rudi would pick from them). Later that night, the researcher was waiting with the younger staff members for the rain to stop. Mas Rehan and Mbak Anisa were grumbling (in Javanese) that Mas Rudi wanted to take his six-year-old daughter to Australia. Rehan asked Anisa whether they were organising family tourism or cross-cultural exchange.[7]

Mas Rehan and the researcher had discussed the opportunity for families to pay for their children to join trips to Australia earlier that month, after a mother had come to the TENGO office with her son to enquire about a replacement trophy for her son. He had been a finalist in the Prince and Princess of the Environment awards and the school had decided to keep his trophy to display in the school trophy cabinet. The mother was quite pushy and asked Rehan if her child could go on the next trip to Australia. She asked if she could reserve a place for her son, complaining that last time they only found out about the trip when the group left and she had felt shocked as her son wanted to go. During this somewhat one-sided conversation, Rehan was polite but made it clear that it was Mas Rudi who arranges the trips (and Mas Rudi was not there at that time). The mother and son left the office and the son reappeared minutes later with a huge bag of traditional snacks for us. Rehan acted like this was no big deal but the researcher saw it as an attempt to win favour. She asked Rehan about the possibility of buying a place on the trip and he explained that this was only possible for the Prince and Princess trip, not the CCE. The victorious Princess and Prince win a free trip, then it is open to others if they can pay their own way. He hinted that there may have to be changes to this set-up for future visits, as Budi had taken a group of primary school children by himself last year. Kit, the CEO of KfC and host to the group, was unimpressed by the lack of adult supervision and told TENGO that if they wanted to do such a trip again they would need more responsible adults, of which at least one must be female.

The judging panel for the selection of students consisted of Mas Rehan and the researcher. She had asked not to be involved in any official sense so that she could observe and take notes. However, due to a lack of available staff and invited judges, she and Mas Rehan were the reluctant MCs and judges for the day. The judging panel for the teachers and principals consisted of Mas Rudi, a representative from the Office of Parks and Sanitation, a representative from PES – the sponsor of the trip, who also went on the trip – and the researcher.

Selection process for students

There were 44 presentations over the course of the day: eight primary school presentations, 25 junior high school (SMP), seven senior high school (SMA) and four vocational senior high school (SMK) presentations, and all were to be in English.

Although the topic was "My Actions", most students presented a list of all the activities that their school had undertaken, not indicating which of these they themselves had been involved with. Mas Rehan reminded the group to highlight their actions for the judges but all of the students had prepared Power-Point presentations and some of them had memorised their presentation in English, so there was no veering off track, despite Mas Rehan's request. The PowerPoint presentations were mostly very similar, with pictures of European landscapes, sparkling rivers and deciduous trees combined with photos of groups of students composting, pictures of handicrafts made from re-used packaging and students making *biopori* holes. Headings were mainly in English and content was mostly in Indonesian. The contestants spoke to the audience, which consisted of other presenters and some teachers, who were seated directly in front of them. (The judges sat to the side.) Very few audience members listened to others' presentations, and most walked around, rehearsing their own presentations while awaiting their turn.

Mas Rehan and the researcher listened to the presentations, taking turns to ask a question at the end of each presentation. The researcher's questions were mostly the same for each student, in an effort to have consistency and to encourage them to try to answer in English. She would ask, "What is your favourite activity?" and would pronounce "favourite" as it is said in Indonesian, *favorit*. She thought that this might be simple enough for the first few students and that the students after that would work out that she was asking the same question for almost all of them. Unfortunately, as no one was paying attention to the questions that others were being asked, each time she asked the question, she was met with looks of horror and had to repeat it in Indonesian. The lack of ability to comprehend this simple question raised her concern about the value of a trip to Perth, if their English communication skills were so limited. Whatever concerns she had for the students' English communication skills on Day One were multiplied at the principals' selection day.

Fajar

There was one presentation that stood out from the others. This was Fajar's presentation. Unlike the other students, Fajar highlighted his personal commitment to improving the environment (where others seemed to only participate in school-based activities), and explained his personal actions at home, with friends and in a leadership role in his school's environmental group. Fajar had clearly put significant effort into preparing his presentation: it had impressive visual effects, pictures of the environmental initiatives that he had introduced to his family, and photos of him undertaking many environmental projects at school and with TENGO. Where others gave the impression that all these environmental activities went smoothly (*lancar*) and without issue, Fajar provided examples of some of the difficulties that he had faced and how he had worked to overcome them. He had memorised his presentation and spoke in English throughout. The researcher congratulated Fajar on his efforts when they were chatting at the end of the day and commented that his presentation was really related to the theme and not just a list of things that his school does. He said that he had worked hard on it and had asked Mas Rehan for feedback. After applying and being unsuccessful the previous year, he said he was determined to get it right this year. Later the researcher asked Rehan what tips he had given Fajar. He said that he advised him to not just provide a list of activities and say that he was the leader, but to explain what the problems were and how he had worked to solve these problems, both at school and at home.

Titis

The presentation of one female, junior high school student highlighted a major shortcoming of TENGO's approach to EE. This student, Titis, gave a 12-minute presentation in Indonesian on climate change. She provided one jam-packed slide after another, describing in great detail the dangers and impacts of climate change. After reading each word on each slide, Titis was required to answer a question from the jury. The researcher asked her (in Indonesian) what actions she does for the environment. She replied that she conserves water and the like. The researcher then asked, "Is the relationship [*hubungan*] between climate change and your actions clear to you?" Titis replied, "No, not clear". She chose to leave it at that and so did the researcher. Titis was not an uninterested, disengaged student at the back of a classroom. She was a high-achieving, enthusiastic environmentalist in the eyes of her school and, presumably, TENGO. She had clearly spent many hours preparing many, many PowerPoint slides full of information and had probably spent hours practising reading every single word so that when her time came, she could read those scientific words and phrases with fluency. What she did not think to learn was the relationships, the connections and the meaning. Her environmental education (and possibly her general education) has failed her by not facilitating the ability to make the connections or think critically. She had a silo-ed, or part-understanding of the issue.

Gruenewald (2004) suggests that one of the central problems with environmental education is the widespread lack of connection between social analysis (analysis of human systems) and ecological analysis (analysis of ecosystems). The example of Titis is clear evidence that the EE these students had been exposed to had failed in helping these bright, concerned and enthusiastic learners to understand the interrelatedness of issues and humankind's active role in these issues. Due to a lack of environmental content in the curriculum, students learn very little about environmental issues. What they do know is whatever information that they have been given to learn for exams, which is always presented in a vacuum, with no context, no linking to other issues or to humankind's role in causing and enhancing issues. It appeared that almost every student knew of the environmental issues and that they must do a certain thing to help it, but there was no thinking beyond that. For example, students in Surabaya knew that flooding was a problem, and that rubbish causes floods, so people need to throw rubbish in the bin, not the river. But the bigger picture – the problems of over-consumption, of consumer waste, of packaging, landfill, upland farming causing erosion – were beyond them. Sometimes students were confused about environmental issues because their knowledge of science was weak or partial. For example, one student explained that he liked growing trees to get more oxygen for the ozone layer. Another senior high school student informed the researcher that if we have lots of trees, our air will get colder with lots of oxygen. In the case of Titis, the knowledge of the issue of climate change was there (copied and pasted on to PowerPoint slides at least) but the reasons for her own behaviour were a mystery.

It was clear that the 44 students were very proud of their environmental endeavours. They were happy to sort waste, make compost and tell others what to do (*sosialisasi*) but hardly any seemed to have given any thought to reducing waste. The researcher asked two or three of the better presenters about reducing waste and they answered by referring to their recycling programmes at school. Only one student made mention in his presentation of bringing a drink bottle to school and trying to reduce the amount of plastic packaging he used. When the researcher spoke about this with Mas Rehan he said that they all do it (reduce waste) but they did not mention it.

Despite her concerns at the presenters' demonstrated lack of understanding of the interconnectedness of environmental issues and human actions, the researcher enjoyed watching students proudly represent their schools and try their best to do so in a second language. She really appreciated these students, as it was the hard work of students such as these that has helped to turn Surabaya into a much more pleasant and healthy city. While the environmental education quality was questionable, there was no question that these students were trying very hard and for that they should be congratulated.

Selection process for principals and teachers

There were eight presentations by school principals, although one was presented by a teacher on behalf of her principal who was unable to attend. There were two male and six female presenters. Three of the principals presented in English, although one of these reverted to Indonesian at some points where she felt that she was struggling to convey her message clearly.

Twenty-two teachers presented on the final day of the selections, 18 female and four male. One male teacher apologised and admitted that he had Google-translated the whole presentation, so it might not make much sense. Most of the teachers switched between the two languages but ten used mostly English, three used about half and half and nine presented completely in Indonesian. A couple of teachers voiced their concern that others had not stuck to the time limits and that those presenting in Indonesian had an advantage as they could communicate more easily. As a judge, the researcher was not willing to select a winner to go to Perth if they were not brave enough to try to present in English, or if they could not stick to the rules (one teacher spoke for approximately 20 minutes before Mas Rudi interrupted her at the urging of the researcher).[8]

A young female principal of an Islamic primary school, Ibu Aliyah, gave a very emotional presentation that highlighted her mission to make her school an Islamic school that really cared about the environment, to represent Islam and to inspire other Islamic schools to join the Surabaya Eco Schools Program. She had tears as she proudly shared the achievements of her school and how in just two years they had become one of the top five Eco Schools in Surabaya. She acknowledged the hard work of her staff and pointed out two who were sitting in the audience. Like Fajar, she acknowledged that it had not been an easy road. When it was question time, the researcher suggested that others could learn from her and asked her what she could teach others if she went to Perth. She replied that, "You can get information from the internet. The difficult thing is how to motivate and include (others). It's like flying a kite, pulling and releasing. I could share that." Ibu Aliyah's presentation was refreshingly honest and heartfelt. She acknowledged the challenges and told how the team were working to overcome them. This open and honest presentation was unlike any of the others, and was perhaps enabled by her gender and comparative youth, and the fact that her school, an Islamic primary school, was at the bottom of the academic ladder. In this presentation, it seemed that, once again, female gender enabled an unconventional, freer approach that was more emotional and honest, though it may have been at the cost of status and regard (Keeler, 1990). Nevertheless, she won the principal's place in the tour.

Another principal took a different approach, presenting in a matter of fact way that made it seem as if bringing about change at the school was easy. She said that when the programme first started, the school had had some problems with parents not supporting it, but then, "We did socialisation [*sosialisiasi*] and now they all understand [*paham semua*]". The facile assumption that if one does socialisation, a program will run smoothly, reflects the New Order approach: in those authoritarian times, messages flowed top-down from the powerful to the powerless, and internalisation and obedience were presumed. In contemporary, democratic times, the term "socialisation" is still often used that way by government officials and teachers, but sometimes, when being "socialised", the recipients of the messages demand proper consultation and fulsome information (Gibbings, 2016).

When the researcher asked this principal what she thought she could learn, or teach others, by doing the CCE to Perth, the principal replied that she hoped that she could represent Surabaya well. Reliance on superficial, oversimplified and well-rehearsed answers was common amongst principals and teachers generally. The aspiration to represent oneself, one's school or even one's family, village, city or country in a positive way was an accepted truism that outweighed the value of an individual, thoughtful answer (Long, 2007). One interpretation of this answer is that recourse to a set phrase obviates the need to think creatively or critically – it is a form of mental laziness. One could compare the two principals' answers and see them as symbolising two value systems, the, latter prioritising the smooth appearance, the concern with the "look" of something, and the desirable avoidance of negatives such as conflict or criticism, and the former valorising individual effort and thought.

For most of the teachers, the idea of highlighting and discussing difficulties and failures would be rather shameful. When the representative from the Office of Parks and Sanitation asked a teacher about the school's efforts to reduce waste, instead of outlining the efforts, or even the future plans, the male teacher lowered his gaze and apologised, "I'm sorry, Sir, it's not optimal." This response was somewhat unusual in the context of a competition, where the participants were supposed to be convincing judges of their achievements and merits in order to win the trip to Australia. The response was, however, very Javanese.

The male teacher was speaking to a man of higher social standing (a high-ranking government official), who had been somewhat belligerent in his questioning of other teachers, often turning away from the speaker to display his dissatisfaction with answers. The researcher was quite horrified when, after asking another teacher a question, the official answered a phone call and held a conversation while the teacher was still speaking. Over the course of the morning, this man had placed himself above everybody else in the room with his way of speaking and body language. The male teacher was acknowledging the government official's position in his answer. By admitting that the school's waste reduction efforts were not optimal, he saved himself the embarrassment of being told so by the government official.

This reluctance to discuss difficulties and failures prevents environmental teachers from being able to work together to provide advice and support to one another. It seems quite likely that the competitive nature of the Surabaya Eco Schools Program makes teachers reluctant to highlight failures and challenges in their schools, and to provide assistance to others. There are models of other, more constructive groups for teachers: for instance, in some districts there are district-wide "teachers' work groups", which are positive opportunities for collegial professional development (Parker & Raihani, 2011). Such horizontal networks stimulate and empower teachers, and refresh those who have been teaching for a long time.

Performing for prestige

The TENGO EE programme, Surabaya Eco Schools, is very popular in Surabayan schools. While the Adiwiyata Programme offers schools the opportunity to gain prestige (*prestasi*) by moving up the Adiwiyata ranks, the TENGO programme offers visibility, and potentially distinction, for schools and individuals through its relentless social media campaigns and strong links to government. In addition to this, it offers cash prizes, environmental resources and the chance to travel to Australia. Mas Rudi understands that most of the adult participants are not motivated by an intrinsic concern for the environment, but by reward and opportunity. For those who are intrinsically motivated by concern for the environment, such as Ibu Aliyah, the reward and opportunity that come with the programme are bonuses. The recognition of one's work is always appreciated. Although the titles (Eco teacher, Eco principal, Eco-preneur) are won as a result of one's environmental merits, Mas Rudi certainly manages the rewards and is very strategic and thoughtful about who wins what when.

Whilst having lunch with Mas Rudi after the CCE Selections, the researcher asked him if he felt bad for a particular principal and teacher who had been long-serving environmental educators but, despite their amazing efforts, were not selected to go to Australia that year. He said, "Yes, of course. They have worked so hard for many years and will be very disappointed." She asked if he would explain to them why they were not selected. "No. No," he said. "There's too many to explain. It'd be never-ending. We don't talk about it. They'll get other rewards." When the researcher had commented earlier in the day that she thought a particular principal had done well in his presentation, Mas Rudi replied, "Yes, but he's Eco Principal of the Year. That's a big reward."

It seemed that the main driver of the Surabaya EE programmes was competition for prizes and rewards. There was a densely-woven net of institutionalised prizes – for student participants, teachers and schools, for each level of school, for individuals as well as groups, for different roles and tasks ("eco-preneur", plastics free, eco-teacher, eco-principal) and with different time scales (weekly, monthly, annual).[9] There were also many minor rewards that were less institutionalised on offer for schools. These included a free visit and incursion from the Eco-mobile; a good word to officials at the Education Office (which may aid one's career progression); selection to speak on television or radio; a visit from the researcher or the American Consul General's wife; and photographs in the local paper. Many of the teachers and principals were not afraid to ask for visits. Many a time a teacher asked the researcher or Mas Rudi when they would come and visit their school.

Arguably the most rewarded and celebrated Eco student in Surabaya was Nadine, daughter of TENGO's biggest private sponsor and supporter, Pak Edo. Nadine's primary school, the *favorit* public primary school in Surabaya, was TENGO's premier Eco School. At the time of fieldwork, Nadine had graduated to junior high school and her younger sister was attending the same primary school. When she attended this school, in 2012, Nadine was crowned Princess of the Environment (*Putri Lingkungan* – see Figure 10.2). She travelled to Australia as the winner, appeared on local television and radio and in newspapers prior to and after her return, and was often featured in TENGO's social media posts. When TENGO was awarded the Kalpataru Award by the President at the Presidential Palace in 2015, Nadine was invited by Mas Rudi to come and meet the President.

Most of these competitions were more acknowledgements of effort and ways to encourage participation, and hence a sense of belonging, than competitions in the traditional sense of "finding the best performer" (Long, 2013, pp. 132–134). Mas Rudi explained that the competitions were very important to keep everyone motivated. We would add that it is not only participation in the competitions that motivates the teachers and principals, but also the opportunity to obtain prestige. When someone wins a competition (even something as small as Eco teacher of the week), they have their photograph posted onto Facebook, Instagram, Twitter and the TENGO websites. Students find themselves being noticed, named and acknowledged by teachers. The Eco teachers and principals know one another and know of each other's accomplishments. Government officials are usually invited to be on the jury of the main competitions and Bu Risma often presents the awards to the winners of the two main competitions. In a city that values environmental action, being known as a green teacher or principal is good for one's career. As Long summarises, winning *prestasi* enhances social capital (Long, 2013, p. 183).

Figure 10.2 Princess of the Environment.
Photo credit: Danau Tanu.

Cross-Cultural Exchange to Perth

The CCE team selected to go to Australia was made up of all of the researcher's preferred applicants from the CCE Selection Competition, with one exception. The exception was a boy from junior high school, Agung. The two teachers and principal who were on the exchange all complained to the researcher at different times and asked how it was that Agung had been selected to come along. The researcher had to agree with them that he was not ready for such an experience and explained that Mas Rudi had made that selection against her recommendation. The other winners were Fajar (senior high school student), Pak Yusuf (male primary school teacher), Ibu Lani (female senior high school teacher) and Ibu Aliyah (Islamic primary school principal) and they were accompanied by Mas Budi and Mas Rehan from TENGO, and Mas Aan, representing the trip sponsor (PES). The researcher arrived in Perth before the group, and met them on their first night at Kit and her partner John's house when they arrived. She spent most days with the group, acting as translator, cultural advisor and participant observer.

There were various cultural difficulties that left Kit and John, and the CCE group bemused and sometimes annoyed at one another, and the researcher often had to explain the position and actions of one to the other. She was horrified on the first night to see Ibu Aliyah refuse to shake hands with John,[10] the owner of the house where she would stay for the week; and equally horrified to learn that the group was not taken to the airport at the end of the stay but instead were given the phone number of a taxi company. There were many small gripes in-between, but Kit and John and the visitors were all very gracious in accepting that these were cultural differences and were to be expected and accepted. Both groups could have done with a few lessons in the basic courtesies of the other's culture, and this would have avoided some of the tension over the week.

Kit had had the expectation that the group would actively participate in the events and activities during their stay. It became evident over the course of the week that some (like Ibu Aliyah) were more enthusiastic than others. Where Kit saw this visit as a learning and training opportunity, the visitors seemed to view it more as a trip to Perth where they could passively observe the work of others. Without question, the lack of English skills amongst the group made participation difficult and Kit's lack of Indonesian language and cultural knowledge did not help either. Mas Rehan and Budi were always busy documenting the group's moves by videoing and taking photographs or sitting away from the group typing up reports and social media posts. Despite these difficulties, the group all reported that they were inspired by the trip and had learnt some important things to take back to Surabaya. Most of the group commented that they were surprised that each project began with a question from a student at the KfC conference. Fajar said in a recorded interview,

> Yeah, this is so, so different to Indonesia. In Indonesia the problem is already defined, we just have to find the solution – that's the hard bit.

We have the problem, but here [in Australia], which we could say is generally good, it's only how to maintain this good situation. So here, we learn
while we discover. In Indonesia we learn what has already been discovered.

(Fajar, interview in Perth, 5 June 2015)

Arguably Fajar gained more insight than anyone else in the group, and was able
to express it clearly. However, we would disagree with him that students in
Indonesia have to find the solution: they are provided with the solution (clean
up the beach, make the compost, sort the rubbish) and "just" have to do the
work. Mas Rehan and Ibu Aliyah said that they had learnt that "Kids make a
change by creating a simple question and finding an adult to help connect their
ideas," and "It becomes a different spirit when it comes from within, not just
from the principal or teacher". The CCE participants certainly saw a different
way of facilitating EE, or better, environmental learning. The challenge for this
group was whether this approach to EE could be adapted to their Surabaya
setting. Ibu Aliyah was very excited by the idea but at the same time felt that it
would be very challenging to convince her staff and parents of the value of it.

A doomsday approach to an uncooperative school

Back in Surabaya, it was Mas Rudi who decided which schools the researcher
would visit, though she had asked to see a variety of schools – those just starting
out, those that had well-established programmes, as well as some *favorit* schools
and some not-so-well-resourced schools, public and private, general senior high
schools and vocational senior high schools. She visited 15 schools during the
course of her field research in Surabaya and spoke to approximately 100 students
at schools and TENGO events. She made some visits in order to conduct FGDs,
and other visits were with the Eco-mobile to observe how TENGO conducted
its school incursions. All but two of the 15 schools that she visited were already
participating in the TENGO programme. The visit to one of the non-
participating schools was particularly insightful because it was a school where
the school administration was not interested.

Prior to arriving at the school, TENGO had not told her that they were not
currently working with the school, nor that TENGO had been finding it difficult to get into the school. Therefore, they had not sought permission to visit
prior to their arrival. That day, the researcher was visiting schools with a
Research Associate from UWA who was also collecting data for the same
project. It soon became apparent that they were not going to be unobtrusively
observing a normal TENGO Eco-mobile visit.

Usually, because visits were pre-arranged, the Eco-mobile was able to drive
straight in and park in a pre-arranged position and the students rushed out to it
with much excitement. In this particular visit, the Eco-mobile was parked
outside the school gates and Mas Rehan invited the two researchers to enter
first with him, leaving the Eco-mobile driver and another TENGO colleague
outside, waiting. Mas Rehan and the two researchers went to the office where

they sat and waited to meet the principal. The male principal and female vice-principal greeted them and enquired why they were there. To their surprise, Mas Rehan started explaining about the PhD research, making it the focus of the visit. He told the principal that they were there to collect data on environmental education and could they please meet with the environmental cadets. The principal quite properly replied that if they were there to collect data, they needed permission. The researcher told the principal that she had permission and showed him the official paperwork which he showed the vice-principal and asked if he could keep. They all made small talk while the vice-principal went to make a photocopy. At this point the researchers were still rather confused as to why they were there and what exactly they would be observing. As the mood seemed to have lightened, Mas Rehan asked permission to bring the Eco-mobile onto the school grounds. The principal agreed and asked which students they needed to talk to. Rehan asked to talk to the student leaders (OSIS) and Emilia, a student who had been involved with TENGO when in junior high school.

Mas Rehan's usual jovial and familiar manner that he used with environmental students was absent and instead he took a sombre approach. He waited silently while the 21 OSIS students plus Emilia gathered in front of him. They sat in a semicircle in the shade while he stood in the sun with a microphone in his hand and the back of the Eco-mobile behind him with the back door open, showing the large television screen and audio system. He started by asking the students about their predictions for the future. The students replied that the situation will get worse, get hotter, and there will be less water. Rehan informed the students that he would show them a video made by a child[11] that predicted the future of the environment. The video was alarmist and portrayed an environmental doomsday. There were mostly images, many artists' impressions of drought and suffering, with plastic trees, and close-up shots of sick and distressed people. There was some text on the screen but no speaking, just emotive music, which was very effectively broadcast though TENGO's audio system. One of the messages that moved across the screen stated that men and women's reproductive organs would be altered. The video ended with the song, "When the Children Cry", by White Lion. Rehan knelt down in front of the silent students and told them, "That is only a prediction". He then asked the students if they were doing enough for the environment. They answered, no, not yet (*belum*), in subdued voices. He told the students that, "We won't tell you what to do, but we will help you, as friends". He told them that, "We can help you, but it's up to you. We can come back and come back, but in-between it's up to you". The students seemed uncomfortable, mostly looking down silently, but clearly listening and thinking about what Rehan was saying. Rehan went on to talk about water cups and plastic before stating that he and the researchers were available to help them. This was simply untrue. The researcher was leaving Surabaya that evening and her colleague was only going to be in Surabaya for a couple more weeks. He showed the students how to set up *takaura* (a form of composting) before having them draw up a brief action plan. He pushed them for dates as to when they would have things done. He told them that their principal was nice, easy to lobby (*enak, gampang dilobby*). Having scared the students with doomsday predictions for the future, Rehan took the mobile phone number of the leader of the group, promised to be in touch, packed up and left.

Rehan explained later that he had to focus on the students as the teachers were not particularly interested in the programme. The researcher emailed Emelia twice to see how they were but there was no positive news. Emilia said that she was frustrated by the lack of action. The researcher asked Rehan to return to the school with her a few months later to find out how they had been progressing and to speak to the students about their experiences in trying to introduce the programme to the school. Mas Rehan said that Mas Rudi had said no, TENGO would not go back there as the school (students) had not done enough to deserve a visit.

The approach taken by TENGO with this school was most unfortunate and irregular. TENGO entered the school under the "cover" of the researcher's government permission to undertake research. There had been no introductions of the researchers, or explanations as to why the students had been pulled out of class; neither the students nor the researchers had been treated respectfully and ethically. The students were not apprised of the purpose of the research, or given the usual Information Sheet and Consent Forms. The session in front of the Eco-mobile had none of the characteristics of EE or EfS: there was no critical thinking, envisioning, systemic thinking, making of connections between social and ecological systems, problem-solving or joint decision-making. The approach was alarmist and burdensome, not unlike that used in the community school in Yogyakarta (Chapter 8), with false promises of support. It is not surprising that it was unsuccessful.

TENGO's lack of success at this school suggests that its approach does require the *paksarela* (forced volunteering) of teachers, or for the school's leaders to co-opt a teacher who would shepherd the students, in cooperation with TENGO, towards full participation in the Eco School programme (or Adiwiyata). In this school, the students were left entirely on their own to initiate action – it is no wonder they floundered. This negative example shows, by contrast, that TENGO's usual approach of working with government agencies and schools is successful. TENGO and the students still need the structure and support of the school institution.

There are many factors that influence engagement or non-engagement in environmental learning and environmental action, and TENGO consistently failed to understand and address this complexity. Two of the main factors that were not addressed by TENGO are enabling students to choose how and when to engage, as Fajar pointed out; and the need to develop the capacity for appropriate and effective action in students (developing reflective, relational and transformative agency) (Stevenson & Dillon, 2010). Many factors must be taken into account: emotions, values, facilities, peer influence as well as levels of knowledge and understanding (Nagel, 2005; Prabawa-Sear & Baudains, 2011; Rickinson & Lundholm, 2010; Watts & Alsop, 1997). There is a significant body of literature that examines the difficulties associated with addressing these factors in North American, Western European and Australian contexts, but almost no literature that examines these issues in Indonesia.

Virtues and vices of "forced volunteering"

Mas Rudi, government officials, teachers and principals all mentioned at various times that Surabaya employed a technique of *paksarela (paksa sukarela)*, forced volunteering. Mas Rudi explained that the idea was that if you make someone do something over and over, it becomes routine and then a natural behaviour for them. Government officials also agreed that this forced volunteering approach would lead to a realisation (*kesadaran*) of the importance and value of environmental actions, which would translate into long-term behaviour change. Until such time as the people of Surabaya achieved this "realisation", the Surabayan government would utilise this approach of *paksarela*.

While this long-term goal might well be seen as valid and useful from the point of view of the environment, and, as mooted in the previous chapter, as the beginnings of environmentality, the fact that most teachers and many students did not volunteer for their environmental activities, and students were never paid for their work, is problematic. However, the idea that this constituted forced labour, or exploitation, was not advanced by any of the student (or teacher) participants. And it must be stated that for many students, working together in mixed-gender environmental clubs was a lot of fun (Tanu & Parker, 2018). The lack of acknowledgement does not mean it is unproblematic – it simply shows the ubiquity and normality of unpaid volunteering in Indonesia.

While there may be some teachers and students in Surabaya who see the "forced volunteering", or forced participation, in the Eco Schools Programme as a negative, most would more likely think to acknowledge the positive outcomes that result from the programme, which benefit the whole community. Indonesia has a long-standing tradition of *kerja bakti* (volunteer service/working together), *koperasi* (co-operatives) and *gotong royong* (mutual assistance) for the betterment of the community and nation (Bowen, 1986). All schools and most government departments still have at least one hour a week allocated to *kerja bakti* where students and employees are expected to clean the premises and gardens. Many neighbourhoods still hold *kerja bakti* sessions when the men come together to cut weeds, paint public buildings and remove dumped rubbish. Most universities make it compulsory for students to serve for two months in community service (KKN, or *Kuliah Kerja Nyata*). They are unpaid but do accrue credits towards their degrees. Further, many kindergartens are staffed by "volunteer" women from the neighbourhood (Newberry, 2014). The idea of working together as unpaid labour is something that is very normal and accepted as an important part of community life and it seems that the *paksarela* approach to EE in Surabaya is a natural extension of this.

The head of the Department for Parks and Sanitation pointed out that the people of Surabaya have a high level of public participation (in the environmental programmes) and a high level of public awareness. He said, "They are not playing around. They want to improve their neighbourhoods, they want to straighten out the rubbish situation, they want to run large-scale activities, and schools are the same. They want everything clean too." A teacher from a lower class school in the far north of the city explained how the school relied on the direction of TENGO to progress their EE programme. The school received instruction from one week to the next which helped them to transform part of their school grounds from a rubbish dumping site to a vegetable garden. This teacher suggested that this guidance was a very important part of the programme and made particular mention of the importance of having young mentors from TENGO, such as Mas Rehan, with whom the students felt comfortable and quickly developed a close relationship.

Over the course of fieldwork the researcher had many conversations with educators and students that suggest that they felt comfortable with and appreciative of the *paksarela* approach. The female principal at Surabaya's most *favorit* junior high school recalled how, when she was principal of another school (in an industrial area), the Mayor, Bu Risma, came to the school and told the principal that her school must become an environmental school. In recounting the story, the principal showed no sign of resentment and in fact spoke in an admiring fashion about Bu Risma who saw an issue and acted on it. The principal was very proud of the excellent progress that the school had made both on the school grounds and with the wider community. While some teachers seemed apathetic and reluctant to be involved in the environmental programs at their schools, we believe this was as a result of seeing themselves as government employees rather than as educators and their reluctance to take on any additional workload or responsibility. Because our interviews and focus group discussions were conducted with teachers and students who were involved in the programme, we did not have much access to those who were not willing participants (such as those students who threw rubbish at their cadet classmates, or taunted them as smelly rubbish-collectors, or those teachers who complained about the high cost of environmental projects or refused to turn lights off). The students, however, seemed completely content to follow the tasks set out for them by TENGO. This is not particularly surprising as Javanese children are taught from birth to do as they are told and to accept the situation for what it is (*terimo*) both inside and outside of the classroom (Geertz, 1959).

Botero, Fediuk, and Sies (2013) argue that in the context of volunteering, individuals will be likely to have future intentions to volunteer only when they feel that volunteering is a positive behaviour; that requiring individuals to volunteer may have negative effects; and requiring students to volunteer can lead to a negative change in attitudes toward volunteering. Contrary to the theory put forward by Mas Rudi and the government officials, Botero *et al.* (2013, p. 313) suggest that "through forcing individuals to engage in volunteerism, the dedication, willingness, even appreciation for service-related behaviour may not be developed, and long-term volunteerism may be negatively impacted as a result". We suspect that in Indonesia, given the normality of *kerja bakti* and other forms of volunteering, the use of "forced volunteering" in and around schools will not in itself have negative effects. Our concern, rather, is that, beyond schools, there is no structure – the governmentality of the environment that Agrawal saw in India – to authorise and motivate environmental work in the community, and that the necessary knowledge and understanding about environmental issues is not there to initiate effective action. If there were, at the level of the local neighbourhood, we hypothesise that the social pressure of shared responsibility for the village or kampung may well kick in, and house-holders would be active in working communally on pollution eradication, waste management, composting and the like.

In her study of mandated service in high schools in Maryland, USA, Helms (2013, p. 308) found that the volunteer mandate did not have the intended effect of promoting lifelong volunteers. She found that while students exhibited higher levels of volunteering early in their secondary school years, after the mandate was introduced volunteering levels of twelfth-grade students dropped, leading her to conclude that one of the major goals of the programme – to inspire a lifetime of service – did not extend through the final years of high school.

At every school that the researcher visited, the Grade 12 students were pro-hibited from joining the environmental programme. This was despite the Sura-baya Eco Schools programme aiming to inspire long-term commitment to the environment. Time and time again she was told that the Grade 12 students had to focus on their exams,[12] even in cases where students wanted to continue to be involved in the programme. It was accepted as a general rule by teachers and principals that students had to focus on academic achievement in Grade 12 (with the assumption that students were not capable of doing both). This banning of participation (which was also the case in Yogyakarta) gave a very strong message that environmental actions are not educationally valuable and that one can pick and choose when one is environmentally responsible. Argu-ably this parallels the idea that the lower class should clean up their neighbour-hoods first and sellers at the traditional market should be targeted to compost waste, while multinational companies and the most elite in society are beyond environmental responsibility.

While visiting one of the best Eco Schools in Surabaya (which is also a *favorit* school), the researcher noted how enthusiastic the students were to show her all of the projects they had in place. They had separated waste, composting, a greenhouse, *biopori*, aquaponics, hydroponics and various other projects going. This impressive show of environmentally-related projects (not necessarily *for* the environment, but *about* the environment) suggested that this school took its environmental actions seriously, yet she was given water to drink in a small plastic single-serve Aqua cup with straw wrapped in plastic, even though she had brought her own re-fillable drink bottle. As they moved about the green-house, she noted to the students that perhaps they might consider not using black plastic (single use) polybags to grow seedlings but could re-use items that they already had at the school (such as Aqua cups). They students showed no interest in the suggestion and did not seem to register that there was a possibility not only to reduce the use of the hundreds of polybags, but also to re-use the plastic cups. It was quite clear that these students were doing their environmental projects exactly as instructed by TENGO and their teachers, and this was just fine by them. This was just one of numerous occasions where it was clear that TENGO (Mas Rudi) did the thinking around environmental projects and the schools carried them out.

All of which shows that the environmental work being carried out in and around schools in Surabaya is not so much environmental *education* as unpaid, community-focused, environmental *labour*. In an article about the difference between activism and volunteering, Eliasoph argues that "Usually, volunteers do not routinely question the roots of the problems they aim to solve, but just try to get in there, hands on, directly, to solve the problem, not necessarily caring about its source" (Eliasoph, 2013, p. 66). This aspect of volunteering is antithetical to the goals of EE, but it does help to explain the apparent lack of interest in and understanding of the environment among the young people of Surabaya. Eliasoph also noted that volunteers' "can-do spirit would collapse if they talked about the bigger issues" (Eliasoph, 2013, p. 83). Our larger project revealed an instance like this. In Palembang, South Sumatra, the university was shrouded in smoke every day; staff and students were finding it hard to breathe, and coughing, with eyes streaming, from the smoke of fires illegally lit to burn native forests to make way for palm oil plantations. Rather than investigate the iniquities of the palm oil industry, the main activity of the university students in the local nature club was picking up litter on campus (Nilan, 2018). The lack of engagement with the major environmental crisis of forest fires was likely due to the feeling that it was "too hard" to tackle the corruption and criminality of the forest fires problem, even though the environmentalists' discourse normalised the problem as "just ordinary life" (Nilan, 2018, p. 337).

Conclusion

This chapter has presented in detail the practices and processes used by TENGO and the Surabayan Government in their efforts to co-opt students into transforming Surabaya into a clean, green city. The dominant approach to EE in Surabaya has been to implement an unwritten policy of forced participation in the Surabaya Eco School EE programme. The Surabaya Eco Schools Programme and the Adiwiyata Programme are both supported by a plethora of competitions and offers of rewards and prizes. This chapter presented details of this competitive approach to EE and some of the benefits and shortcomings of such an approach. The abundance of contests and prizes is not restricted to our field of EE – they are to be found in fields as diverse as beauty pageants and Quranic recitation. The many competitions share a preoccupation with participation and competition in order that participants accumulate both prestige (*prestasi*) and social capital. The result is unfortunate for EE, because it means that the focus of these two programmes is not on learning or producing environmental awareness, understanding or responsible environmental behaviour, but on winning (Prabawa-Sear, 2018).

The chapter presented some possible negative aspects and effects of forced volunteerism that have been identified in Western research, and considered them in light of the data and the normality of unpaid volunteering in this non-Western, Indo-Javanese context. The negative example of the school where *paksarela* by teachers and students alike has not been implemented, suggests that, given the poor quality of the education system, the *paksarela* approach has been realistic. Many of the issues around a lack of critical thinking and lack of understanding of the interconnectedness of environmental issues and human behaviour are not issues specific to EE but are issues related to the wider education system in Indonesia: the curriculum and textbooks; commonly used pedagogies that emphasise passive regurgitation and not active or deep learning; insufficient training of teachers; teachers' perceptions of themselves as civil servants first and teachers second; and a preoccupation with examination results.

While Surabayans want a clean, green city, and appreciate the efforts made so far, the *paksarela* approach will need to alter norms and beliefs as well as actions if it is to achieve a realisation of the importance of environmental behaviours. If it failed to achieve this, the need for *paksarela* would never end. So far, the efforts of students, as targets of compulsory EE, have been successful as environmental labour rather than as environmental education. Students – even those who are eco-champions in the eco-champion schools – are not yet capable, responsibilised environmental citizens.

In the previous chapter, we discussed the Surabayan Government and TENGO's approach of placing the responsibility for behaviour change on those with the least voice and power in society (children and, in the newest project, the urban poor in traditional markets). This approach was found to be problematic both in regard to internationally promoted principles and characteristics of EE/ESD/EfS and Javanese societal expectations around the position of children and the importance of respect for elders and those of higher social standing. Nevertheless, Surabaya is yet to see engagement with waste management and environmental restitution from groups outside of the city government, schools and NGOs.

These two very different ideas about the role and position of children not only do not fit but clash when brought together. This clash was voiced by Fajar and other students who felt that adults and their peers did not listen to them or support their environmental actions. Unlike the students, the uncritical teachers, TENGO staff and government officials all reported that TENGO's approaches were working well.

It remains to be seen whether the hoped-for responsibilisation for the environment is taken up by teachers and students once the forced volunteering structure is no longer present.

Despite the focus on processes rather than outcomes in this and the preceding chapter, there were some very obvious positive environmental outcomes that resulted from the Eco Schools and Adiwiyata Programme and these were acknowledged. Surabaya and Surabayan schools are an example to other cities across Indonesia in the way that they have brought about positive environmental change.

We asked whether the approach in Surabaya might be seen as "environmentality". As described above, environmentality is the academic term, coined by Agrawal and derived from Foucault's notion of governmentality, wherein the environment is made conceptually separate from human society and constructed as an object to be measured and regulated. The objectification of the environment thus renders it manageable, by governments, through laws and regulations, but also by individuals, and environmental discourses, such that "technologies of self and power are involved in the creation of new subjects concerned about the environment" (Agrawal, 2005b, p. 166). We can see how this happened with anthropogenic climate change: at first, a phenomenon known only by a few scientists and environmentalists, climate change has come to be an issue of global concern and measurement, to dominate news broadcasts and election campaigns, to the extent that we now have huge class actions wherein populations sue their governments for not taking the action required to meet UN-required climate change commitments.[13] Thus, environmentality refers to the "creation of environmental subjects along with the emergence of the 'environment' as a domain that requires regulation and protection" (Bauer, 2005, p. 116).

The green movement in Global North countries – First World environmentalism – is usually said to have begun at the grass-roots level, with motivated "greenies", usually from the educated middle classes, applying pressure to government to act via environmental organisations and lobby groups. Inglehart's "post-material" thesis, based on a survey of values in 43 countries, is one of the earliest and best-known expositions of this evolution: that people only become environmentalists once basic material needs have been met (Inglehart, 1995). There is considerable evidence that the pattern is different in Global South countries, where quite often conflicts over natural resources foster environmentalism, but it is a local and ephemeral environmentalism based on the interests of local farmers, peasants and workers, sometimes led by urban activists (e.g. Gadjil & Guha, 1994; Meek, 2015).

Various scholars have argued that in the Global South, environmentalism tends to be issue-based, local and ephemeral, as Development encroaches on poor people's livelihoods and resources (e.g. Kalland & Persoon, 1998, pp. 6–18). Gadjil and Guha argue that India's environmental movement includes poor individuals and communities that have become involved in the green movement as a result of their own problems of land and resource depletion (notably of forests and water), pollution, and the decimation of biological diversity, often protesting *against the state*, from colonial times onwards (Gadjil & Guha, 1994). In contrast, Agrawal has argued the environmentality thesis: that wholesale participation of rural villagers in government-imposed programmes focused on environmental protection transforms people into "environmental subjects" in rural populations (Agrawal, 2005a, 2005b).

Referring back to Birkenholtz and his study of the management of groundwater resources in Rajasthan, we would argue that in Surabaya, as in Rajasthan, two precepts of environmentality have largely not materialised: (1) these new ideas, knowledge and rules about the environment are most effective when produced and enforced at the local level; and (2) there is a need for these to be implemented through the simultaneous creation of environmental subjects who willingly monitor and enforce. After our study of EE in Surabaya, we conclude that students have not yet become "responsibilised environmental subject citizens". The way that EE is being done in Surabaya, the environmental problem and its solution are pre-defined: all the students have to do is the "grunt work". Educationally, as Fajar noted, "we learn what has already been discovered".

Surabaya is a unique example in Indonesia, in that the government set the green agenda. Surabaya has enjoyed a good run of >7 per cent per annum growth in GDP since 1998; there are many wealthy residents, and the middle class, whose material needs have largely been satisfied, is burgeoning. But there has been no interest in the environment among this cohort in Surabaya. The desire to make Surabaya green was not born of affluence, education and "greenie" environmentalists, nor of environmental destruction through development, nor of green governmentality, but because the Mayor and the government saw value in having a clean, green city. Pushed by Bu Risma, government began working with NGOs and schools to bring about change in the community. So while "the environment" has definitely become a subject of interest and management by the government, as have groundwater resources in Rajasthan, knowledge about how to manage it has been top-down. The Rajasthan and Surabaya governments share a paternalistic "government knows best" attitude, implying that the current situation "is the people's fault and if the state can just get people to become conscious and take responsibility then the problem can be solved." "But on the other hand, [in Rajasthan] the state does not integrate people into the regulatory decision making process" (Birkenholtz, 2008, p. 91) and in Surabaya the government must force people to implement the policies that it, and TENGO, have designed.

While Agrawal acknowledges "variations in the transformation" of subjects (Agrawal, 2005a, p. 185), Surabaya's approach to EE and to addressing environmental issues is largely to place the burden of change on those with the least voice: children and people in the lower socioeconomic classes. The hoped-for movement *upwards* from the powerless to the powerful is radical and contrary to the traditional flow of power and authority in Java and Indonesia. It is not obvious that the modelling of good behaviour by the powerless will motivate those of higher social standing to change their environmental behaviours, in order not to be embarrassed or shamed.

With this strategy, TENGO and the Environmental Agency are avoiding taking on the challenge of the wealthy and powerful in society, and thereby avoiding confrontation. Instead, they focus on the least powerful in society who are not in a position to question or resist directives (whether it be from government or a school principal).

Surabaya's approach to environmental change seems not only to perpetuate a system that creates inequality, but to place the burden and blame on those most vulnerable to the negative consequences of continual development that simultaneously benefits the city's wealthiest.

Notes

1 These competitions are independent and not associated with the Adiwiyata Programme.
2 The researcher was unclear about how her criterion of "music" differed from Mas Rudi's criterion of "compilation". The criteria did not seem of much importance to Mas Rudi.
3 This is an important aspect of any EE programme as it is often the case that environmental teachers and students feel unsupported in their efforts.
4 While this would not be a criterion of success in EE generally, it appeared to be something that was very important in this competition. Students and teachers chanted over and over again the same messages.
5 While TENGO called it a "cross-cultural exchange", this title referred to travel to Perth, not to any ethos, values or principles of a cross-cultural exchange. There was no exchange of people. Also, there was no pre-trip cultural training and no post-trip debriefing and little evidence of cross-cultural learning by either side during the trip.
6 In Indonesian competitions, judging and judges are widely held to be suspect. Long (2013) reports, for example, one student who claimed that "the judges are all corrupt or incompetent anyway – it's because they're Indonesian judges – so who wins is really quite random" (Long, 2013, p. 182).
7 Mas Rudi's daughter did not come on the trip that year.
8 At the end of the teacher selection day, the judges' table was filled with items brought in by the teachers for the judges to try: mostly food and drink made by the schools out of produce grown in the school gardens. Items included fried spinach crackers, fruit syrup, ice-cream and bananas. There were also bags of compost and various items made from re-used materials.
9 One can get an idea of this from the TENGO website. For reasons of Ethics commitment to anonymity, this website cannot be cited. The website is kept up-to-date with all the "success stories", e.g. a primary school wins a prize for being "plastic free" (though there is plenty of plastic visible in the accompanying photographic documentation); there is a "best poster" competition; there is a clean-up-rubbish hand-puppet competition (though the connection between puppets and cleaning rubbish in the park is hard to fathom).
10 While shaking hands is an everyday practice for most Muslims in Indonesia, some will refuse to touch (including shaking the hand of) a non-family member of the opposite sex as it is said to be not permissible according to the teachings of Islam (as it may lead to temptation). The researcher later spoke to Ibu Aliyah about this cultural difference and suggested that if she was going to refuse to shake hands with men, she would need to explain to them why. No doubt this was an uncomfortable proposition for her and she was careful to wear gloves thereafter, thereby avoiding the uncomfortable situation. She shook hands with a male principal at a school they visited a few days later, wearing gloves.
11 There was nothing about this film to suggest a child was involved in its production. It was a very polished, adult-like video.
12 This is the case with non-environmental extras too. For instance, the second author conducted research in schools on adolescents, in West Sumatra, Jakarta, Yogyakarta and Bali, and invariably was directed away from Grade 12 classes, as, it was said, the students would be distracted from their exam study.

13 The first case was one in which 886 Dutch citizens sued their government for contributing to dangerous climate change. In 2015, the District Court of The Hague ruled that the government must cut its greenhouse gas emissions by at least 25 per cent by the end of 2020 (compared with 1990 levels); the ruling was appealed and is still being heard at time of writing (www.urgenda.nl/en/themas/climate-case/). Similar cases have been brought in countries as diverse as New Zealand and Pakistan. See www.climate liabilitynews.org/2018/03/07/climate-accountability-uk-paris-agreement-plan-b/).

References

Agrawal, A. (2005a). *Environmentality: Technologies of Government and the Making of Subjects*. Durham: Duke University Press.

Agrawal, A. (2005b). Environmentality: Community, intimate government, and the making of environmental subjects in Kumaon, India. *Current Anthropology*, 46(2), 161–190.

Bauer, J. (2005). Environmentality: Technologies of government and the making of subjects. *Ethics & International Affairs*, 19(3), 116–118.

Birkenholtz, T. L. (2008). "Environmentality" in Rajasthan's groundwater sector: Divergent environmental knowledges and subjectivities. In M. K. Goodman, M. Boykoff, & K. Evered (Eds), *Contentious Geographies: Environmental Knowledge, Meaning, Scale* (pp. 81–96). Aldershot, UK: Routledge.

Botero, I. C., Fediuk, T. A., & Sies, K. M. (2013). When volunteering is no longer voluntary: Assessing the impact of student forced volunteerism on future intentions to volunteer. In M. W. Kramer, L. K. Lewis, & L. M. Gossett (Eds), *Volunteering and Communication: Studies from Multiple Contexts* (pp. 297–318). New York: Peter Lang.

Bowen, J. R. (1986). On the political construction of tradition: Gotong Royong in Indonesia. *The Journal of Asian Studies*, 45(3), 545–561. doi:10.2307/2056530.

Creese, H. (2014). The Utsawa Dharma Gita competition: The contemporary evolution of Hindu textual singing in Indonesia. *The Journal of Hindu Studies*, 7(2), 296–322. doi:10.1093/jhs/hiu026.

Eliasoph, N. (2013). *Politics of Volunteering*. Wiley ebook.

Gade, A. M. (2004). *Perfection makes Practice: Learning, Emotion, and the Recited Qur'an in Indonesia*. Honolulu: University of Hawai'i Press.

Gadjil, M., & Guha, R. (1994). Ecological conflicts and the environmental movement in India. *Development and Change*, 25(1), 101–136. doi:10.1111/j.1467-7660.1994. tb00511.x.

Geertz, H. (1959). The vocabulary of emotion: A study of Javanese socialization processes. *Psychiatry*, 22(3), 225–237.

Gibbings, S. L. (2016). Sosialisasi, street vendors and citizenship in Yogyakarta. In W. Berenschot, H. G. C. H. Schulte Nordholt, & L. Bakker (Eds), *Citizenship and Democratization in Southeast Asia* (Vol. 115, pp. 96–122). Leiden, Boston: Brill.

Gruenewald, D. A. (2004). A Foucauldian analysis of environmental education: Toward the socioecological challenge of the earth charter. *Curriculum Inquiry*, 34(1), 71–107.

Helms, S. E. (2013). Involuntary volunteering: The impact of mandated service in public schools. *Economics of Education Review*, 36, 295–310. doi:http://dx.doi.org/10.1016/j. econedurev.2013.06.003

Inglehart, R. (1995). Public support for environmental protection: Objective problems and subjective values in 43 societies. *Political Science and Politics*, 28(1), 57–72.

Kalland, A., & Persoon, G. (1998). An anthropological perspective on environmental movements. In A. Kalland & G. Persoon (Eds). *Environmental Movements in Asia* (pp. 1–43). Richmond, Surrey: Curzon Press and Nordic Institute of Asian Studies.

Keeler, W. (1990). Speaking of gender in Java. In J. Atkinson & S. Errington (Eds), *Power and Difference: Gender in Island Southeast Asia* (pp. 127–152). Stanford: Stanford University Press.

Long, N. (2007). How to win a beauty contest in Tanjung Pinang. *Review of Indonesian and Malaysian Affairs, 41*(1), 91–117.

Long, N. (2013). *Being Malay in Indonesia: Histories, Hopes and Citizenship in the Riau Archipelago.* Singapore: Asian Studies Association of Australia in association with NUS Press and NIAS Press.

Meek, D. (2015). Learning as territoriality: The political ecology of education in the Brazilian landless workers' movement. *The Journal of Peasant Studies, 42*(6), 1179–1200. doi:10.1080/03066150.2014.978299

Nagel, M. (2005). Constructing apathy: How environmentalism and environmental education may be fostering "learned hopelessness" in children. *Australian Journal of Environmental Education, 21*, 71–80.

Newberry, J. (2014). Women against children: Early childhood education and the domestic community in post-Suharto Indonesia. *TRaNS: Trans-Regional and National Studies of Southeast Asia, 2*, 271–291. doi:10.1017/trn.2014.7.

Nilan, P. (2018). Smoke gets in your eyes: Student environmentalism in the Palembang haze in Indonesia. *Indonesia and the Malay World, 46*(136), 325–342. doi:10.1080/1363 9811.2018.1496624.

Parker, L. (2003). *From Subjects to Citizens: Balinese Villagers in the Indonesian Nation-state.* Copenhagen: Nordic Institute of Asian Studies Press.

Parker, L., & Raihani, R. (2011). Democratizing Indonesia through education? Community participation in Islamic schooling. *Educational Management Administration & Leadership, 39*(6), 712–732. doi:10.1177/1741143211416389.

Prabawa-Sear, K. (2018). Winning beats learning: Environmental education in Indonesian senior high schools. *Indonesia and the Malay World, 46*(136), 283–302. doi:10.108 0/13639811.2018.1496631.

Prabawa-Sear, K., & Baudains, C. (2011). Asking the participants: Students' views on their environmental attitudes, behaviours, motivators and barriers. *Australian Journal of Environmental Education, 27*(2), 219–228.

Rickinson, M., & Lundholm, C. (2010). Exploring student learning and challenges in formal environmental education. In R. B. Stevenson & J. Dillon (Eds), *Engaging Environmental Education: Learning, Culture and Agency* (pp. 13–29). Rotterdam: Sense Publishers.

Sears, L. L. (1989). Aesthetic displacement in Javanese shadow theatre: Three contemporary performance styles. *TDR (1988-), 33*(3), 122–140. doi:10.2307/1145992.

Stevenson, R. B., & Dillon, J. (2010). Introduction to issues in learning, culture and agency in environmental education. In R. B. Stevenson & J. Dillon (Eds), *Engaging Environmental Education: Learning, Culture and Agency* (pp. 3–10). Rotterdam: Sense Publishers.

Tanu, D., & Parker, L. (2018). Fun, "family", and friends: Developing pro-environmental behaviour among high school students in Indonesia. *Indonesia and the Malay World, 46*(136), 303–324. doi:10.1080/13639811.2018.1518015.

Watts, M., & Alsop, S. (1997). A feeling for learning: Modelling affective learning in school science. *The Curriculum Journal, 8*(3), 351–365.

11 Young people as environmental subjects?

Identity, behaviour and responsibility

The previous four chapters reported on EE in and around schools in Yogyakarta and Surabaya, as observed by Prabawa-Sear during ethnographic fieldwork. While we did find some inspirational individual environmentalists, and in Surabaya there is an impressive amount of environmental work happening, overall we found the formal and informal education about human–environment interactions disappointing. In the next, final chapter, we suggest some ways forward for schools in Indonesia.

In this chapter, we explore the self-identification of young people as environmentalists, their reported environmental behaviour and their ideas about who is responsible for the environment. We used a different methodology – basically a written survey – and the results are quantitative.

Researchers have addressed how, when and why individuals become environmentalists in the Global North, but countries in the Global South are much less studied. Chawla studied the life histories ("life paths") of individual environmental activists in the US and Norway, and identified these influences: childhood experiences in nature; experiences of environmental destruction; environmental values held by the family; environmental organisations; role models (friends or teachers); and education (Chawla, 1998, 1999). Her study is often too cavalierly encapsulated in the acronym SLE (Significant Life Experiences); it was controversial and raised many questions for further research (Chawla, 1998, 2001; Gough, 1999). The larger point for us here is that with one or two exceptions, such influences have not been identified for environmentalists in Indonesia.[1]

However, our goal in this chapter is not to seek out individual environmentalists to find their motivations. Rather, we are interested in what we might call practising pro-environment citizens, who share an environmentalist subjectivity. Chawla noted that she believes that besides individual environmental activists who variously demonstrate, petition, lobby and participate in direct action to save wilderness, stop fracking, etc., environmentalism requires

> a large population of citizens who support the protection of the environment in other ways as well: through their voting records on state and local referenda, through holding politicians accountable for their environmental

positions, through recycling, reducing consumption and other day-to-day behaviors.

(Chawla, 2001, p. 455)

The processes of creating new environmental subjectivities and practice *en masse*, and thus a new social movement for environmental sustainability, are complex, but all point to the value of education, whether formal or informal. Many hundreds of studies have been done in the countries of the Global North, using different disciplines and various theoretical frameworks. For many of these countries, we have small- and large-scale survey data on attitudes towards the environment, e.g. many using the NEP (New Environmental Paradigm) (Dunlap, 2008; Dunlap & Van Liere, 2008 [1978]; Dunlap *et al.*, 2000). Kollmuss and Agyeman (2002) usefully surveyed this literature, noting the ubiquity of the "gap" between environmental sensibilities and environmental action. Work has begun in a few non-Western contexts, as identified in Chapter 3. However, as far as we are aware, no-one has conducted a large-scale survey of attitudes towards the environment in Indonesia,[2] nor has there been other research about how education can contribute to creating an environmental subjectivity among peoples in Indonesia. This chapter addresses this dearth of studies from Indonesia.

The survey (Survei Pendidikan Lingkungan 2014–2015)

The team designed a set of questions (detailed below) that were common to all surveys, and the survey was completed by 1000 students in senior high schools in Yogyakarta and Surabaya.[3]

There was one overarching question about participants' identity as an environmentalist: do you identify as an environmentalist? This was followed up with questions about activities and behaviour that aimed to show this self-identification in action. Of course, a survey can only report on respondents' reported behaviour: we were not able to compare actual behaviour with reported behaviour. The other common team questions asked respondents to identify the most important environmental issues, at local, national and global levels; to identify whose responsibility it is to address the problem; to say how they (the respondents) can act on the problem; and to identify the main constraints that hinder efforts to address the environmental issues.[4]

In Surabaya, Prabawa-Sear worked through TENGO, which approached the schools for permission to survey students. In Yogyakarta, she approached the schools directly. The schools in Yogyakarta were all participating in the national Adiwiyata Programme, an EE programme run by the national Ministry of Environment and Forests, as described above in Chapter 6.

Thus, all of the selected schools participated in either the Adiwiyata Programme or TENGO's Eco Schools Programme and so all the students in our survey had been exposed to EE programmes at school. In this they were, by definition, not representative of schools in Indonesia. Most students in most schools in Indonesia are not exposed to EE. For this reason, participants in our survey are *more environmentally aware and active than most students in Indonesian senior high schools*. Schools were selected to ensure a range of public and private schools; non-religious, Catholic and Islamic schools; high-achieving and high socioeconomic status schools and lower-achieving and lower socioeconomic status schools; vocational schools and general schools.[5]

The number of schools participating in the survey was much higher in Surabaya than in Yogyakarta because there are many more schools participating in environmental programmes there, and it was much easier to get access to the schools through TENGO as a result of their working partnership with the Surabayan government. Once schools accepted our request to survey students, we asked for a range of classes (science, social science and language streams),[6] ages and genders. In some cases these requests were met and in others they were not. Teachers often commented that they would make "good" students available despite our requests for a representative sample of students. It was particularly difficult to get access to year 12 students because they were supposed to be focusing on their national exams, and we were interrupting class time.

Approximately half of the surveys were completed in classrooms, with the other half completed in the grounds of the schools. The students were all environmental club members or student leaders (i.e. members of the OSIS, Organisasi Siswa Intra Sekolah, School Student Council). The researcher or a TENGO colleague was present in all cases. Prior to handing out surveys, the researcher or TENGO representative provided a brief introduction, explaining that this survey was for an Australian research project on environmental education in Surabayan and Yogyakartan high schools and that participation was voluntary and anonymous. Teachers were invited to take a break while the surveys were being completed and most teachers took up the offer. Students were asked to direct any questions to the researcher or TENGO representative and to complete every question in the survey. Where possible, we checked that surveys were complete as the students handed them in, and returned any with missing sections to students for completion. The researcher and TENGO representatives walked around while surveys were being completed in an attempt to limit the sharing of answers.[7] In instances where the Australian researcher was present, students were invited to ask questions of the researcher after the survey was complete.

The respondents

The characteristics of the respondents are tabulated in Table 11.1. The median age of students was 16, and most students were in Grade 11, the middle grade of senior high school. Two-thirds were girls. This gender imbalance was not intended, but we think it is because girls are disproportionately active in environmental organisations.[8] The religious breakdown of our respondent group follows the national census statistics quite closely: 90.8 per cent of our respondents were Muslim; 5.4 per cent Protestant; 2.9 per cent Catholic; 0.2 per cent Hindu and 0.7 per cent "Other".[9]

Table 11.1 Characteristics of respondents (n = 1000)

	n (%)	Median (IQR)
Age		16 (16)
Sex:		
Male	332 (33.2)	
Female	668 (66.8)	
Religion:		
Islam	908 (90.8)	
City:		
Yogyakarta	173 (17.3)	
Surabaya	827 (82.7)	
Year level:		
10	286 (28.6)	
11	523 (52.3)	
12	191 (19.1)	

Source: Survei Pendidikan Lingkungan 2014–2015.

The last census, in 2010, showed that 52.78 per cent of young people aged 16–18 attend school (BPS (Badan Pusat Statistik), 2010b). These are young people from more socioeconomically secure families, but it is too broad a generalisation to say that they are, by definition, middle class, as some have intimated. What we can say is that these young people come from families on at least adequate monthly incomes, and they are at least aspirational middle class. Students at these schools expect to find "white collar" work, in offices, clinics, businesses, schools and universities – virtually none aim to become farmers or fishers (Nilan *et al.*, 2011, pp. 716–620)

Although there has not been a lot of research into the question of how subjects come to assume an environmentalist identity, Payne has suggested that, "For the vast majority of younger people, identity issues and options are now utterly entangled in the lifestyle preoccupations and consumptive imperatives of a technologically-replete, image-driven postmodernity" (Payne, 2001, p. 74). While Payne is probably talking about young people in countries of the Global North, we can see something of the entangled identities of the students reported on in this study. Perhaps most noteworthy is the fact that almost all of the students owned mobile/cell phones, albeit not all of them were "smart phones" connected to the internet. Sometimes teachers asked students to use their phones for research in class. While some of the students owned the newest iPhones, most tended to use older versions of the cheaper brands such as ASUS or Huawei, or older versions of the more expensive Samsung. This symbol of participation in the global media world is the single most visible status symbol that students in schools in Indonesia can display, as they all wear school uniforms.

Fourteen schools participated in the survey: only two in Jogja and 12 in Surabaya. Seven schools are state senior high schools, with 55.8 per cent of our student respondents; four are vocational schools, with 28.2 per cent of our respondents, and one is a state *madrasah*.[10] State senior high schools are generally considered the most desirable schools; typically the children of public servants, professionals and other members of the middle class attend these schools. However, richer people will often send their children to academically superior private schools, particularly the Christian schools. In Surabaya we tried to enlist Year 11 students, as it was the beginning of the school year and students in Year 10 had not yet experienced much exposure to the programmes we were studying.

Identity as an environmentalist

Our understanding of personal "identity" derives from the now large body of literature on identity studies that grew out of the postmodern critique as well as the "culture wars" over history and race/ethnicity, gender/sexuality and youth studies from the 1960s (e.g. Calhoun, 1994; Clarke *et al.*, 1976; Erikson, 1968; Giddens, 1991; Hall, 1990, 1996). We acknowledge that identity is a hybrid of external ascription and individual development, and is fluid – changing according to context. Identity is historically produced and historically specific, so female identity is different in nineteenth-century Batavia to female identity in twentieth-century Jakarta, and also diverse, because a person's identity is formed within their particular milieu: their family, neighbourhood and school, and the perceptions within that milieu of what "a girl" or "a boy", a "Chinese" or a "Javanese", should be like. Individuals develop their individual identity over time, adapting to the surrounding expectations, constrictions and opportunities, in unique ways, but commonalities are produced because of group expectations and common experiences. Humans are social animals and are almost never purely autonomous. Identity is an expression both of social structure (in Giddens' sense) and of agency, as individuals "ad lib", perform, resist, negotiate and comply. Because of this, identity is always "in process". While there has been considerable research conducted on how people become environmentalists, this work has not commonly engaged with the identity literature.[11]

Given the reported low level of environmental awareness in Indonesia, perhaps the most startling statistic in the survey was that *81.9 per cent of our respondents self-identified as environmentalists.* We can explain this very positive association with an environmentalist identity by reference to the selection of schools: all schools in the survey participated in either the Adiwiyata Programme or TENGO's Eco School Programme. We should also mention that the researcher was known to be researching environmental education, and to the extent that students understood what that was, and wanted to give her what she wanted, the students were probably anxious to present their school in a good light.

Another team member, Pam Nilan, was researching environmental sensibilities among university students. She did not select the universities, the courses nor the respondents with reference to any environmental attributes or environmentalist affiliations. She surveyed 804 undergraduate students in four different cities: Jakarta, Bandung and Yogyakarta in Java, and Palembang in Sumatra. In contrast to the strong identification with environmentalism from our school students, more university students said that they were not an environmentalist (52.2 per cent) than said they were (47.8 per cent).

Environmental behaviour

So our sample is atypically "environmentalist" in orientation for high-school students in Indonesia. Bearing in mind the oft-remarked "gap" between environmental sensibilities and environmental action, as highlighted by Kollmuss and Agyeman (2002), our next task was to try to work out what our respondents meant by ticking the environmentalist box. We first asked students if they have ever participated in environmental activities, without suggesting what these might be. Over 75 per cent of students said they had (76.8 per cent). Given the high percentage who self-identified as environmentalists, perhaps it is surprising that 23.2 per cent said "no". Then we presented them with a range of environmental care activities that are common in Eco and Adiwiyata Schools and asked them if they "usually" do those things: making *bio-pori* (cylindrical holes stuffed with leaves and organic matter, dug into the compacted ground in cities to improve water absorption), participating in clean-ups, making compost, growing plants for herbal remedies, managing waste through activities such as a recycling bank, and re-using containers such as drink bottles and lunch boxes (Table 11.2). And then we asked them an open question for miscellaneous other environmental activities.

Clean-ups are the most common activity and that is not surprising: Indonesian students usually have to clean up their school rooms and yards on Fridays or Saturdays, whether or not their schools are Eco or Adiwiyata Schools. Perhaps in other schools this would not be seen as an environmental activity. It should be noted that not all the students would be involved in all the activities at each school: a student might be on the "compost team" and therefore not be involved in digging *biopori*. Re-using containers is the only behaviour that could be freely chosen, though would usually need the cooperation of mothers.

Table 11.2 Environmental "care" activities conducted by senior high school students

Activity	% of students who say they usually do the activity
Clean-ups	62.8
Waste management such as recycling bank	41.7
Re-using containers such as drink bottles and lunch boxes	40.2
Making *biopori*	34
Making compost	31.4
Growing plants for herbal remedies	14.7

Source: Survei Pendidikan Lingkungan 2014–2015.

The open-ended question about "other" environmental care activities elicited 21 individual responses, ranging from praiseworthy activities such as using waste water from air-conditioners and *wudhu* (water used for washing before prayers) to watering plants, and cleaning up rubbish from mountain tops, to more dubious ones such as burning rubbish. It was interesting that two students listed "not smoking" as an environmental activity – an answer that might not have occurred to students in Global North countries.

The next questions were about environmental activities that aimed to advance knowledge or educate about the environment: through environmental lessons, learning in groups or with friends, through workshops or training, or campaigning such as through social media, or on the street (Table 11.3). Clearly these young people are engaging in a range of environmental learning, training, communication and advocacy activities – more than a quarter of our sample say they are engaged in campaigning, e.g. via online media as well as on the streets.

In the final question of the survey we returned to the question of the students' environmental behaviour. We provided a list of behaviours, as in Table 11.4, and asked students how often they carried out this behaviour, on a four-point scale ranging from "never" to "always".

Comparing Tables 11.4 and 11.5, and our other survey answers, we can identify some inconsistencies and commonalities. Turning off taps when not in use is something that is in the power of young people and easy to do, so this frequency and consistency is understandable. Similarly, turning off the television when not in use, is something these young people can, and do, do.

There is some inconsistency when we look at behaviour around litter. In Table 11.2, we saw that most students said that they are usually engaged in "clean-ups" (62.8 per cent). If litter is perceived as a problem (see below), the fact that ~500 of 1000 mainly "environmentalist" students will "always" throw litter on the ground if there are no bins, and that 80 per cent of them "often" throw litter on the ground, is a concern. And it is surprising that only one-quarter "always" pick up litter. While throwing litter is not perceived as "dirty", picking up litter is. We hypothesise that the concept of "the clean-up" actually contributes to littering practice, because it implies that it is someone else's job to clean it up later. In other words, young people don't have to be responsible for their own waste. It is encouraging that so many students bring food containers from home – and Tables 11.2, 11.4 and 11.5 are consistent in this. This should help to minimise rubbish, as well as reduce use of materials. However, it

Table 11.3 Environmental "knowledge" activities conducted by senior high school students

Environmental lessons	Study groups/learning with friends	Training	Campaigning, e.g. via online media, on the streets
453	293	335	261

Source: Survei Pendidikan Lingkungan 2014–2015.

Table 11.4 Frequency of everyday pro-environment behaviours

Always——Never				Variables	Variables	Never——Always			
4	3	2	1			1	2	3	4
143	241	420	196	Refuse plastic bags when shopping (6)	Turn off the taps when not using water (2)	3	27	97	873
166	304	393	137	Print on both sides of the paper (4)	If no one else is watching TV, turn it off when you leave the room (1)	16	94	163	727
272	318	309	101	Encourage friends to recycle at school (13)	Bring a drink bottle or lunch box to school (5)	58	147	214	581
420	322	172	86	Go by car or motorbike even when it's not far to go (8)	If there are no bins, just throw rubbish on the ground (7)	50	141	317	492
263	371	296	70	Tell friends about environmental things that you have learnt (15)	Go by car or motorbike even when it's not far to go (8)	86	172	322	420
313	382	241	64	Tell your parents/older friends about environmental lessons (11)	Walk or ride a bicycle when going somewhere close (3)	62	264	262	412
412	262	264	62	Walk or ride when going somewhere close (3)	Recycle at school (12)	40	235	370	355
253	369	319	59	Pick up litter (16)	Tell your parents/older friends about environmental lessons (11)	64	241	382	313
581	214	147	58	Bring a drink bottle or lunch box to school (5)	Encourage friends to recycle at school (13)_	101	309	318	272
492	317	141	50	If there are no bins, just throw rubbish on the ground (7)	Tell friends about environmental things that you have learnt (15)	70	296	371	263
355	370	235	40	Recycle at school (12)	Pick up litter (16)	59	319	369	253
727	163	94	16	If no-one else is watching TV, turn it off when you leave the room (1)	Print on both sides of the paper (4)	137	393	304	166
873	97	27	3	Turn off the taps when not using water (2)	Refuse plastic bags when shopping (6)	196	420	241	143

Source: Survei Pendidikan Lingkungan 2014–2015.

Table 11.5 Frequency of everyday pro-environment activities (grouped answers, "rarely" and "often", arranged according to declining frequency of "often")

Variable	Rarely	Often
VII.2. Turn off the taps when not using water.	30	970
VII.1. If no-one else is watching TV, turn it off when you leave the room.	110	890
VII.7. If there are no bins, just throw rubbish on the ground.	191	809
VII.5. Bring a drink bottle or lunch box to school.	205	795
VII.8. Go by car or motorbike even when it's not far to go.	258	742
VII.12. Recycle at school.	275	725
VII.11. Tell your parents/older friends about environmental lessons.	305	695
VII.3. Walk or ride a bicycle when going somewhere close.	326	674
VII.15. Tell friends about environmental things that you have learnt.	366	634
VII.16. Pick up litter.	378	622
VII.13. Encourage friends to recycle at school.	410	590
VII.4. Print on both sides of the paper.	530	470
VII.6. Refuse plastic bags when shopping.	616	384

Source: Survei Pendidikan Lingkungan 2014–2015.

is disappointing that students do not generally think to print on both sides of the paper or refuse plastic bags – again, these are individual actions that are in their power. Reducing the use of paper, and minimising the use of such problematic materials as plastic, will help the environment in many ways.

There is also internal inconsistency over mode of transport: Three-quarters (74 per cent) say they go by car or motorbike even if it's not far to go, while over two-thirds (67.4 per cent) say they walk or ride by bicycle when going somewhere close.

When it comes to telling friends and family pro-environment messages, there is some consistency in the survey answers, and we are encouraged (and a bit surprised) that so many (313) reported "always" telling their parents and seniors about environmental matters.

Perceptions of environmental problems

The team devised some common team questions around perceptions of environmental problems locally, nationally and internationally (Table 11.6); perceptions of what our respondents could do about these problems; the barriers to their solution; and finally perceptions of who should be responsible for solving the problems. First we report on the students' perceptions of the most important environmental problem "in the area where you live". This was an open-ended question and students were free to enter as many words as they liked. We have grouped them into rubbish/waste[12] (76.5 per cent), water pollution (9 per cent), air pollution (7.2 per cent) and other/don't know (7.3 per cent), realising that rubbish/waste can be encompassed by pollution.

Given that all our respondents live in urban areas, it is not surprising that waste was the answer provided by more 75 per cent of our 1000 participants.

Table 11.6 In your opinion, what is the most important environmental issue where you currently live? (n = 1000)

Issue	Frequency	%
Rubbish/waste	765	76.5
Water pollution	90	9.0
Air pollution	72	7.2
Other/no answer	73	7.3

Source: Survei Pendidikan Lingkungan 2014–2015.

Breaking down the figures by city (Yogyakarta versus Surabaya), a larger percentage of students in Surabaya (78.96 per cent) considered waste the most important problem than in Yogyakarta (64.7 per cent). We put this down to the fact that the Surabayan government, TENGO and the students involved in environmental activities in Surabaya were very active in waste management, whereas the Yogyakarta government is yet to act on this problem.

When asked to identify the most important issue nationally (Table 11.7), "waste" disappeared. Most students (52 per cent) identified "pollution" as the most important environmental issue nationally.

It may be that students are not well informed about environmental matters nationally, and that they just extrapolated from their local experience and knowledge. It is also not clear if their identification of "pollution" included waste (as this was the most serious issue locally) as well as smoke pollution from forest fires. The coding is important here: students were asked an open-ended question and they answered in words – there were no pre-defined categories. We have broken down "the exploitation of natural resources" into "the exploitation of natural resources generally/in the environment" and "the exploitation of natural resources in the forest" as these were common categories. If combined, they still only comprise 27.8 per cent of the total.

The identification of natural disasters as the most important environmental problem facing Indonesia (although only 12.8 per cent made this claim), raises some interesting points. Indonesians have only recently begun to understand that their country suffers an unusually high occurrence of natural disasters,

Table 11.7 In your opinion, what is the most important environmental issue nationally? (n = 1000)

Issue	Frequency	%
Pollution	522	52.2
Exploitation of natural resources/the environment	157	15.7
Natural disasters	128	12.8
Exploitation of natural resources in the forest	121	12.1
Other/no answer	72	7.2

Source: Survei Pendidikan Lingkungan 2014–2015.

especially those due to tectonic movements (earthquakes, and resultant tsunami, and volcanism) around the so-called Pacific "ring of fire". Alongside these naturally-caused disasters are a host of human-induced disasters, such as floods and smoke haze. Much of the devastation wrought by disasters can be traced to unwise land use activities and settlement patterns, overpopulation, weak implementation of laws, and corruption, which tend to magnify the effects of natural hazards, causing human disasters.[13] It is also noteworthy that climate change is not seen as a national problem. Given Indonesia's dubious distinction as the world's fourth largest emitter of carbon pollution, that lack of awareness is a source of consternation.

Climate change (often called global warming) does make an entry at the international level: when asked what was the most important environmental issue internationally (Table 11.8), over one-quarter of the 1000 students identified global warming. More students, though, identified pollution (34.3 per cent) as the most important problem globally. It is hard to know if students saw pollution as separate from global warming, or if one was seen as a subset of the other. Number three position went to the exploitation of natural resources and the environment.

Who is responsible for the environment?

We turn now to the group of questions around responsibility. The question looks forward, not backward: i.e. it is not a question about who caused an environmental problem, but rather, who is going to be responsible for solving the problem. For each level – local, national and global – students were asked an open-ended question: in your opinion, who is responsible for addressing this most important issue? Remembering that over 75 per cent of students had identified rubbish/waste as the most important environmental problem at the local level, it is perhaps not surprising that the vast majority (91.5 per cent) identified society (*masyarakat*) as being responsible (Table 11.9).

This coded term, "society", was mainly made up of responses such as "all people" (*semua orang*, 10.3 per cent), "all citizens" (*semua warga*, 9.4 per cent), "we ourselves" (*kita dirinya, diri sendiri*), all of us (*kita semua*, 4.2 per cent), society (*masyarakat*, 3.8 per cent), all of society (*semua masyarakat*, 3.7 per cent),

Table 11.8 In your opinion, what is the most important environmental issue internationally? (n = 1000)

Issue	Frequency	%
Pollution	343	34.3
Global warming	258	25.8
Exploitation of natural resources/the environment	169	16.9
Natural disasters	38	3.8
Other/no answer	192	19.2

Source: Survei Pendidikan Lingkungan 2014–2015.

Table 11.9 In your opinion, who is responsible for addressing this (most important) issue (locally)? (n = 1000)

Issue	Frequency	%
Society	915	91.5
Society and government	30	3.0
Government	29	2.9
Other	26	2.6

Source: Survei Pendidikan Lingkungan 2014–2015.

the whole of society (*seluruh masyarakat*, 1.9 per cent), citizens (*warga*, 1.8 per cent), all of the citizens (*seluruh warga*, 1.5 per cent), and variants of these. Students often identified themselves in concert with local citizens or inhabitants, as responsible for their own environment, and this was pleasing. Dividing the 915 "society" responses into those who explicitly identified themselves as being responsible, and those who identified only some group such as society, residents or citizens, 152 indicated that they included themselves as being responsible (all of us, we ourselves, etc.). However, we should not put too much meaning onto this, as there is nothing that excludes the writer in answers such as "all people" or "all citizens".

What is surprising is the few students who identified "the government" as responsible for addressing the environmental issue. This assumption of rubbish/waste as a social responsibility rather than a government responsibility reflects the reality in Yogyakarta, where the provincial government has been uninterested in environmental matters and citizens are left largely on their own to cope with their rubbish. Almost 91 per cent of Yogyakartan students identified "society" as responsible. However, 82.7 per cent of our respondents were from Surabaya, where the city government has assumed responsibility for garbage and composting services, and just over 91 per cent of Surabayan students also identified "society" as responsible.

When we turn to what the students themselves perceived that they could do to address the issue that they had identified as most important locally (Table 11.10), unsurprisingly there were some who said they could put rubbish

Table 11.10 In your opinion, what can you do to address this (local) issue? (n = 1000)

Issue	Frequency	%
Contribute by protecting the environment	342	34.2
Throw rubbish in the bin/manage and provide rubbish facilities	341	34.1
Campaigning and enriching knowledge about environmental issues	226	22.6
Greening the environment	40	4.0
Other/no answer	51	5.1

Source: Survei Pendidikan Lingkungan 2014–2015.

in the appropriate receptacle (34 per cent). Other than this, the answers were distressingly vague: 34 per cent suggested they could protect the environment, 22.6 per cent suggested they could strengthen knowledge or awareness, and 4 per cent suggested greening the environment. The paucity of these responses suggests the inadequacy of the EE that is being conducted by TENGO and the Adiwiyata Programme: these programmes focus on group activities, the nature of which is decided not by students themselves but by NGO workers and the government. There is no education about the complexity of environmental problems, nor is there training in problem-solving.

The next question – again, an open-ended question to be answered in students' own words – asked students what they considered the main barriers to addressing the main environmental issue that they had identified at the local level (Table 11.11). Again, the students focused on society rather than government: almost 75 per cent identified factors such as limited environmental awareness and knowledge and lack of care for the environment as the main barrier; only 9.6 per cent identified limited facilities and services. Interestingly, lack of discipline was identified by 7.6 per cent of students. "Discipline" has been identified as a ubiquitous key word in education in Indonesia (Parker & Nilan, 2013, pp. 95–99), and is a much-desired value of students. In a different survey of senior high students in Indonesia, laziness was identified by middle-class students as the main barrier that they anticipated in achieving their life goals (Nilan *et al.*, 2011, pp. 721–724). These answers all reflect a neoliberal, responsibilisation discourse that traces responsibility for addressing problems to the individual.

Moving to the national level, where "pollution" was identified as the most important issue by 52 per cent of students, the "exploitation of natural resources" by 27.8 per cent and "natural disasters" by 12.8 per cent of the total, we go through the same set of questions: who is responsible for the issue, what can you do to address this issue, and what are the barriers to addressing this issue. We were surprised that for the national level, more than two-thirds of students still considered that "society" is responsible for addressing these problems. Again, students saw the role of government as very limited: only 9.4 per cent identified the government alone as responsible, and another 17.1 per cent identified the government as responsible "with society" (Table 11.12).

Table 11.11 In your opinion, what are the barriers to addressing this (local) issue? (n = 1000)

Issue	Frequency	%
Limited knowledge/awareness/care about the environment	740	74.0
Limited facilities	96	9.6
Lack of discipline	76	7.6
Other/no answer	88	8.8

Source: Survei Pendidikan Lingkungan 2014–2015.

Table 11.12 In your opinion, who is responsible for addressing this (most important national) issue? (n = 1000)

Issue	Frequency	%
Society	675	67.5
Society and government	171	17.1
Government	94	9.4
Other	60	6.0

Source: Survei Pendidikan Lingkungan 2014–2015.

Also notable is the silence around industry and business. Indonesia does have a reasonably strong framework of environmental legislation, including planning regulations that require new developments to have environmental impact assessments. However, it is an open secret that implementation is weak, with decentralisation fostering corruption and nepotism at devolved levels, rendering national and international protections and commitments weak, if not useless (McCarthy & Robinson, 2016; Setiawan & Hadi, 2007).

At the national level, students again suggested rather vague actions that they could do to address the issue: contribute by protecting the environment (68.1 per cent) and run a campaign and strengthen understanding and knowledge about environmental issues (20.8 per cent) were the main responses (Table 11.13).

Students identified the two main barriers to addressing the most important environmental issue at national level (Table 11.14) as

a   the low level of awareness, care, knowledge and capacity among society (79.1 per cent), and
b   the government's failure to prioritise the handling of environmental issues (10.2 per cent).

At the international level (Table 11.15), students had identified "pollution" as the most important problem (34.3 per cent), followed by "global warming" (25.8 per cent), which might refer to the same thing, and the exploitation of the environment (16.9 per cent). In answering the question, who is responsible

Table 11.13 In your opinion, what can you do to address this (national) issue? (n = 1000)

Issue	Frequency	%
Contribute by protecting the environment	681	68.1
Campaigning and enriching knowledge about environmental issues	208	20.8
Advocating about the environmental problem to government	22	2.2
Other/no answer	89	8.9

Source: Survei Pendidikan Lingkungan 2014–2015.

Table 11.14 In your opinion, what are the barriers to addressing this (national) issue? (n = 1000)

Issue	Frequency	%
Society's limited knowledge/awareness/care/capacity about the environment	791	79.1
Government doesn't prioritise handling environmental issues	102	10.2
Other/no answer	107	10.7

Source: Survei Pendidikan Lingkungan 2014–2015.

Table 11.15 In your opinion, who is responsible for addressing this (most important international) issue? (n = 1000)

Issue	Frequency	%
Society – humankind	677	67.7
All sides	96	9.6
Governments/states	62	6.2
Other/missing	165	16.5

Source: Survei Pendidikan Lingkungan 2014–2015.

for addressing this issue, again, what is striking is the low expectation of government: over two-thirds of students answered "humankind" or "society" or similar (67.7 per cent); only 6.2 per cent answered "the government" or "the state/s"; and 9.6 per cent replied "all sides". And again, there was silence around industry as responsible for cleaning up the obvious pollution.

Answers to the "What can you do?" question for the international level (Table 11.16), were again vague: 61.2 per cent replied that they could contribute to the protection of the environment, and 13.9 per cent said they could participate in campaigns to enhance awareness among society.

Again, in answer to the question about the main barriers to addressing the issue at international level (Table 11.17), most students responded that the low level of awareness, care, and capacity in society was the main barrier (63.3 per cent) and only 7.4 per cent saw this as a failure of governments.

Table 11.16 In your opinion, what can you do to address this (international) issue? (n = 1000)

Issue	Frequency	%
Contribute by protecting the environment/ecosystem	612	61.2
Campaigning about environmental issues to society	139	13.9
Advocating about the environmental problem to government	5	0.5
Other/no answer	244	24.4

Source: Survei Pendidikan Lingkungan 2014–2015.

Table 11.17 In your opinion, what are the barriers to addressing this (international) issue? (n = 1000)

Issue	Frequency	%
Society's limited knowledge/awareness/care/capacity about the environment	633	63.3
Government doesn't prioritise handling environmental issues	74	7.4
Other/no answer	293	29.3

Source: Survei Pendidikan Lingkungan 2014–2015.

Conclusion

We began this chapter by noting the serious lacuna in knowledge about environmental awareness and sensibilities in Indonesia, explaining that this paper was an attempt to begin to fill the hole. The students reported on in this article are unusual in Indonesian education in that they have been exposed to (varying levels of) environmental education. We were pleased to find that a strong majority of students did self-identify as environmentalists. However, after looking at their reported environmental behaviours and their perceptions of environmental problems, we began to ask, "What does it mean to self-identify as an environmentalist in Indonesia?"

We did not expect to find the romantic answers that often appear in such surveys in richer countries, where respondents not infrequently talk about life-changing experiences in the woods or the importance of saving wilderness or iconic species such as whales, baby seals and orangutan. We did expect to find concern at rubbish/waste and indeed this issue was the issue of most concern at the local level. But we were surprised by the fact that in all the answers to our questions about environmental issues, perceptions and behaviours, students never once raised the issue of consumption or of consumerist culture. There was never any attempt to link the issues of waste, pollution (the issue of most concern at the national and international levels) or global warming (which only appeared as the No. 2 problem at international level) to consumption of material goods or of carbon-based, non-renewable energy. There was not a single mention of loss of biodiversity, urbanisation or human population growth. We think this failure to identify the complex interactions among environmental problems and human behaviour reflects the shallow and activity-dominant form of EE conducted in these schools.

It is important to discover how senior high school students in Indonesia see their world and their place in it. They will be the next generation of teachers and parents, business leaders and politicians in the fourth most populous country in the world. It is a country that is developing quickly, and will therefore not only be depleting its globally-important resources of forests, mangroves and its rich marine and coastal biodiversity at breakneck speed for (non-renewable) energy, transport and material goods, but also it will be pumping out vast amounts of carbon and other warming and toxic pollutants.

We have found that student participants in our research seemingly agree with the World Bank and the Government of Indonesia, that "environmental values are not deeply embedded in society, leading to undervaluation of natural resources and environmental services". However, students go much further than merely noting a low level of awareness of environmental problems in society. This survey has shown that, overwhelmingly, young people believe that "society" – rather than governments or industry – is responsible for addressing environmental problems. They tend to include themselves as being responsible for addressing local problems, but are vague about what they can actually do to ameliorate or solve other environmental problems. Further, they barely mention the role that governments must play, and do not think to mention that industry or consumers should be responsible. Thus, it would seem that while young people are happy to self-identify as environmentalists, they have absorbed the neoliberal message of small government and assigned responsibility to "society" – those who are least aware, most ignorant, and most poorly equipped to meet the challenges of environmental destruction.

Notes

1 Exceptions include Crosby (2013); Nilan and Wibawanto (2015).
2 In this chapter we are not talking about what we can call ethno-environmental attitudes and knowledge, i.e. local knowledge and "traditional" understanding of local ecologies. While we recognise that these are important in the context of natural resource management at the local level (Laumonier, Bourgeois, & Pfund, 2008), such understandings are outside the scope of this chapter.
3 A somewhat different version of this chapter has already been published in (Parker, Prabawa-Sear, & Kustiningsih, 2018).
4 The questions were grouped according to scale. The first set of four questions was about the local level: (A) What is the most important local environmental issue? (B) Whose responsibility is it to fix this problem? (C) What do you think you can do about it? (D) What are the problems/constraints that hinder those efforts? The next set of four questions was about the national level, and the third set about the global level. Piloting of the surveys revealed that we had to change the wording for "local". We settled on "Apa masalah lingkungan yang paling penting di tempat tinggal anda saat ini?" What is the most important environmental issue where you currently live? We would be the first to acknowledge that identifying "the most important environmental problem" at any level is a difficult task. Indeed, we would expect that answers would vary among scientists and environmentalists. We were not seeking "the right answer": rather, we wanted to explore students' understandings of what they might consider constitute "environmental problems" at different scales.
5 The structure of the Indonesian education system is described in Chapter 5.
6 In senior high school, students must choose one of three streams. The science stream is the most academically prestigious and consists of subjects such as Physics, Chemistry and Mathematics. Next comes the social science stream, with subjects such as Geography, History and Economics, and the least prestigious is the languages stream. Not all schools offer all three streams; the most common streams are science and social science. Of our respondents, 55.1 per cent were from the more academically prestigious science stream. We tested the significance of subject stream for the survey questions addressed below, and found there was no significance.

7 It is common in schools in Indonesia for students to complete work communally, sharing ideas and answers.
8 The gender aspects of environmentalism in Indonesia are very interesting. Our field-workers noted that although girls are disproportionately active as the "foot soldiers" of environment clubs and organisations in schools, the leaders of ENGOs are almost always young men, and this was the pattern found in Blora also (Crosby 2013).
9 According to the 2010 census, 87.18 per cent of the population follow Islam; 6.96 per cent Christianity; 2.91 per cent Catholicism; 1.69 per cent Hinduism and the reminder Buddhism, Confucianism and "other" (BPS 2010a).
10 Vocational schools can be public or private: we had two of each in our sample. *Madrasah* are administered by the Ministry of Religion (MOR), rather than the Ministry of Education and Culture (MOEC). However, 70 per cent of their curriculum comes from MOEC; the additional curriculum (30 per cent) comes from the MOR.
11 Some exceptions include Dillon, Kelsey, and Duque-Aristazabel (1999), and Payne (2001).
12 The term in Indonesian, *sampah*, does not distinguish between waste and rubbish.
13 See Blaikie *et al.* (2004), Cannon (1994) and Warren (2016).

References

Blaikie, P., Cannon, T., Davis, I., & Wisner, B. (2004). *At Risk: Natural Hazards, People's Vulnerability and Disasters* (2nd edn). Oxford: Routledge.

BPS (Badan Pusat Statistik). (2010a). Sensus Penduduk 2010 (2010 Census). Retrieved 24 August 2016 http://sp2010.bps.go.id/index.php/site/tabel?search-tabel=Penduduk+Menurut+Wilayah+dan+Agama+yang+Dianut&tid=321&search-wilayah=Indonesia&wid=0000000000&lang=id.

BPS (Badan Pusat Statistik). (2010b). Sensus Penduduk 2010 (2010 Census). Education. Retrieved from http://sp2010.bps.go.id/index.php/site/topik?kid=6&kategori=Pendidikan.

Calhoun, C. (1994). Social theory and the politics of identity. In C. Calhoun (Ed.), *Social Theory and the Politics of Identity* (pp. 9–36). Cambridge: Blackwell.

Cannon, T. (1994). Vulnerability analysis and the explanation of "natural" disasters. In A. Varley (Ed.), *Disasters, Development and Environment* (pp. 13–29). Chichester, New York, Brisbane, Toronto and Singapore: Wiley.

Chawla, L. (1998). Research methods to investigate significant life experiences: Review and recommendations. *Environmental Education Research*, 4(4), 383–397.

Chawla, L. (1999). Life paths into effective environmental action. *Journal of Environmental Education*, 31(1), 15–26.

Chawla, L. (2001). Significant life experiences revisited once again: Response to Vol 5(4) "Five critical commentaries on significant life experience research in environmental education". *Environmental Education Research*, 7(4), 451–461.

Clarke, J., Hall, S., Jefferson, T., & Roberts, B. (1976). Subcultures, cultures and class. In S. Hall & T. Jefferson (Eds), *Resistance Through Rituals: Youth Subcultures in Post-War Britain* (pp. 9–74). London: Hutchinson.

Crosby, A. (2013). Remixing environmentalism in Blora, Central Java 2005–2010. *International Journal of Cultural Studies*. doi:10.1177/1367877912474535.

Dillon, J., Kelsey, E., & Duque-Aristazabel, A. M. (1999). Identity and culture: Theorising emergent environmentalism. *Environmental Education Research*, 5(4), 395–405.

Dunlap, R. E. (2008). The new environmental paradigm scale: From marginality to worldwide use. *The Journal of Environmental Education*, 40(1), 3–18.

Dunlap, R. E., & Van Liere, K. D. (2008 [1978]). The "New Environmental Paradigm": A proposed measuring instrument and preliminary results [Reprint of original 1978 article]. *The Journal of Environmental Education, 40*(1), 19–28.

Dunlap, R. E., Van Liere, K. D., Mertig, A. G., & Jones, R. E. (2000). Measuring endorsement of the new ecological paradigm: A revised NEP scale. *Journal of Social Issues, 56*(3), 425–442.

Erikson, E. H. (1968). *Identity, Youth and Crisis*. New York: Norton.

Giddens, A. (1991). *Modernity and Self-Identity: Self and Society in the Late Modern Age.* Cambridge: Polity Press.

Gough, S. (1999). Significant life experiences (SLE) research: A view from somewhere. *Environmental Education Research, 5*(4), 353–363.

Hall, S. (1990). Cultural identity and diaspora. In J. Rutherford (Ed.), *Identity: Community, Culture, Difference* (pp. 222–237). London: Lawrence & Wishart.

Hall, S. (1996). Introduction: Who needs identity? In S. Hall & P. du Gay (Eds), *Questions of Cultural Identity* (pp. 1–17). London: Sage.

Kollmuss, A., & Agyeman, J. (2002). Mind the gap: Why do people act environmentally and what are the barriers to pro-environmental behavior? *Environmental Education Research, 8*(3), 239–260.

Laumonier, Y., Bourgeois, R., & Pfund, J.-L. (2008). Accounting for the ecological dimension in participatory research and development: Lessons learned from Indonesia and Madagascar. *Ecology and Society, 13*(1), 15.

McCarthy, J. F., & Robinson, K. (Eds). (2016). *Land and Development in Indonesia. Searching for the People's Sovereignty.* Singapore: ISEAS Publishing.

Nilan, P., Parker, L., Bennett, L., & Robinson, K. (2011). Indonesian youth looking towards the future. *Journal of Youth Studies, 14*(6), 709–728.

Nilan, P., & Wibawanto, G. R. (2015). "Becoming" an environmentalist in Indonesia. *Geoforum, 62*(2), 61–69.

Parker, L., & Nilan, P. (2013). *Adolescents in Contemporary Indonesia.* New York: Routledge.

Parker, L., Prabawa-Sear, K., & Kustiningsih, W. (2018). How young people in Indonesia see themselves as environmentalists: Identity, behaviour, perceptions and responsibility. *Indonesia and the Malay World, 46*(136), 263–282. doi:10.1080/13639811.2018.1496630.

Payne, P. (2001). Identity and environmental education. *Environmental Education Research, 7*(1), 67–88.

Setiawan, B. B., & Hadi, S. P. (2007). Regional autonomy and local resource management in Indonesia. *Asia Pacific Viewpoint, 48*(1), 72–84.

Warren, J. F. (2016). Typhoons and the inequalities of Philippine society and history. *Philippine Studies: Historical and Ethnographic Viewpoints, 64*(3–4), 455–472.

12 Conclusion, and a way forward

The need for EE

Human inaction in preventing and ameliorating global environmental destruction is one side of a coin; the other side is human greed, materialism and economic development. We would argue that both of these are present in Indonesia but, in addition, there is a prevailing lack of awareness and knowledge about both the rate and extent of environmental destruction, and the role of human beings in it. Although there is a dearth of research into environmental understandings in the Global South, we think these conditions are not unique to Indonesia, and indeed characterise much of the "majority" world. There is a crying need for Environmental Education in the Global South.[1]

While global ecological decline is a "slow violence" (Nixon, 2011), the problems are urgent. Environmental education is also a slow process. We know it can only be one component of a broad social movement that galvanises large portions of a population to change their consuming ways and somehow exert pressure on powerful corporations and self-interested governments. The authors' own country of Australia is an example where the citizens are ahead of their conservative government in environmental awareness and preparedness to take pro-environment action. So we are well aware that the political sphere (where actors are in cahoots with powerful economic players, e.g. in the coal industry) is a great challenge. Nevertheless, we cannot do without a strong social movement of people motivated by a responsible environmental subjectivity. We think that an effective way to build such a movement is to encourage people to think about how they are connected to other people in the world, and how they rely on natural resources to sustain themselves.

I. M. Young's theory of connection, responsibility and social justice (Young, 2011) can be extended to become a useful framework of environmental justice and responsibility. Looking forward to how we must take responsibility for our environmental consumption and behaviour (rather than backwards to attribute blame and liability for the mess we are in), this theory urges both serious reflection and considered action, and has the advantage of a global scope. To many, it is a new thing to think about the source of the items we buy and use, the energy used to extract, produce, transport, package, sell and use them, the sometimes iniquitous wages paid to workers and the unhealthy conditions in which they work, the unconscionable over-consumption of "fast fashion" and disposable goods, the deliberate design of devices, furniture and buildings to have a quick life, and the everyday "lifestyle" of using motorcars, air-conditioners and bottled water. In Indonesia, this lifestyle is not yet questioned: these are assumed "goods" that are universally desired and never critiqued. And we argue that it is into this space that EE must step, to help young people connect this desired lifestyle with the questionable sustainability of the natural world (the earth's life-support system).

But we recognise that there are many challenges. In setting out below the recommendations for EE in Indonesia, we explain how each addresses problems that are already described in the ethnographic chapters (Chapters 7–10), and in the survey chapter (Chapter 11). But over and above these is a broad economic, political, social and religious national landscape, and an even larger global socioeconomic context that we have barely mentioned: the world of popular culture and social media, the apparently unassailable power of transnational corporations, the increasingly fundamentalist Islamic *ummat* and rising tide of anti-Westernism, the unstable and increasingly threatening international relations of competing nation-states, and the growing economic gap between the haves and have-nots in the world. After discussing our recommendations we turn to the issue of EE in the Global South.

Recommendations

Berryman and Sauvé (2016) caution readers of the need to be realistic and consider the historical, cultural and ecological context when designing environmental education. As discussed above, Indonesia experiences the challenges of being an emerging middle-income nation, having a deeply religious population, a strong top-down education culture, and a long history of colonisation and strong leadership. It is developing rapidly, and consistently achieves growth in GDP of the order of 6 per cent per annum. Its population is the fourth largest in the world, and growing quite fast. The education system has grown rapidly and while Indonesia has produced an "education revolution" in terms of access to schooling, the quality of that education is lamentable. Competencies such as critical thinking, imagining future scenarios, problem solving and making decisions in a collaborative manner do not fit well within the current education system. There are many factors that could impede the pace of change in the Indonesian education system. These include: an enormous education system servicing almost 50 million students (Sekretariat Jenderal, 2017); geographical, cultural, religious and linguistic diversity; a strong culture of bureaucracy; low levels of teacher training (although it is constantly improving); and, despite calls by educationists, a persistent focus on national exams. We argue that Indonesia needs its own unique approach to EE that meets its complex cultural, educational and environmental needs, rather than trying make internationally promoted approaches to EE fit where they simply do not. Here are our suggestions.

(1) *The creation of environmentally responsible citizens should become one of the principle objectives of education.* The objectives of education are perhaps not much known at the level of the school, but policy documents shape the content of curricula and curricula shape textbooks, and it's textbooks that are taught in schools. So an explicit commitment to transformative environmental education and the objective to create environmentally responsible citizens would have an impact in the classroom. By the phrase "environmentally responsible citizens", we mean not just knowledge of environmental concepts and issues (as that too easily becomes conventional Science subjects) (Hollweg *et al.*, 2011), but also "the principle that ecosystems have evolved to sustain the web of life, attended by an understanding of the interdependence between natural processes and human ways of living" (Eames, Barker, & Scarff, 2018, p. 192), an emotional connection to nature and the willingness to act in its defence and preservation, as well as the skills and capacities to identify and solve social-environmental problems.

This recommendation aims to undermine and transform the dominant discursive constructions of the environment in Indonesia. In Chapter 6, we highlighted the persistent religious and educational discourse that portrays the environment as a resource provided by God, to be exploited for humankind's benefit. This is complemented by a strongly nationalist discourse about the wealth of Indonesia's natural resources. This confidence in the infinitude of Indonesia's natural resources is coupled with ignorance about the disastrous loss of species and biodiversity in Indonesia, through resource extraction, over-exploitation (e.g. of marine resources), urbanisation and the loss of habitat. Students have to be taught that there is an environmental crisis, that Indonesia's natural resources are threatened, but also that they can do something about this.

Although by personal inclination we would not locate the creation of environmentally responsible people within the frame of the nation-state of Indonesia, we think that in Indonesia the education system has been such a dominant force in uniting and creating the nation-state, it is unavoidable. The introduction of the new curriculum in 2013 was done through nationalist appeals to unify the nation against the claimed threat of religio-ethnic conflict leading to national fragmentation as well as the immoral and violent behaviour by young people (for instance in street gangs). The strength of nationalism and the unquestioning commitment to national identity are such that they could be mobilised in a discourse of "saving Indonesia's environment".

Obviously, to add such a radical new objective would require strong support from the Ministry for Education and Culture, as well as inter-ministry co-ordination among the Ministries of Education, Environment and Religion. Currently, both at the central apex in Jakarta and in regional offices of the Ministry of Education and Culture, there is no interest in the environment. Aside from the fieldwork detailed in Chapters 7–10, we spent a week in Jakarta talking to officials, and found no one in this Ministry who indicated a spark of interest, even when we were offering partnerships and sponsorships to work on new environmental initiatives. We therefore feel it would require extraordinary intervention, from influential outsiders, such as the President, or the World Bank (the major external funder of education in Indonesia). While this might seem wildly optimistic, there are precedents of the government taking dramatic pro-environmental action, e.g. when a ban on logging was introduced in 2011, under the sponsorship (and funding assistance under REDD+) of Norway, when leaded petrol was banned and when Presidential Instruction 3/1986 banned the use of pesticides on rice.

The addition of this major objective to the role of education would help with the problem of leadership. Above, we identified that a major reason for the success of the "clean and green" movement in Surabaya was the commitment and initiative of the Mayor. The downside is that her term will come to an end, and, as far as we can see, there is no ongoing policy commitment or plans to continue her good work. Concomitantly, the absence of environmental action in Yogyakarta – despite it being a city of university students and NGOs – was at least partly due to the absence of interested leadership by the Sultan or the Mayor. If we can make the environment more important structurally, regional leaders will have no choice but to become more active in environmental protection, and environmental education and action will be more sustainable because they will be mandated and made "core programmes" of the education system.

Further, if the Ministry of Education and Culture were publically committed to Environmental Education, that problem of teachers identifying first as public servants and only second as teachers, might be alleviated. As Bjork noted, many teachers' commitment to teaching is mediated by their motivation to advance their professional status as civil servants (Bjork, 2005). If the status of the new teachers of the new subject of EE were tied to the promotion of Environmental Education, they would have an incentive not only to improve their knowledge of EE but also to develop their pedagogical skills.

The Ministry of the Environment and Forests is historically much more pro-environment, and, if this "green" objective were initiated sensitively, with their full involvement, could be a valuable partner. At the moment, this Ministry is a "second-class citizen" ministry and does not have much clout.

The Ministry of Religion, which is responsible for many Islamic schools, is an unknown quantity in regards to EE. While there is no doubt that Islamisation is becoming increasingly fundamentalist and strident in the public sphere in Indonesia, there is considerable action being taken by Islamic scholars and leaders in the environmental space.[2] The environmental discourse in Islamic circles can be highly anthro-centric, but there is increasing talk of environmental crisis and of human responsibility for that. Gade has identified a strong moral ethos in Islamic environmentalism in Indonesia, e.g. a statement put out by Nahdlatul Ulama in 2007, described

> … [t]he development of an attitude of disbelief and mistrust toward the government, weakness of the rule of law, disaster of moral decadence, growth of "individualism", disappearance of authenticity and integrity and erosion of national pride, increase of characteristics of "konsumtif

hedonisme", the disease of corruption, sickness of laziness and the desire to take the "easy road", that is, the path of least resistance that allows for anything and which leads to the greatest destruction on the face of the earth.

(Cited in Gade, 2015, p. 170)

This statement is worth quoting at length, because it captures a sense of ethical crisis as well as social injustice as it describes

> those who are responsible for the environmental crisis, [who] carry out illegal logging, provoke conflicts over land tenure, and carry out [forest] over-cutting, create monopolies and privatize the resource of water, pollute water and water sources ... [and] develop hillside areas with a land slope greater than 40% or land in a water catchment area.
>
> (Cited in Gade, 2015, p. 170)

The description becomes more judgemental as it identifies those who

> would exterminate all forms of life with [their] "confident excuses" ... discard waste indiscriminately, and are heedless in caring for natural resources ... who endanger the lives of the people, ... agents of *over exploitasi* ... those who will not carry out redistribution ... in order to improve the condition of the local people ... and [the] many who have wrought environmental evil and acted with a great cruelty that disturbs the overall peace and destroys life itself.
>
> (Cited in Gade, 2015, p. 170)

This recommendation is an attempt to address the national-level lack of policy on EE. While Indonesia has signed up to international EE documents, in Jakarta, even UNESCO has stopped talking about the environment.[3] Currently, the Ministry of Education and Culture leaves it up to the Ministry of Environment and Forests, but that Ministry has no authority within the education system and EE is neither part of the curriculum nor embedded in the education system in any serious way. The result is that no one is taking responsibility for Environmental Education. In order to locate itself within the formal education system, EE needs to be part of the bureaucratic system and hierarchy, which is overseen by the Ministry of Education and Culture.

(2) (i) EE *should become a standalone, examinable subject in the curriculum*, across all levels of schooling. This recommendation goes against conventional wisdom in UNESCO literature and the curricula of some Global North countries, such as Australia, New Zealand and the UK, where a whole-school approach is advocated, with EE integrated across the curriculum. Another, potentially more serious criticism might be that the curriculum is already overfull – particularly for those in Islamic schools, where 100 per cent of the curriculum in general schools constitutes only 70 per cent of the curriculum and teachers and students add several hours on to the school day in order to squeeze

in the additional 30 per cent of curriculum content. Further, to change the curriculum again will not be a popular move. The disastrous way the 2013 Curriculum was introduced was largely to blame for its unpopularity, as in reality it is not a huge change from the 2006 Curriculum. Both of these criticisms would need to be sensitively pre-empted by government in introducing both a new subject and new emphases.

We advocate this move for several reasons. First, it will establish the environment as a serious priority for the government, and for the country. As we have seen repeatedly, environmental awareness, knowledge and understanding in Indonesia are low. Second, it will establish EE as a serious subject with academic value. Making it an examinable subject strengthens this recommendation. We have seen in Yogyakarta that schools that are not strong academically follow the Adiwiyata Programme because it gives them an identity, a possible claim to status. Year 12 students are exempted from environmental activities because the environment is not considered as important as studying for exams. Third, it will give EE a solid identity. At the moment, insofar as teachers have any inkling what the term "EE" might connote, it is usually inferred to be something to do with Science. This means that teachers in other subjects think that it does not apply to them. Currently, even in national-level Adiwiyata schools, which, in theory, have achieved integration of environmental sustainability across the curricula, teachers have no idea what "environmental sustainability" might mean or how to do it. Fourth, because this new subject would be compulsory for all students in all grades of school and in all streams, it will normalise and universalise thinking and action for the environment. Currently, not only are Year 12 students not allowed to participate in environmental activities, but also it is only a small percentage of students who are involved in most of the environmental activities conducted by students in programmes such as Adiwiyata and Surabaya's Eco Schools. Although we have not discussed it here, students who do not participate in such activities often deride those who do with stigmatising labels such as "smelly" or "dirty" (because they deal with compost), or "*kuli*" (coolie) or "*pasukan kuning*" (literally, yellow brigade, a reference to lowly street sweepers) (Tanu & Parker, 2018). Such derision indexes the strength of the dominant paradigm that places no value on caring for the environment. We propose that normalising care for the environment will undermine the power of such discourses.

Like any other subject in the curriculum, the EE subject teachers would be required to understand the content, and there would be textbooks to facilitate the teaching and learning. This subject would provide a dedicated time for students to learn the science of the environment (*about* the environment) and, with the right textbook content and questions, could facilitate deeper considerations about the complexities of the environment and environmental issues, aiming at students being able to identify humankind's role in environmental problems and solutions *for* the environment.

(ii) *The academic subject of EE would have a partner subject or organisation, such as a revamped Adiwiyata, that would provide the shell for outdoor, place-based, practical, problem-solving activities.* Research from the Global North shows conclusively that spending time in nature provides opportunity for connection to it, seeing oneself as part of it, and increasing the desire to protect it (Braun & Dierkes, 2017; Nisbet, Zelenski, & Murphy, 2009; Schultz, 2000). In Indonesian cities, where green spaces are often hard to find, there may have to be excursions, e.g. to outside the city or to botanical gardens or to rivers, lakes and beaches where remediation activities could be conducted (Suárez-López & Eugenio, 2018).[4] Ideally these would be student-led activities, but informed by the knowledge and skills developed in the more academic EE subject. It is imperative that these activities are not just environmental labour (as we saw so often in Surabaya) (Tanu & Parker, 2018); and that they are not just meaningless performances (as described in Chapter 8). Also see below re pedagogy.

A much-neglected cohort of young people with reference to EE is rural youth – and not only in Indonesia. Some have been active in social issues such as local land reclamation or dam protests and we think there is much scope with them to extend understanding of social justice issues to environmental justice. In West Java, a recent experimental programme with mixed rural and city young people in an environmental justice campaign has been very effective for EE (Tillah & Rahman, 2017).

(iii) The establishment of EE as a separate school subject would not obviate *the need to make a whole-school approach to environmentalism a priority.* The two are not inimical. Indeed, proper teaching and learning of EE would highlight the need for schools to instantiate environmentally-friendly, energy-renewing, environmentally sustainable programmes in schools, paying attention to procurement of facilities, equipment and supplies such as plastic water cups, food and photocopying paper, relations with service providers, energy use, waste management, as well as student clubs, staff professional development, community relations, and a host of other school programmes. We think that the new academic subject, coupled with the habits of environmentally responsible behaviour as members of the school community, will be an effective starting-place for young Indonesians who can then move on into the larger world, sensitive to environmental problems, and equipped with the tools to analyse and work towards solving them. In this way, they will have become responsibilised environmental subjects.

(iv) The establishment of a whole-school approach implies *the need for schools to develop "green" connections with their local community and government,* including, of course, service providers (e.g. of electricity, internet, waste disposal), neighbours, parents and the local civil community. A picture that has stayed in our minds comes from one of the better schools we visited. Students in this school sorted their rubbish into organic and non-organic bins. When the external rubbish collector came to the school, he collected the differently-coloured bins and just outside the school gate he dumped all the contents on the ground, mixing up the sorted organic and non-organic waste, before taking it all away in his truck, all mixed up. In full view of the students, this was an explicit statement of the significance of the school's fences: in this instance, the school was a bounded entity, and beyond the school fence, the students' pro-environment work was futile. We cannot wait for the children currently in school to grow up. This sort of community engagement is important for flow-on effects. If schools demand that garbage collectors practise waste separation and send organic waste to composting facilities, that will force them to seek appropriate end-users; if parents see that schools are serious about issues such as waste management, they might join in too – reducing the pressure on children to somehow convince their parents to "go green". Recent work on community waste initiatives in Java shows that community-based efforts, such as *bank sampah* (waste banks), are much more effective than top-down projects, because they mobilise the strong pressure of community-mindedness in Java (Schlehe & Yulianto, in press). To repeat a point made in Chapter 1,

> The highly idealistic notion – which assumes that we just need to change the way we educate our kids and students in order to make sustainability fall into our lap – is both horribly naïve *and* utterly unfair to the younger generation.
>
> (Jucker, 2002, p. 9, emphasis in original)

It is also important that students be involved in these community inter-actions: we note that EE can only be one (albeit interrelated) element in a movement towards environmental sustainability, and in Indonesia, the sphere of government and politics is sadly lacking in environmental awareness. Involv-ing students in community relations – developing the confidence to see how to discuss issues with more powerful people – can help them move from silence and invisibility to visibility and influence. It would be great for students to learn how to reach out to and negotiate with local politicians and government offi-cials, businesses, and the media.

(3) *A radically innovative teacher pre- and in-service training programme for EE would need to be developed.* With regard to selection of existing teachers, it may be useful to note that at the moment, at senior high school level, the two sub-jects whose content is the closest to the EE subject we have in mind are Geo-graphy and Biology. Another possibility is the Citizenship subject, as environmental citizenship is a concept that we think would resonate in nation-alistic Indonesia. There could be a role here for international educators and ENGO activists to contribute as trainers.

We think it important that teachers of the new subject be supported once they are (back) in schools. The formation of Teachers' Groups (Kelompok Guru) (Parker & Raihani, 2011) and partnerships with local ENGOs seem likely avenues.

(4) *The take-up of new pedagogies and forms of assessment is a matter of urgency.* It is important to mention again that many of the shortcomings that we have identified for EE in Indonesia are not special to EE: they are characteristics of the education system in general. There are many interconnected problems – for instance, the reliance on rote learning pedagogy and student memorisation is

related to teacher insecurity about their knowledge, the continuing importance of exams, and the fact that many teachers are public servants first and teachers second, and so are less committed to curriculum development, pedagogic experimentation and innovation, etc., than might otherwise be expected. And it is only fair to note that the MOEC has been pushing the new active learning, student-centred pedagogy that we are advocating for EE in its 2006 and 2013 Curricula. The problem of poor support for existing teachers with the introduction of new curricula, and the low quality of their education in both the subject content and pedagogies, combined with the insecurity of teachers, poor quality of textbooks, etc., are all to blame.

We advocate critical eco-pedagogy, outlined in Chapter 3, and suggest that it be taken up first in the new, standalone subject of Environmental Education. We think that focusing on this new pedagogy and forms of assessment in the one new subject might be a more effective way to begin, rather than trying to bring them in across the board all at once. We would like to see the new EE subject incorporate systemic thinking, critical thinking, the positive valuation of questioning and curiosity, the imagination of future scenarios, the ability to identify social inequality, the connections among social, economic and environmental problems, participative problem-solving, active research projects in the community and the natural environment, and collaborative decision-making (team work). It will take confident and capable teachers to take up these new pedagogies, but it will solve the problem that Fajar identified: "In Indonesia we [only] learn what has already been discovered" (Chapter 10).

Ramifications for EE in the Global South

Of course the "Global South" is not one thing. Indonesia is a rapidly developing country, but there are some Global South countries that languish at the bottom of the "human development" league tables and are actually dysfunctional nation-states, rent by conflict and war, where people cannot live in dignity and safety. In between are many, "developing", mostly postcolonial states, often with growing populations and threatened environments. It is simply not possible to address this diversity properly, so here we identify some salient commonalities and issues.

"Global South" status usually implies a range of common features: a marginal or inferior position in the world hierarchy of nations; vulnerability to the power of richer countries and international patron organisations such as the World Bank and IMF, which wield great economic power; compromised independence of action; growing populations; and long histories of colonisation by other, often European, powers. Colonisation has often left them economically, politically and social weak, with concomitant poor infrastructure, such as waste management systems, transport, markets, and energy supply, inadequate expertise in the bureaucracy, and sometimes unfamiliarity with modern structures and processes associated with the template of the nation-state. Often there are diverse populations (cultures, religions, languages), and national unity is contested. Politically, they are often unstable and volatile, and the mass of people is voiceless. Their education status is also often weak, sometimes with low rates of access to education, particularly for girls, and in terms of the quality of education. Reasons for the low quality of education include the newness of the mass education system; the need to develop a national system of education using teachers whose own education was meagre; lack of resources; poor infrastructure; the low socio-economic background of students; and so on. Typically, school education is a matter of rote learning facts, en masse, and passing exams that show that one is educated.

Thus, we think that many of the practical difficulties of introducing EE into Indonesia would be encountered in other Global South countries. Many are problems of "developing" education systems and are not unique to EE.

More significantly, it behoves us to consider the globalisation of EE, and in particular the power relations involved when concepts and discourses such as that of "Education for Sustainable Development" (ESD) are disseminated and imposed from the metropole upon Global South countries. The dominance of the "sustainable development" discourse, with its emphasis on economic development, rather than the sustainability of the environment, is a matter of great concern (Griggs et al., 2013). And strangely, not only in Indonesia but also in Australia, when teachers and students talk about the environment in sustainable development, there is a silence about economics and the role of material development in destroying the planet (Dyment, Hill, & Emery, 2015).

Both ESD and the International Environmental Education Programme (IEEP) of UNESCO-UNEP, introduced in 1974, have "cultivated a neo-colonialist discourse in environmental education by systematically privileging Western (and especially US) interests and perspectives" (Gough, 2008 [2003], p. 55). Many in EE have assumed the universality of science knowledge, when it is only one of many types of knowledge. People in different parts of the world see their natural environments through their own worldviews, epistemologies and cosmologies. Science knowledge of the environment might sit side-by-side with a view of the environment in which natural events might occur for religious or moral reasons, e.g. an earthquake happens because God is displeased or the country has become immoral. In Indonesia, many people go to medical doctors trained in Western medicine when they are sick, but find explanations such as bacteria or viruses inadequate for answering the "why me?" question, and seek further answers from *balian* or traditional healers who are likely to be able to identify a displeased ancestor or jealous in-law who is the cause of the problem.

Our discussion of the possibility of religious environmental education (Chapter 6) is one little foray into the possibilities of a more inclusive curriculum. Generally, the so-called "cosmopolitan" discourse of EE is actually a specifically Western, post-Enlightenment one that defines what constitutes knowledge and learning, and often (implicitly and unintentionally) privileges Western knowledge and learning over that of others. Science knowledge is a local knowledge (Gough, 2014, p. 132, following Turnbull), specifically developed during the era of industrialisation and now post-industrialisation, in the era of the Anthropocene. It has often spearheaded that same economic development that has destroyed the Earth's biosystem. We might do well to "provincialise Europe", as Chakrabarty (2007) enjoins. This is not to say that science knowledge is not valid or not useful. It is, but we need to acknowledge that there are other ways of knowing the world, and that these too are valid and useful for citizens where they prevail.

In African nations there tends to be more talk about postcolonial power relations and also more talk about Indigenous knowledge systems to do with the environment, than we hear in Indonesia. Ketlhoilwe has shown how the technologies of governmentality and international power relations – starting with peak global bodies such as UNESCO – shape EE policy, discourse and teaching programmes in Botswana (Ketlhoilwe, 2013). Postcolonial discourses are also prevalent in India. But white, European, scientific knowledge is both desired and rejected – desired as prestigious, glamorous and the path to a decent job, but also rejected as imperial and therefore unacceptable. "Indigenous" is a tricky term in Indonesia, and has multiple meanings that are mostly quite different to those attached to the term in African nations, as well as in predominantly white settler countries such as Australia. In Indonesia, the complexity of prehistorical and historical waves of immigration and settlement, as well as contemporary racism, and the strength of the First Nations discourse globally, mean that Indigeneity is sometimes an advantageous status and sometimes an object of parody and vilification. All of which means that it is no simple thing to incorporate "local knowledge" or "local wisdom" into EE.

However, we are not arguing for education that does not change things, or that is so accommodative of existing practices that it is "business-as-usual", with blissful ignorance about global environmental destruction. We want transformative education, which wakes people up to the sad loss of species and ecosystems, and beauty, and galvanises them to action. We are looking for education that will help young people deconstruct global capitalism, and critique entrenched power, and "out" corrupt leaders, and learn how they can fight for their local environment as well as for the world's biosphere. The most effective way to do this will be to develop culturally sensitive, locally relevant, environmental education programmes that lead young people to become environmentally responsible citizens.

Notes

1 Suárez-López and Eugenio (2018, p. 1103) have recently noted that "the development of EE remains incipient in Latin America". There seems to be more EE activity in southern Africa, and more research into it, but much of the latter is quantitative (Velempini *et al.*, 2018).

2 For instance, the two largest Islamic organisations, Nahdlatul Ulama and Muhammadiyah, both now have Environmental Councils. Majelis Lingkungan Hidup of Muhammadiyah has declared that in its second century of existence, Muhammadiyah is to be devoted to environmental jihad (http://lingkungan.muhammadiyah.or.id/

berita-14461-detail-penyelamatan-lingkungan-jihad-muhammadiyah-di-abad-kedua. html). Gade has described how ulama and Islamic legal scholars have been developing new Islamic environmental jurisprudence (*fiqh*) (Gade, 2015); the HMI (Himpunan Mahasiswa Islam, Association of Islamic University Students) is active in environmental education and research; and there are pesantren (Islamic boarding schools) that are famously "green".

3 Interview with the UNESCO National Programme Officer for Education, Jakarta, 31 January 2017.

4 In Indonesia, the ENGO, Rimbawan Muda Indonesia, has been running the Repling programme of EE via excursions to Bogor Botanical Gardens for many years. One PhD study of senior high school Chemistry students and their response to different approaches to EE in Java showed that active learning in community action or in the field outdoors was more effective than classroom education (Kusmawan, 2007).

References

Berryman, T., & Sauvé, L. (2016). Ruling relationships in sustainable development and education for sustainable development. *The Journal of Environmental Education, 47*(2), 104–117. doi:10.1080/00958964.2015.1092934.

Bjork, C. (2005). *Indonesian Education: Teachers, Schools, and Central Bureaucracy*. New York: Routledge.

Braun, T., & Dierkes, P. (2017). Connecting students to nature – how intensity of nature experience and student age influence the success of outdoor education programs. *Environmental Education Research, 23*(7), 937–949. doi:10.1080/13504622.2016.1214866.

Chakrabarty, D. (2007). *Provincializing Europe: Postcolonial Thought and Historical Difference*. Princeton, NJ: Princeton University Press.

Dyment, J., Hill, A., & Emery, S. (2015). Sustainability as a cross-curricular priority in the Australian Curriculum: A Tasmanian investigation. *Environmental Education Research, 21*(8), 1105–1126.

Eames, C., Barker, M., & Scarff, C. (2018). Priorities, identity and the environment: Negotiating the early teenage years. *The Journal of Environmental Education, 49*(3), 189–206. doi:10.1080/00958964.2017.1415195.

Gade, A. M. (2015). Islamic law and the environment in Indonesia. Fatwa and Da'wa. *Worldviews, 19*, 161–183.

Gough, N. (2008 [2003]). Thinking globally in environmental education: Implications for internationalizing curriculum inquiry. In W. F. Pinar (Ed.), *International Handbook of Curriculum Research* (pp. 53–72). Mahwah, NJ: Laurence Erlbaum Associates.

Gough, N. (2014). Globalization and curriculum inquiry: Performing transnational imaginaries. In N. P. Stromquist & K. Monkman (Eds), *Globalization and Education Integration and Contestation across Cultures* (pp. 87–101). Lanham: R & L Education.

Griggs, D., Stafford-Smith, M., Gaffney, O., Rockström, J., Öhman, M. C., Shyamsundar, P., … Noble, I. (2013). Sustainable development goals for people and planet. *Nature, 495*(7441), 305–307.

Hollweg, K. S., Taylor, J. R., Bybee, R. W., Marcinkowski, T. J., McBeth, W. C., & Zoido, P. (2011). Developing a framework for assessing environmental literacy. Washington, DC: North American Association for Environmental Education.

Jucker, R. (2002). Sustainability? Never heard of it!" Some basics we shouldn't ignore when engaging in education for sustainability. *International Journal of Sustainability in Higher Education, 3*(1), 8–18.

Ketlhoilwe, M. J. (2013). Governmentality in environmental education policy discourses: A qualitative study of teachers in Botswana. *International Research in Geographical and Environmental Education*, 22(4), 291–302. doi:10.1080/10382046.2013.82 6545.

Kusmawan, U. (2007). *An Analysis of Student Environmental Attitudes and their Impact on Promoting Sustainable Environmental Citizenship: A Multi-Site Study in Indonesian Secondary Schools*. PhD, The University of Newcastle.

Nisbet, E. K., Zelenski, J. M., & Murphy, S. A. (2009). The nature relatedness scale: Linking individuals' connection with nature to environmental concern and behavior. *Environment and Behavior*, 41(5), 715–740. doi:10.1177/0013916508318748.

Nixon, R. (2011). *Slow Violence and the Environmentalism of the Poor*. Cambridge, MA: Harvard University Press.

Parker, L., & Raihani. (2011). Democratizing Indonesia through education? Community Participation in Islamic schooling. *Education Management, Administration and Leadership*, 39(6), 712–732.

Schlehe, J., & Yulianto, V. I. (in press). An anthropology of waste. Morality and social mobilisation in Java. *Indonesia and the Malay World*.

Schultz, P. W. (2000). Empathizing with nature: The effects of perspective taking on concern for environmental issues. *Journal of Social Issues*, 56(3), 391–406.

Sekretariat Jenderal, Pusat Data dan Statistik Pendidikan dan Kebudayaan (2017). *Ikhtisar Data Pendidikan Tahun 2016/2017*. Jakarta Retrieved from http://publikasi.data. kemdikbud.go.id/uploadDir/isi_FC1DCA36-A9D8-4688-8E5F-0FB5ED1DE869_.pdf.

Suárez-López, R., & Eugenio, M. (2018). Wild botanic gardens as valuable resources for innovative environmental education programmes in Latin America. *Environmental Education Research*, 24(8), 1102–1114. doi:10.1080/13504622.2018.1469117.

Tanu, D., & Parker, L. (2018). Fun, "family", and friends: Developing pro-environmental behaviour among high school students in Indonesia. *Indonesia and the Malay World*, 46(136), 303–324. doi:10.1080/13639811.2018.1518015.

Tillah, M., & Rahman, F. (2017). Fighting for existence. *Inside Indonesia*, 127.

Velempini, K., Martin, B., Smucker, T., Randolph, A. W., & Henning, J. E. (2018). Environmental education in southern Africa: A case study of a secondary school in the Okavango Delta of Botswana. *Environmental Education Research*, 24(7), 1000–1016. doi:10.1080/13504622.2017.1377158.

Young, I. M. (2011). *Responsibility for Justice*. Oxford and New York: Oxford University Press.

Index

2013 Curriculum in Indonesia 100–7; relative importance of environment in 107–8; religious EE in 108–15

academic learning: perceived as valuable and examinable while practical learning is less valued 163; versus practical learning 167–9; student understandings of academic or theoretical knowledge is limited 1, 168–9; students prefer practical learning to 167–8

Adiwiyata Programme: absence of in Ministry of Education and Culture 129; aims of 99; associated with academically weaker schools 135; audit culture in, as symbolic of lack of trust 140–1; burden of documentation in 139; coverage of 99; and ersatz accountability 138–41; false documentation in 140; history of 98–9; importance of documentation in 139; importance of high participant numbers for point-scoring in 162; importance of *prestasi* in 140; meaningless performance of EE events 162–3; motivation of teachers to participate in 140, 141; organisation of in schools 136–8; problematic administration of in Yogyakarta 128–31; schools join for marketing 135; schools join because forced 136; schools join by accident 135–6; teachers as decision-makers, students as cadre 137; trapped in administrative silo 128

affect: desired among students as expressed in 2013 Curriculum 112–13

agency: of animate and inanimate phenomena 21; silence of 2013 Curriculum about human agency with respect to the environment 112–13, 119; of students with regards to environmental behaviour 23, 39, 45, 46, 50, 156, 160, 223, 238

Agrawal and environmentality 14, 27, 175, 201, 226, 229, 230

Anthropocene age 5, 15n2, 263

anthropo-centric worldviews 4, 23, 111

banning of Year 12 students from participation in EE 226

Biology lesson 150–2

burden of change towards environmental responsibility in Surabaya: as against the traditional top-down flow of power and authority in Java 230; placed on children and poor market-sellers 230–1

causation: different cultural understandings of 20–1; neglected in 2013 Curriculum 112–13

children 2, 3, 14, 21, 30, 34, 43, 45, 47, 58, 59, 61, 80–2, 84, 96, 134, 155–6, 159–60, 161, 183, 195–7, 201, 212, 222, 225, 228, 230, 238, 260

Chawla and Significant Life Experiences as triggers for environmentalism 234

citizenship 1, 9, 11, 14, 18–38, 84, 103, 260

civil servants 3, 69–70, 86–7, 139, 140, 164–5, 228, 256; *see also* public service in Indonesia and teachers

classes in environment 3; examples 147–55; Biology as example of 150–2; Craft and Entrepreneurship (mushroom fiasco) as example of 148–52; Geography as example of 152–5

climate change: in Indonesia 63; perceptions of among populace in Indonesia 65–6; perceptions of among students in survey 244; students' lack of understanding as related to their environmental behaviour 169, 214; as taught in the curriculum 105, 112; as taught in class 153

competitions 207–17; in Adiwiyata 139; creating community and maintaining hope 198; for *prestasi* 163, 207–17, 218, 227; in Rubbish Day event 157; TENGO's 181, 190–1, 207–8; ubiquity of in Indonesia 140, 192, 207, 227; *see also* Yel-Yel Eco Schools Competition

competitiveness in EE: means teachers reluctant to discuss failures or problems 217; winning as motivation in Eco Schools (rather than environmental concern) 218, 228

consumerism and lack of understanding re connection to environmental destruction in Indonesia 66, 184, 214, 249, 250, 253–4

Craft and Entrepreneurship lesson (mushroom fiasco) 148–52

creationism: as basis of environmental education 13, 109–11, 115

critical eco-pedagogy 48–50, 261

critical thinking: in Dewey's thinking 45; as necessary in EE 51, 168, 261; lack of in Indonesian education 92, 163, 198, 223, 228, 254

"Cross-Cultural Exchange" tour organised by TENGO to Perth 220–1; differing expectations of by hosts and guests 220; inter-cultural difficulties during 220; no parents attended selection tests 211; selection of students 210–215, 220; selection of teachers and principals 215–17, 220; weakness of English competence by candidates 213, 215, 220

curricula: as product of the state 115

decentralisation: of education 82, 88–9, 100, 135, 165, 180; in Indonesia 12, 27, 59–60, 128, 247

Deep Ecology 23–4

Dewey, John 45

"doomsday" approach to EE 222; lack of success by TENGO shows need for *paksarela* approach 223–4; at an uncooperative school 221–4

eco-centric worldviews 4, 8, 23

ecological citizenship 9, 32–3

education: traditionalism and progressivism in 45

education system in Indonesia: access to schooling 81–3; history of 2; objectives of 83–4; quality of 84–90; pedagogy in 90–1; size of 3, 81; structure of 80–1; as unifier of nation-state 2

Education for Sustainable Development (ESD) 1, 11, 39–42, 44, 69, 95, 98, 119, 194, 198, 262

Education for Sustainability (EfS) 11, 40–2, 44, 194

"the environment": absence of in Indonesian national discourse 2; in Indonesian education system 3; meanings of 4–7; as "natural resources" in Indonesia 4, 5; as social construct 5–6; as something to be managed (precursor to environmentality) 228; *see also* "the environment" in 2013 Curriculum

"the environment" in 2013 Curriculum 107–8: in Biology 104–5; in Physics 105; in Chemistry 105–6; in Geography 106; in Economics 106; relative importance of in 2013 Curriculum 107–8; creationism 109–11; instrumentalism 111

Environmental Agency: problems with MOEC in Yogya 128–32; as responsible for training Adiwiyata teachers 129; in Surabaya 179–83, 185, 192, 196, 197, 208, 230

environmental awareness: lack of in Indonesia 2–4, 10, 22, 29–4; lack of in the Global South 253

environmental behaviour: as goal 3, 10, 20, 160, 163, 228, 230; gap between pro-environmental attitudes & knowledge and 11, 46, 47, 65; meaningless performance of pseudo- 170; as reported by survey respondents 14, 239–42

environmental citizenship 14, 18, 30–3, 103

environmental education: critical ecopedagogy in 48–50; dominance of Global North countries in 1, 24, 39, 262–3; dominance of science in 1, 43–4, 262–3; Eurocentrism of 24, 262–3; history of internationally 6; incorporation of Indigenous and local knowledge into 263; lack of knowledge about in Global South 1, 234; lack of

environmental education *continued*
knowledge about in Indonesia 234;
location of discipline in school curricula
43–4; must be political 42–3; needs
large movement of people, political will
and action to be truly effective 253;
objectives of 42–3; pedagogy in 44–8; is
a slow process 253; as transformative
education, not transmissive 42–3, 263
environmental education in Indonesia:
history of 95–9; positioned outside of
education system in Indonesia 7; should
be enabling rather than discouraging 8;
as meaningless ritual 162–3; does not
involve identification of environmental
problems 221, 227; involves
environmental labour rather than
education 221, 227; recommendations
for the way forward for 254–61
Environmental Education Network (JPL)
71, 97
environmental movement in Indonesia
70–3
ENGOs (Environmental Non-
Government Organisations) in
Indonesia *see* TENGO and Kids for
Change
environmental responsibility 18–30;
example of with blocked river in Jakarta
19–20; backward-looking theories of
(liability approach) 21; forward-looking
theories of 21; introduction of idea of
7–9
environmentality 9, 14, 24–7; in Surabaya
175–6, 186, 201, 207, 224, 230–1;
requires structure and responsibilised
citizens which are missing in Surabaya
226, 228, 229–31
environmentally responsible behaviour
(ERB) 22, 23
environmentalism 1, 7, 9, 14, 47, 50, 64;
in the Global North 1, 229; in the
Global South – as local, ephemeral 7,
229; in Indonesia 70, 71, 184, 234, 239,
256, 259; as sometimes against the state
and with the state under
environmentality 229
examinations 45, 46, 47, 80, 88, 89, 101,
102, 108, 115, 120n12, 137, 146, 163,
167, 214, 226, 228, 236, 254, 258, 261,
262

Fajar (exceptional EE student) 197, 213,
223, 228, 230, 261; discovers differences

between EE in Perth and Surabaya
220–1
Foucault, Michel 24–5, 112, 175, 190, 229
Freire, Paolo 48–9

Geography class 152–5
gender: in access to schooling 79, 80, 81,
82; among teachers 216; in
environmental organisations 236; in
leadership 178–9; in school and in the
classroom 138, 155
Global North: EE in 1, 234–5;
environmental awareness in 1, 4;
privileging of in EE 263; as responsible
for environmental destruction 4, 263
Global South: aspirations for prosperity in 4;
characteristics of 261–2; lack of
knowledge about EE in 1, 47, 234–5;
ramifications of this study for EE in 261–3
governmentality 18, 24–7, 175, 226,
229–30, 263
Gruenewald, David A. 42, 108, 159, 214

hope, the importance of in EE 160–1

identity as an environmentalist 238–9
Indonesia: demography of 57–8; economy
of 58–9; environment of 62–4;
environmental awareness in 64–70;
government attitudes towards the
environment in 67–9; government
capacity re the environment in 69–70;
growing middle class in 58; NGOs' and
universities' environmental awareness
in 70–3; politics and government of
59–60; religion and culture of 60–2;
urbanisation in 58
Islam 5, 11, 18, 60–2; Islamic education in
Indonesia 80, 83, 89; in 2013
Curriculum 104, 107, 109–11, 113–15;
and environmentalism in Indonesia
256–7

Kids for Change (ENGO in Perth,
Western Australia) 190–2, 194–5,
197–8, 211–12, 220
knowledge about the environment: in the
Global South 253; lack of in Indonesia
21–2, 234–5

lack of science knowledge among
Indonesian population, teachers and
students 3, 2, 21–3, 34n7, 41, 71, 85,
92, 185, 193–4, 214, 262–3

lessons in environment *see* classes

national park, reading and comprehension questions about in textbook 116–18

objectives of schooling vs objectives of EE 42–3
OSIS (Organisasi Siswa Intra Sekolah, Student School Council) 137, 156, 222, 236

paksarela (forced volunteering) in EE 224–8; apparent need for in Surabaya shown by an uncooperative school 223–4; common in Indonesia 224; not expressed as exploitation by students or teachers 224–6; unpaid labour 224; usual response to - was appreciation of the positive outcomes of environmental labour 225–6; will have to continue as long as people are not "responsibilised" with environmental norms, beliefs and knowledge 228
Pancasila (Indonesia's national ideology) 2, 7, 60, 62, 83
parents 3, 82, 196; child-parent relationships 195, 196, 197; and EE 185, 208, 211, 216, 241–2, 259–60; and schools 82, 88, 89, 135, 216
pedagogy, critical ecopedagogy 48–50
pedagogy in Indonesia, straying from traditional pedagogy 8
performing for prestige *see prestasi*
prestasi (rank, prestige, status): importance of in Surabaya Eco Schools programme 217–9; performing for 217–9; prestige rather than environmentalism as motivation for teachers and schools 217–18
problem solving 45, 49, 51, 163, 168, 223, 246, 254, 259, 261
public service in Indonesia: lack of capacity to deal with environment 70; size and characteristics of 69–70; *see also* civil servants

recommendations for the way forward for EE in Indonesia 254–61
religious EE in 2013 Curriculum 108–15
reluctance to admit failures and difficulties in EE 217
responsibility: as aim of Adiwiyata 99; in EE 40, 47; for the environment 30, 32, 33, 41, 51; for the environment in

Yogya 7; for environmental problems 19–20; for pro-environment behaviour change 228; for justice 27–30; history of concept of 18–19; human, for the environment 114; lack of, for the environment in the Curriculum 95, 112–13, 119; lack of, for the environment in schools 99; liberal environmental 20–4; no one taking, for EE in government or in the education system 125; social connection model of 7–8, 27–9, 253; student perceptions of, for environment in survey 244–9
responsibilisation 18, 246; Northern Territory Intervention as example of neoliberal responsibilisation 26; in Surabaya 176, 228; under environmentality 24–7
rote learning: as major pedagogy 11, 45–6, 87, 90, 91, 101, 146, 150, 163, 169, 260, 262
Rubbish Day event 156–8; creation of waste in 158; missed opportunities in 158, 160–3, 170; student-led 157–8

Significant Life Experiences (SLE) literature 234
siloed knowledge: example of 213; lack of connection between social analysis and ecological analysis 213
"slow violence" of environmental destruction 6, 253
social (in)justice: as environmental (in)justice 7, 8, 253–4; as more potent than environmental (in)justice in Indonesia 7, 29
socialisation (*sosialisasi*): assumed effectiveness of 216; as top-down teaching 216
stewardship: human, for nature 4, 23, 68, 114
student leaders: as agents of change 156–60; of Rubbish Day event 156–9
student perceptions: of barriers to overcoming these problems 244–9; of local, national and international environmental problems 242–4; of who is responsible for local, national and international environmental problems 244–8; of what they can do to overcome these problems 244–8
student voice: in EE 155–61; in Rubbish Day event 156–8; vis-a-vis lack of power and authority for children 159–60, 197–8

sungkan (value of docility, respectful politeness, submission and of knowing one's place in a hierarchy) 14, 147, 166–7

Survey (Survei Pendidikan Lingkungan Hidup 2014–15) 235–50; characteristics of respondents of 236–8; perceptions of barriers to overcoming these problems 244–9; perceptions of what respondents can do to overcome these problems 244–8; perceptions of who is responsible for local, national and international environmental problems 244–8; responses re environmental behaviour 239–42; responses re identity as an environmentalist 238–9; responses re perceptions of environmental problems 242–4

sustainable development 1, 11, 13, 27, 39, 40–2, 67, 69, 95, 98–9, 106, 108, 119, 127, 155, 161, 194, 198, 262

Tbilisi Declaration of 1977 11, 40, 98

teachers: as civil servants 69, 86, 163–6; their lack of training and awareness in EE 196–7, 225; motivation of teachers 164–6; teaching as desirable job, especially in rural and remote areas 86

TENGO (environmental NGO in Surabaya) 14, 175–231; *see also* Yel-Yel and Cross-Cultural Exchange

textbooks 115–18; as basis of most classes 148, 150; example of English reading about a national park 116–18; funding arrangements 115; "textbook culture" and importance of in Indonesia 90, 101, 115, 255; well known by teachers 115

time, cross-cultural concepts of 20–1

transmissive education 42–3

transformative education 40, 42, 43, 263

UNESCO: lack of UNESCO EE goals in Indonesia 198; role of in EE 40, 41, 44, 262–3; role of in EE in Indonesia 95, 98–9, 257

universities: EE in 44, 119; environmentalism in in Indonesia 48, 66, 67, 71, 72, 238–9; environmental knowledge, attitudes and behaviour in 65

values in the 2013 Curriculum 113

WALHI 67; and EE 96–7

waste management: lack of in Yogyakarta 128, 162; need for social pressure for collective action in 226, 228; as part of whole-school approach to EE 259–60; practised by students 239, 243; successful implementation in Surabaya 201

Yel-Yel Schools Competition 208–11; community-building in 209; creation of waste in 210; indoctrination in 210; shallow messages in 210

Young, Iris Murdoch: her theory of social justice and its connections to environmental justice 7–8, 27–9, 253–4

young people 1–4, 10, 14, 29, 45–6, 58, 79, 91, 92, 125, 160–1, 164, 170, 194, 195, 196, 197, 198, 234–52, 254, 255, 259, 263

For Product Safety Concerns and Information please contact our EU
representative GPSR@taylorandfrancis.com
Taylor & Francis Verlag GmbH, Kaufingerstraße 24, 80331 München, Germany

9 7 8 0 3 6 7 7 8 4 2 4 9